MOON LAUNC

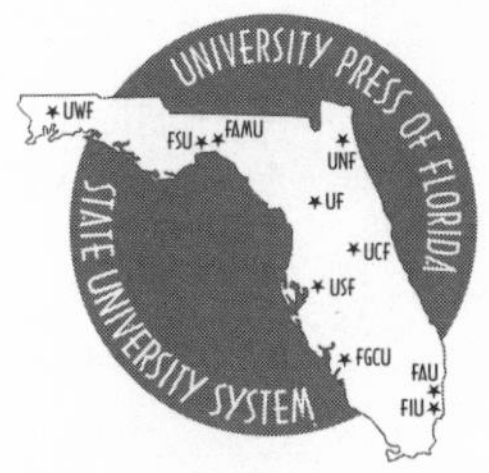

Florida A&M University, Tallahassee
Florida Atlantic University, Boca Raton
Florida Gulf Coast University, Ft. Myers
Florida International University, Miami
Florida State University, Tallahassee
University of Central Florida, Orlando
University of Florida, Gainesville
University of North Florida, Jacksonville
University of South Florida, Tampa
University of West Florida, Pensacola

USA

MOON LAUNCH!

A History of the Saturn-Apollo Launch Operations

Charles D. Benson and William B. Faherty

University Press of Florida

Gainesville · Tallahassee · Tampa · Boca Raton

Pensacola · Orlando · Miami · Jacksonville · Ft. Myers

First cloth printing 1978, as first half of *Moonport: A History of Apollo Launch Facilities and Operations,* by National Aeronautics and Space Administration, Scientific and Technical Information Office
NASA SP-4204

First paperback printing 2001 by University Press of Florida
Printed in the United States of America on acid-free paper

06 05 04 03 02 01 6 5 4 3 2 1

Library of Congress Cataloging-in-Publication data are available.
ISBN 0-8130-2094-8

The University Press of Florida is the scholarly publishing agency for the State University System of Florida, comprising Florida A&M University, Florida Atlantic University, Florida Gulf Coast University, Florida International University, Florida State University, University of Central Florida, University of Florida, University of North Florida, University of South Florida, and University of West Florida.

University Press of Florida
15 Northwest 15th Street
Gainesville, FL 32611-2079
http://www.upf.com

CONTENTS

Gateway to the Moon: Building the Kennedy Space Center Launch Complex was originally published in 1978 by the National Aeronautics and Space Administration as part of the NASA History Series. The original volume was a single book titled *Moonport: A History of Apollo Launch Facilities and Operations*. This volume corresponds to chapters 15–24 and pages 317–551 of that publication.

Foreword

By now the grandeur of the achievement of landing men on the moon and returning them to earth has taken its place in our language as a yardstick of human accomplishment—"If we could send men to the moon, why can't we do so-and-so?" The most imposing artifact of that achievement is the Apollo launch facilities at Kennedy Space Center.

When the national objective of landing men on the moon was dramatically announced in May 1961, it quickly became apparent within NASA that the remainder of the decade was little enough time to design, build, and equip the extensive and unprecedented facilities required to launch such missions. Indeed, time was so pressing that for many months the planning, designing, even initial construction of launch facilities had to go forward without answers to some essential questions, such as: How big would the launch vehicle(s) be? How many launches would there be, and how often?

Intense effort by a rapidly growing team of people in government, industry, and the universities gradually filled in the grand design and answered those questions. Land was acquired, ground was broken, pipe was laid, concrete was poured, buildings rose. When the launch vehicles and spacecraft arrived, the facilities were ready and operations could begin. Seldom was the pressure off or the path smooth, but the end of the decade saw the deadline met, the task accomplished.

This history tells the story of the Apollo launch facilities and launch operations from the beginning of design through the final launch. You will meet many of the cast of thousands who took part in the great adventure. You will read of the management techniques used to control so vast an undertaking, of innovation in automation, of elaborate, repetitive, exhaustive testing on the ground to avoid failures in space. You will also learn something of the impact of the Apollo program on the citrus groves and quiet beaches of Florida's east coast.

It is fitting that, as this manuscript was being prepared, these same facilities were being modified to serve as the launch site for Apollo's successor, the Space Shuttle, for at least the remainder of this century.

August 1977

Lee R. Scherer
Director
Kennedy Space Center

Preface

On 28 July 1960, the National Aeronautics and Space Administration (NASA) announced a new manned spaceflight program. Called Apollo, its aim was to put three astronauts into sustained earth orbit, or into a flight around the moon. The timing of the announcement was not auspicious. The next day, NASA's first Mercury-Atlas (MA-1) disintegrated and fell into the ocean 58 seconds after takeoff from Cape Canaveral. This disaster ushered in a bleak four months during which the test rocket Little Joe 5 joined the MA-1 in the ocean, and the first Mercury-Redstone lifted a fraction of an inch and settled back on its launch pad. The last failure, on 21 November, marked the absolute nadir of morale for the engineers working on Mercury. The people at the new NASA headquarters in Washington, coping with financial and administrative problems and facing a change of administration after the national election, were only a little less dispirited than the workers in the field. But the fledgling space agency had an asset that made its announcement of an ambitious Apollo program more than an exercise in wishful thinking—it had the support of the American people.

If there is an American psyche, it had been shaken 4 October 1957 by the news that Russia had launched the first man-made earth satellite—Sputnik 1. To those apprehensive of anything Soviet, the news was a red flag. The military and the President played down Sputnik's significance, but a layman could not but wonder if Sputnik was one of those scientific breakthroughs that could alter the balance of power. The average American was perhaps most concerned because someone else was excelling in technology—an area in which the U.S. was accustomed to leading.

There was an almost unanimous determination to get into the space race and win it. Three Presidents, with firm support from Congress, channeled the public will into an answer to the Russian challenge. Lyndon B. Johnson, the Senate majority leader, pushed the Aeronautics and Space Act through Congress in 1958. Under its authority, President Eisenhower set up NASA and transferred the armed services' non-military space activities to the new civilian agency. The following year NASA received a vital asset—the Army team of former German V-2 experts who were working up plans for Saturn, a large rocket. Assigned the task of manned spaceflight, NASA's immediate goal was the successful orbiting of a man aboard a Mercury spacecraft. NASA's Ten Year Plan of Space Exploration, revealed to Congress in early 1960, called for nearly 260 varied launches during the next decade, with a manned flight to the moon after 1970. The House Committee on Science and Astronautics considered it a good program except that it did not move ahead fast enough.

Meanwhile, the Russians were not idle. On 12 April 1961, they put Major Yuri A. Gagarin into orbit around earth. The Soviet Union and the United States were locked in a confrontation of prestige in Cuba, in Berlin—and in space. Convinced it was necessary to show the world what America could do, President Kennedy told Congress on 25 May 1961:

> Now it is time to take longer strides—time for a great new American enterprise—time for this nation to take a clearly leading role in space achievement which in many ways may hold the key to our future on earth . . . . I believe that this nation should commit itself to achieving the goal, before this decade is out, of landing a man on the moon and returning him safely to earth. No single space project in this period will be more exciting or more impressive to mankind or more important for the long-range exploration of space and none will be so difficult or expensive to accomplish . . . . In a very real sense, it will not be one man going to the moon—it will be an entire nation. For all of us must work together to put him there.

If President and people were agreed on the end, what about the means? Kennedy's proposal was not made lightly. Before coming to a decision, he had taken counsel with advisors who believed that the moon project was feasible, largely because it could be accomplished without any new scientific or engineering discoveries. It could be done "within the existing state-of-the-art" by expanding and extending the technology that existed at that time.

What was the "existing state-of-the-art" as of 25 May 1961? Since December 1957, when the first Vanguard orbital launch attempt had collapsed in flame before a television audience, the United States had tried to put 25 other scientific satellites into earth orbit; 10 had been successful. Two meteorological satellites had been placed into orbit, and both had operated properly. Two passive communications satellites had been launched, but only one had achieved orbit. Nine probes had been launched toward the moon; none had hit their target, although three achieved a limited success by returning scientific data during flight. After its 1960 failures, NASA had put a Mercury with Alan B. Shepard aboard into suborbital flight on 5 May 1961.

Just 18 months before the Kennedy recommendation, the Atlas military missile, at that time America's most powerful space booster, had made its first flight of intercontinental range—some 10 000 kilometers. Not three years had gone by since the smaller intermediate range ballistic missiles, Jupiter and Thor, had made their first full-range flights. Yet by May of 1961 none of these military rockets had reached a high degree of reliability as space carriers.

When the President laid his proposed goal before the Congress, the spacecraft that would carry man to the moon existed only as a theoretical concept tentatively named Apollo.

The powerful rocket that would be necessary to launch the spacecraft with sufficient velocity to escape earth's gravity was only a few lines on an engineer's scratch pad. Conceivably, it would be one of a family named Saturn: specially designed space carrier vehicles, each generation larger and of greater power than the preceding one. The first Saturn would not make its maiden flight for another six months.

The vast support, checkout, and launch facilities of the earthbound base whence men would launch other men on their journey did not exist. The moonport had yet to be located, designed, built, and activated—and this book tells that story.

Other books now being prepared for NASA deal with the other aspects of the program—the Saturn launch vehicles, the Apollo spacecraft, astronaut training and the missions. Another volume, a history of NASA administration, 1963–69, will include the headquarters story of Apollo.

The central feature of this book is launch complex 39 (LC-39), where American astronauts were launched toward the moon. Its story begins in early 1961 with the earliest plans for a mobile launch complex and proceeds through design and construction to the launching of Apollo 11 and subsequent lunar missions. The construction story is a big one—the building of the Apollo launch facilities was the largest project of its time. In many ways, however, the operations at LC-39 were an even greater challenge. As an Apollo program manager has noted, the Kennedy Space Center was at the "tail end of the whip." There all the parts of the Apollo program came together for the first time. The launch team ensured that the space vehicle would work.

While LC-39 is the principal focal point, it is not the only one. Two other Apollo-Saturn complexes on Cape Canaveral, LC-34 and LC-37, launched the program's early flights; at LC-34 the program's great tragedy occurred. The Apollo spacecraft were tested in the operations and checkout building in the Merritt Island industrial area. Vital telemetry equipment was located nearby in the central instrumentation facility. Moreover, the size and shape of the launch facilities were largely determined by the Saturn family of launch vehicles, which were produced under the direction of Marshall Space Flight Center at Huntsville, and by the Apollo spacecraft, under the Manned Spacecraft Center at Houston. An understanding of launch facilities and operations requires, to some degree, an appreciation of program-wide activities.

The history is complicated because planning, construction, and launch operations were conducted concurrently during much of the program. Three topics take up most of the first ten chapters: the construction of launch complexes 34 and 37 and the subsequent Saturn I tests; the planning of a moonport on Merritt Island and the purchase of that area; and the buildup of the launch team. Chapters 11–15 relate the design, construction, and activation of launch complex 39. Chapters 16–23 describe the Apollo launch operations from early 1966 through the launch of Apollo 17 in December 1972. Chapter 24 is a tentative summing-up.

The work comprehends three kinds of history: official, contemporary, and technological. The technology of the moonport crossed many scientific and engineering disciplines from microelectronics to civil engineering; expertise was needed in telemetry, fluid mechanics, cryogenics, computers—even lightning strikes. Although NASA engineers gave us a great deal of help, it was our task to make the technical terms comprehensible. Another problem stems from NASA's requirement that its authors use the new international system of units. One obvious way to comply, without losing most of our readers, would have been to give all measurements in both international and old-fashioned units. Unfortunately, with that solution the prose immediately bogs down. We have therefore proceeded as follows. First, where physical units were not essential, we have eliminated them. Second, the more familiar of the international units, such as meters and kilograms, we have used alone. Third, only the more esoteric terms, such as newtons, have we translated in the text.

The contemporary historian's task is to walk into a virgin forest of unsifted materials, with no clearings made by destruction of the unimportant and no trails blazed by prior researchers. Yet the journey can be propitious: we were able to interview hundreds of eyewitnesses who told it as they saw it. They recalled personality conflicts that sometimes affected major decisions. They narrated events never put down in writing and reached into personal files for documents not available in the archives. The use of eyewitnesses naturally required the resolution of some conflicting evidence, and their additional material increased the problems of selection. The insights gained, however, more than compensated for the trouble.

The great weakness of contemporary history, a want of perspective, is irremediable. Until the Russian story is on the record, our view of the space race is limited. Future judgments of the Apollo program will reflect further developments in space exploration. Thus, with respect to the launch facilities, the wisdom of building the moonport in the way it was done depends in part on the programs to be launched henceforth. The moonport was funded, designed, and built on the assumption that the lunar landing was only a beginning. With these considerations in mind, we defer to 21st-century historians a definitive evaluation of the effort.

Under the contract with the University of Florida, NASA enjoyed the rights to final review and publication of this book. We worked largely from NASA documents and with NASA officials. This may have tempered some of our conclusions, consciously or not, but we are satisfied that this is not a court history. Criticisms directed at the Kennedy Space Center (KSC) team and mistakes in the launch operations are treated in detail. Contrary to the wishes of some participants, conflicts within the program are aired. A greater fault may lie in our dependence on NASA documents. Although we tried to balance the account with corporation documents and interviews, the history inevitably focuses on NASA's direction of Apollo launch operations. The Apollo contractors and other support agencies, such as the Air Force, may receive less than their due.

Understandably, our treatment of certain events will not satisfy everyone. For example, too much controversy still surrounds the Apollo-Saturn 204 fire. We have largely avoided two other controversial questions. Was the KSC operation more or less efficient than other governmental projects of the 1960s? There was undoubtedly waste in the construction of the Apollo launch facilities and in the launch operations, but we are not in a position to judge the cost efficiency of the KSC team against similar projects, such as a large defense contract. The second question—the worth of the Apollo program—will be, as previously stated, left to future historians. In our personal view it was a noble goal, nobly achieved.

A word is in order with regard to Kennedy Space Center speech usages, especially acronyms. The scientists and engineers at KSC do not use a peculiar tongue to mystify the layman—but as a matter of fact, that is one result. When an LCC man says "the crawler is bringing the bird back from pad 39 to the VAB," he is understood by anyone at the space port. Every discipline has its technical language, which sometimes goes too far. We believe we reached the nadir in space jargon when we uncovered the record of a "Saturn V Human Engineering Interstage Interaction Splinter Meeting of the Vehicle Mechanical Design Integration Working Group."

Apollo scientists and engineers were establishing a terminology for new things; no one had defined them in the past because such things did not exist. *Module* is an example. As late as 1967, the *Random House Dictionary of the English Language* gave as the fifth definition of *module* under *computer technology:* "A readily interchangeable unit containing electronic components, especially one that may be readily plugged in or detached from a computer system." The space world was well ahead of the dictionary because, as every American television viewer knew, a module—command, service, or lunar—was a unit of the spacecraft that went to the moon. *Interface* is another word that was recast at the space center. Defined in the dictionary as "a surface that lies between two parts of matter or space and forms their common boundary," it grew to encompass any kind of interaction at KSC. Perhaps this was subliminal recognition that Kennedy Space Center was the Great Interface where the many parts and plans that went into the moon launch had to be fitted together.

Like all government agencies since 1950, NASA made extensive use of acronyms. In February 1971, the Documents Department of the Kennedy Space Center Library compiled a selective list of acronyms and abbreviations. It contained more than 9500 entries. We have tried to avoid acronyms as much as possible; when used, the acronym is coupled with its full and formal terminology on its first use.

The astronauts were quick to acknowledge that Apollo was a team effort. Appropriately enough, the same can be said for this history of the Apollo launch operations. We drew extensively upon the work of previous researchers. Dr. James Covington and Mr. James J. Frangie prepared material on the design and construction of the launch facilities. Dr. George Bittle and Mr. John Marshall performed helpful research on launch operations. Mr. William A. Lockyer and Mr. Frank E. Jarrett of the KSC historical office provided much reliable criticism. Dr. David Bushnell, the University of Florida's project director for the history, rendered administrative and editorial assistance. Finally, thanks are due to scores of KSC personnel who provided recollections, documents, and patient explanations on the workings of Apollo.

15

Putting It All Together: LC-39 Site Activation

The Site Activation Board

In 1965 KSC officials prepared to put it all together at LC-39. After two years of construction, and midway through President Kennedy's decade of challenge, Kennedy Space Center approached a milestone known in NASA parlance as "site activation." Two parts of the task were complete: the brick and mortar construction of the facilities, including installation of the utility systems for power, water, heating, and air conditioning; and the electrical and mechanical outfitting such as propellant piping and intercommunications systems. Now came the installation, assembly, and testing of ground support equipment. Earlier chapters have dealt with the first two phases. The third phase in some ways constituted Apollo's greatest challenge. Hundreds of contractors sent nearly 40 000 pieces of ground support equipment to the Cape for installation at LC-39. On Merritt Island, KSC's Apollo Program Office had to integrate the activities of more than two dozen major contractors. Engineering and administrative interfaces numbered in the thousands. At NASA Headquarters, Gen. Samuel Phillips, a veteran of the Minuteman site activation program, and his boss, George Mueller, doubted that KSC would have LC-39 ready in time for Saturn V.[1]

Rocco Petrone took the first step toward site activation in September 1964 by appointing Lt. Col. Donald R. Scheller "Staff Assistant for Activation Planning." Scheller, a B-17 pilot in World War II, had just completed four years with the Atlas Missile Project Office. An October 1964 memo from William Clearman's Saturn V Test and Systems Engineering Office listed the responsibilities of Scheller's new position. He was to analyze:

- Construction schedules of facilities under the cognizance of the Corps of Engineers.
- Delivery schedules of all ground support equipment to be installed on complex 39 regardless of the source of the equipment.

- Tests to be performed on the facilities by contractors prior to release to KSC as well as the tests on utilities, subsystems, and systems to be performed after these facilities are accepted by KSC.
- Tests that are to be performed under the direction of KSC personnel after ground support equipment is installed.[2]

Drawing upon the support of KSC's various design and support elements, Scheller was to develop a work schedule for site activation. His plans would become the management tools to accomplish the task efficiently. After KSC began implementing the site activation plans, Scheller would prepare facilities description documents for LC-39.

Although Clearman was projecting no mean task, he underestimated the job. Scheller took several months to review the situation before organizing a Site Activation Board in March 1965. At the board's first meeting, he outlined his plans to 40 NASA and contractor representatives. The Site Activation Board, under the aegis of the Apollo Program Office, would work at the management level of KSC and the stage contractors; subordinate groups would handle daily site activation problems. The board was not to usurp other organizations' responsibilities.

Scheller's subordinates presented a performance evaluation and review technique (PERT) for the LC-39 activation. PERT schedules would provide three levels of control. At the A level, PERT would focus on the major control milestones for the Saturn V program, e.g., the first facility checkout of the Saturn V test vehicle, SA-500-F. B networks would track each major element required to support the key milestones, e.g., firing room 1 for 500-F. Level C networks, providing a further breakdown of B-level networks, would follow the progress of all subsystems within each major facility, e.g., the propellants loading panel in firing room 1. The Schedules Office, supporting the board, would maintain the A and B levels, while NASA line organizations and stage contractors prepared the C networks.[3]

The PERT networks brought order to LC-39 site activation. PERT defined each task, performer, and deadline in a descending and expanding level of detail. The top or A network also served as the site activation master schedule, establishing major milestones. This master schedule, prepared by Scheller's office, was divided into segments or "flows." Flow 1 charted the activation of the minimum facilities and equipment necessary for the checkout and launch of the first Apollo–Saturn V vehicle, AS-501. A preliminary objective was the arrival and erection of the facilities checkout vehicle, SA-500-F, which would be used for testing and validation of launch facilities and operating procedures. Flow 1 listed as minimum facility requirements:

- Mobile launcher 1
- Crawler-transporter 1

- High bay 1 in the assembly building
- Firing room 1 in the launch control center
- Mobile service structure
- Launch pad A
- Propellants and high-pressure gas facilities
- Related mechanical equipment, electrical-electronic support equipment, and other ground support equipment.

Flow 2 charted the activation of additional facilities to support AS-502, including launcher 2, high bay 3, firing room 2, and related ground support equipment. Flow 3, originally intended for AS-503, tracked the activation of pad B and related facilities such as the crawlerway. A fourth flow covered the remaining LC-39 facilities.[4]

B and C networks supported each of the flows. The B networks eventually listed over 7400 events, e.g., completion dates for equipment installation. These events covered all facilities and some major components within the facilities. The C networks, largely a contractor responsibility, listed 40 000 activities in sequence and set the dates by which one contractor would have to complete a job to make way for the next operation. A numbering system facilitated the transposition of data between C and B levels, matching as many as 15 C-level activities with their B counterparts.

Following the PERT description, the Site Activation Board discussed a second management tool, the equipment records system. NASA was compiling in Huntsville a list of 40 000 pieces of ground support equipment, the data coming from the engineering divisions of the three spaceflight centers. The lists provided: a name and number; an estimated-on-dock date, the expected delivery date at KSC; and a required-on-dock date, when KSC needed the item for installation. The board intended to use the equipment records system as the communications medium between KSC users and equipment suppliers. Representatives of the Facilities Engineering and Construction Division reported on the current status of key construction milestones, including rack and console installation. All agencies were asked to review the construction status in terms of their organizational needs for access, and to report any problems to the Site Activation Board. Scheller requested comments on the board charter and PERT networks within a week and an early submission of level C data.[5]

Under its charter, the board was responsible for ensuring that all facilities and support equipment comprising the Apollo–Saturn V operational launch base were "constructed, outfitted, installed, interconnected, and tested" in preparation for subsequent operations. This included equipment modifications during site activation. KSC division chiefs quickly expressed concern about these broad powers. In his comments Dr. Hans Gruene asked:

"Will decisions of the Board be made at the discretion of the chairman or by some other method? . . . Can a decision be appealed by the director of an operating unit [such as the Assistant Director for Launch Vehicle Operations] to the Director of Plans, Programs, and Resources [Petrone]?"[6] Col. Aldo Bagnulo, acting Assistant Director for Engineering and Development, not wanting the board to assume any of his responsibilities for facility development, said: "The recent emphasis on performing work through normal procedures rather than by committee action should be followed."[7] Raymond Clark, Support Operations Chief, raised the same sensitive issue: "Additional clarification is needed as to the depth of management control anticipated by the Site Activation Board."[8] Despite these objections, Petrone had his way; when the board began operations, it enjoyed a wide-ranging authority.

The preparation of PERT schedules monopolized the board's attention for several months. The fashioning of the detailed C-level networks proved time-consuming, and when submitted, data from the contractors forced revisions in the B networks. KSC also had trouble bringing the equipment records system under its control. By August 1965, however, both that system and PERT were computerized and operational. Early that month Scheller initiated biweekly board meetings.[9] In early October the board moved to new quarters in firing room 4 of the launch control center. Even desks and telephones were in short supply, but KSC got on with the installations, and the following month the board had work space, conference areas, and a management information display and analysis room. The display brought home the immensity of the board's task. Magnetic devices on a 21 × 5-meter metal wall revealed the status of each PERT chart and told the story of site activation. Tiered seats accommodated 90 people with standing room for another 50. Four rear-projection screens, above the metal wall, provided simultaneous or selective viewing of activation data. Apollo officials could view level A networks, milestone event charts, and major problem summaries on two rear-lighted display areas located to the right and left of the four screens. As activation moved into high gear, the display room was used to brief visiting dignitaries on program goals and progress. Level C networks had an area of their own behind the huge display wall. The Site Activation Board laid out the 40000 events of the C networks on 418 square meters of metal wall space. A Boeing team, responsible for updating the network, worked at nearby desks. Offices and a graphics section occupied the rear section of the firing room.[10]

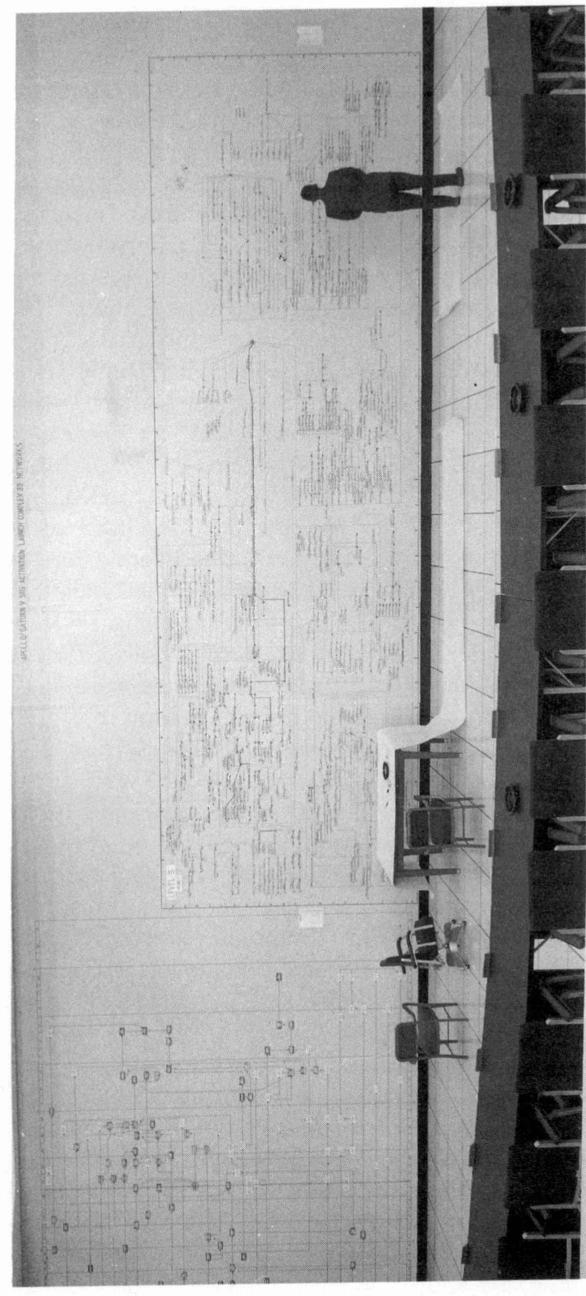

Fig. 105. Site activation schedules displayed in firing room 4 of the launch control center, January 1966.

Site Activation Working Groups

During its first months, the Site Activation Board functioned more as a working group than a management team, but in September Scheller activated several subordinate groups. Lt. Col. Richard C. Hall took command of the Site Activation Working Group, set up to resolve technical interface problems and devise methods of accomplishing new requirements within a given facility. At its first meeting in December, Hall introduced a typical problem. The Communications Service Branch had not received telephone requirements for firing room 1 from the operating organizations. Hall asked the group to submit all requirements at least 60 days prior to the "need date" (the date on which an item was required). The following month Hall's group assumed formal responsibility for daily site activation matters.[11]

James Fulton, Launch Vehicle Branch chief in Clearman's office, recruited Donald Simmons to handle LC-39's electrical cable problems. Simmons's experience on Atlas served him well as the first chief of the Cable Working Group. The group's mission in September 1965 involved preparation of a cable accounting system, the Site Activation Board's third essential management tool. This tracking system kept tally on more than 60000 cables including all connectors by part number, the length of cable, cable makeup, procurement action and date, the agency furnishing the cable, the need date as assessed from the PERT schedules, "from and to" locations, and the installation contractor. Communication and instrumentation from the launch control center to pad B alone required nearly 160 kilometers of cable. KSC let a $2 million contract for the job in October 1965; the work included the installation of 142 kilometers of coaxial, video, telephone, and instrumentation cables plus terminal equipment. The group managed network configuration through computer printouts and network diagrams, with a General Electric team in Huntsville preparing the cable interconnect drawings.[12]

A contractor and two working groups played important logistical roles in site activation. The Boeing Company, primary integration contractor on the Minuteman program, gathered, processed, and reported data for the Site Activation Board. While much of Boeing's effort involved the PERT schedules, its management systems staff at the Cape effected major improvements in the equipment records system. During the fall of 1965, few engineers relied on the system. When someone needed information about ground support equipment, he normally went to the designer. In early December the equipment records system lacked nearly 33% of its essential data; 79% of the support equipment did not correlate with a PERT activity. Boeing initiated a four-month search and classification program that reduced the respective figures to 5% and 7% and made the equipment records system an effective tool.[13]

The Equipment Tracking Group benefited from the resulting improvements. This group resolved differences between the estimated-on-dock and required-on-dock dates and tracked all items until installation and final testing. The group reflected Colonel Scheller's belief in management by exception, concentrating on items that failed to meet schedule dates or arrived in the wrong configuration. When this happened, team members scrambled to devise acceptable "work-around" measures.[14]

Interface Control Documentation

Interface control documentation, an essential activity during site activation, was another responsibility of KSC's Apollo Program Office. Apollo configuration control dated from February 1963, when the manned spaceflight centers had agreed to consolidate and store interface control documents. During the next several years, Apollo-Saturn subpanels placed hundreds of such documents in a Huntsville repository. Through the interface control documents, Apollo managers made sure that thousands of items, built in many different places, would fit and work together. The documents provided design requirements and criteria for hardware and software interfaces, describing the parameters and constraints under which the interfacing items functioned. The information in the documents varied and might include physical and functional design details and operational and procedural requirements. Where an interface involved two NASA centers, a level A document applied—for example, the interface between a command module (Houston responsibility) and the mobile service structure (KSC responsibility). Level B documents pertained to intra-center interfaces such as the S-IVB–Instrument Unit interface covered by Marshall's Saturn ICD, 13M06307 (October 1965). When changes affected performance, cost, or schedule accomplishment, the centers prepared interface revision notes.[15]

Although the Panel Review Board (established in August 1963) gave NASA Headquarters limited control over configuration decisions, General Phillips provided the centers with detailed directions in his May 1964 *Apollo Configuration Management Manual*. The manual, patterned after Air Force procedures, included a requirement for Configuration Control Boards at each center. KSC had difficulty fitting Phillips's management scheme onto a program already under way. In September 1965 however, Petrone announced plans to implement it. Maj. Andrew Reis's Configuration Management Office would "interpret the requirements of [the manual] and define the degree of flexibility necessary to integrate KSC operations consistent with the requirements of Configuration Management."[16] Petrone's directive also established a series of Configuration Control Boards, or change boards as they were usually called. Edward Mathews chaired the Saturn IB board; William Clearman, the Saturn V board; and Hugh McCoy, the spacecraft board.[17]

Apollo-Saturn subpanels continued to prepare interface control documents and notes. When inter-center panel representatives reached technical agreement on an interface requirement, the proposal would go to an appropriate change board. The board would circulate a "request for impact" through KSC to ensure that the proposed document had no adverse impact on any center function. Other details solicited by the change board included the cost of modifications and the "need dates" of operations and maintenance groups. The Configuration Management Office served as a secretariat for the change boards. When a proposal proved acceptable, the board would notify the other centers to implement the document.[18]

Since unapproved interface control documents left open the possibility of an unsatisfactory interface, program offices made strenuous efforts to coordinate their work. Nevertheless a backlog of "open" documents had developed by 1968 that gave NASA officials much concern. A Boeing investigation in May 1968 found two weaknesses in KSC's program: the documents contained extraneous material that made inter-center coordination difficult, and the complicated processing wasted time. KSC's program office overhauled its procedures during the next six months and closed out all control documents before the first manned launch of an Apollo-Saturn in October 1968.[19]

500-F—A Dress Rehearsal

The Site Activation Board focused its attention in the fall of 1965 on the 500-F test—a dress rehearsal for the new Saturn V rocket and launch complex 39. Plans for the 500-F test vehicle dated back to early 1962 when LOD engineers were still studying the use of barges to move the giant Saturn V to its launch pad. As the facilities checkout vehicle, 500-F would test the mating of the stages in the assembly building, the fit of the service platforms, the launcher-transporter operation, the propellant loading system, and the test connections to the mobile launcher and support equipment. Each dummy stage would duplicate the flight configuration, ordnance, and umbilical connections of its live counterpart. Although inert, the retrograde rockets, ullage rockets, and shaped charges would have the dimensions of the live ordnance. This allowed the launch team to practice ordnance installation. Facility checkout would culminate with a "wet test" to verify the storage and transfer of the propellants. The wet test would involve hundreds of components: pneumatic valves, liquid sensors, time delay relays, pressure switches, circuit breakers, pumps, motors, fans, vaporizers, vents, and the burn pond. The launch team scheduled the delivery of the 500-F stages at the Cape nine months before the first Saturn V flight. The Office of Manned Space Flight translated this into a tentative July 1965 test date.[20]

This was not to be. When George Mueller revised the Apollo schedule in November 1963, erection of the SA 500-F stages on the mobile launcher slipped back to 1 February 1966. Marshall would deliver an S-IVB-F stage (used in the Saturn IB checkout of LC-34) in May 1965, and the S-IC-F and S-II-F stages in January 1966. General Phillips announced the Apollo launch schedule in February 1965, as follows:

January 1966:	AS-201, first Saturn IB launch, from LC-34
February:	Start of 500-F test, checkout of LC-39
October:	AS-204, first manned Saturn IB, from LC-34
January 1967:	AS-501, first Saturn V launch, from LC-39

Although planners were dubious about meeting the AS-501 launch date, two more Saturn V launches were scheduled for 1967. From the start of the 500-F test, KSC would have nearly a year to prepare for the first Saturn V launch.[21]

There was little margin for error. In December 1964, Dr. Arthur Rudolph, Saturn V Program Manager in Huntsville, asked KSC to agree to a delay in delivery dates for the 500-F and AS-501 launch vehicles. After reviewing the schedules for equipment installation and checkout, the 500-F test, and AS-501, Petrone replied that there was no room for delay. KSC had already eliminated the detailed receiving inspection for the 500-F and 501 vehicles. Although Marshall's contract with Douglas omitted the digital data acquisition system test for 500-F propellant loading, KSC would not waive this check. The schedule did include several weeks of learning time, primarily in crawler operations with a space vehicle aboard the mobile launcher. Petrone, however, considered the 500-F schedule "optimistic since it does not allow time for resolution of major difficulties which may occur."[22]

The Crawler-Transporters Begin to Crawl

Events were to reveal a little slack in the LC-39 activation schedule, just enough to recover from a near disaster. The crawler was the prima donna of the Site Activation Board drama of 1965. This gargantuan tractor, designed to carry the 36-story Apollo–Saturn V space vehicle from the vehicle assembly building to the launch pad, caught the public eye; no other facility, excepting the assembly building, got like publicity. Perhaps on account of the public interest, the crawler engendered a series of labor and political disputes, as well as mechanical problems, that nearly disrupted the site activation schedule.

The Marion Power Shovel Company built the two crawlers in Ohio and then took them apart for shipment to the Cape. Under its contract, Marion intended to reassemble the crawlers on Merritt Island with an Ohio work crew, members of the AFL–CIO United Steelworkers. The Brevard (County) Florida Building and Construction Trades Council, citing the Davis-Bacon Act, insisted that on-site construction fell under its jurisdiction. The local unions won a Department of Labor decision in August 1964, but agreed to a compromise that let the Marion crew remain on the job. Although the labor dispute simmered throughout the winter, W. J. Usery and the Missile Site Labor Commission managed to avert a major shutdown. On the basis of the labor difficulties, Marion won a delay in the crawler testing date from November 1964 to late January 1965.[23]

The crawler moved under its own power for the first time on 23 January. NASA officials observed that "the initial crawler-transporter was not in a state of complete assembly ready for joint testing" and forwarded a list of deficiencies to Marion.[24] Additional runs in April tested the propulsion and steering systems. On the 28th Gunther Lehman of Marion drove the crawler about 900 meters at a speed of 1.1 kilometers per hour; this was a "press day" ride with Debus, Petrone, and other KSC and Marion Power executives aboard. The hydraulic jacking and leveling system was ready for testing on 22 June when the crawler picked up its first load, a mobile launcher. Although the test was labeled a success, the launch team noted high hydraulic pressures when the crawler trucks scuffed on the crawlerway during turns. The treads also chewed up large portions of the macadam surface.[25]

For Want of a Bearing

On 24 July the crawler moved a launch umbilical tower about 1.6 kilometers to test the crawler on two short stretches of road, one surfaced with washed gravel ("Alabama River rock") and the other with crushed granite. Preliminary data on steering forces, acceleration, vibration, and strain pointed to the gravel as the better surface. While the crawler was making its run, members of the launch team found pieces of bronze and steel on the crawlerway—the significance of which was not immediately recognized. The transporter was left out on the crawlerway over the weekend because of problems with the steering hydraulic system. On the 27th more metal fragments were discovered and a thorough search disclosed pieces of bearing races, rollers, and retainers from the crawler's traction-support roller assembly. After the transporter was returned to its parking site, a check of the roller assemblies revealed that 14 of the 176 tapered roller bearings were damaged. KSC engineers attributed the failure primarily to thrust loads encountered during steering; the anti-friction support bearings, about the size of a can of orange juice concentrate, were underdesigned for loads exerted during turns. For want of a bearing, the crawler was grounded indefinitely. And for want of a crawler the site activation schedule and the entire Apollo program would be seriously delayed.[26]

A reexamination of Marion's design calculation indicated some other significant facts. The designers had assumed an equal load distribution on all traction support rollers; perfect thrust distribution over the entire bearing, i.e., an axial thrust equivalent to the radial load; and a coefficient of sliding friction of 0.4 (meaning it would take four million pounds of force to move a ten-million-pound object). During the early crawler runs, KSC engineers discovered an unequal load distribution on the traction support rollers. At times as many as four of the eleven rollers on one truck were bearing no load. The thrust, or side load, proved greater than expected. Finally, the crawler tests revealed that the estimated coefficient of sliding friction was far below the actual resistance experienced on the crawlerway. At a crawlerway conference on 27 June 1963, NASA engineers had insisted on a minimum design coefficient of 0.6. In the first runs on the crawlerway's macadam surface, the coefficient reached nearly 1.0.[27]

Troubles with the crawler had not been unforeseen. Prior to the roller bearing crisis, M. E. Haworth, Jr., chief of the KSC Procurement Division, upbraided Marion for making difficulties about the tests:

> KSC has tolerated innumerable delays in the assembly, tests and checkout operations of CT-1. These delays are to the definite detriment of Apollo facilities readiness and Marion's position as to the testing operations, will, if carried out, likely cause even further delays which will have a definite and substantial dollar impact on other projects directly and indirectly connected to the crawler transporter concept. The failure of Marion to fulfill its delivery obligations is in itself costing the government substantial sums which were not contemplated.[28]

On 14 October 1965 Haworth wrote Marion, expressing grave concern over the inactivity at the erection site consequent on a new labor dispute (the unions stayed off the job for nearly six weeks). The roof fell in on both NASA and Marion when the bearing story reached the press and television. Walter Cronkite told his evening newscast audience that the crawler was sitting on wooden blocks under the hot Florida sun, with a top Washington official stating privately that it might never work. The press and Cronkite revived the controversy over the award of the contract to Marion. Politics, they hinted, was involved; and in any case the low-bid procedure might prove penny wise and pound foolish.[29]

NASA and Marion could answer that the design and construction of a land vehicle expected to carry 8000 metric tons was without precedent. Its very size, as the Corps of Engineers had pointed out, ruled out preconstruction tests of the coefficient of friction in its moving components. A more pertinent answer was to develop a new bearing, a hydraulically lubricated sleeve bearing made of Bearium B-10. KSC selected the bronze alloy after testing a half-dozen materials at Huntsville. The new design provided separate bearings for axial thrust and radial loads. KSC retained in the design the original supporting shafts that housed the bearings. Although the sleeve bearings would not reduce the amount of friction, they would eliminate the possibility of a sudden, catastrophic failure. Periodic inspection could determine the rate of wear and need for replacement. The disadvantages of the sleeve bearings—lubrication difficulties, the inability to predetermine useful life, and a need for more propulsive power because of increased friction—were acceptable. Fortunately, while the crawler design had underestimated friction, there was a considerable reserve of power. At KSC and Marion, engineers designed a new bearing system. A parallel effort modified the crawler's steering hydraulic system, almost doubling the operating pressure. At KSC, the burden of the bearing crisis fell principally on Donald Buchanan's shoulders. In Marion, Ohio, Phillip Koehring directed the redesign.[30]

Marion reinstalled the support roller shafts in early December. A prototype of the sleeve bearing arrived on the 14th. After cooling it in dry ice and alcohol, the assembly crew placed the bearing in its housing. The fit proved satisfactory, and the remaining bearings were installed by mid-January. On 28 January 1966, the crawler transported a mobile launcher approximately 1.6 kilometers to the assembly building. Bearing measurements indicated an acceptable heat factor. Fortunately, KSC had initiated the crawler contract early enough to allow for both labor disputes and redesign of the bearing.[31]

"Negative Slack" in "Critical Paths"

The nerve center for site activation lay in John Potate's scheduling office. Potate, a young engineer from Georgia Tech, had previously worked on site activation for LC-34 and LC-37 and brought that experience to his new

The bearing problem

Figure 106

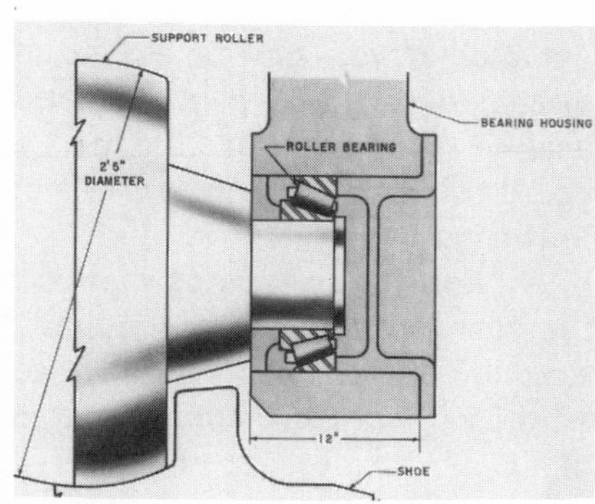

Figure 107

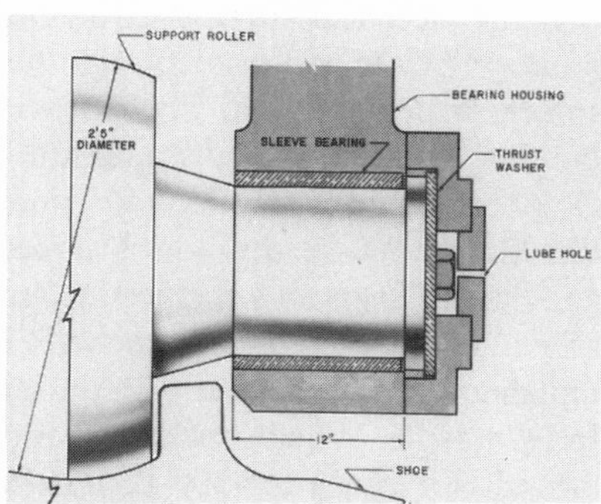

Figure 108

Fig. 106. Crawler carrying a mobile launcher 24 July 1965, the day the bearing trouble was discovered. Fig. 107. The original design, which failed in early tests of the crawler. Fig. 108. The sleeve bearing, which solved the problem.

and much bigger assignment. Supported by a Boeing team, Potate put the B PERT network to use. The scheduling office had little control over the A networks; the Apollo Program Offices at NASA Headquarters and KSC set the key milestone dates. The line organization level-C networks provided valuable data, but were too detailed for quick program evaluation. KSC officials, in large part, based their program decisions on the B networks—the level where Potate reconciled C-level capabilities with A-level deadlines. Potate relied heavily on PERT to identify problems. A computer, after processing all available network data, printed out "critical paths," which traced the controlling chain of events leading to a goal. For each critical path, the computer sheet also indicated the "negative slack." That curious term indicated how far a facility's development lagged behind its readiness date; or, put another way, how far behind schedule the Site Activation Board was in meeting a particular goal. The critical paths, consolidated into the PERT analysis reports, became the focal point of board meetings. Potate and the board examined the negative slack in each critical path and searched for ways to eliminate it.[32]

Some critical paths showed more than a year's negative slack when the board started work in August 1965. One involved the mobile service structure, under construction for only six months and nearly a year behind schedule. Since there would be no spacecraft tests during the initial checkout of LC-39, the absence of the service structure would not affect the 500-F schedule. If the service structure continued to lag, however, it would delay the AS-501 mission, the first Saturn V launch. At the board meeting on 5 August, the engineering directorate reported new efforts to speed up development of the service structure. A Marshall representative acknowledged that the electrical support equipment was a week behind schedule. At the moment LC-34 had priority, but the LC-39 electrical equipment dates would somehow be improved. The Corps of Engineers disclosed that the assembly building's first high bay would not be ready for use by its scheduled date, 1 October, without accelerated funding. Since the October date impinged directly on 500-F, the board agreed to spur construction. In one piece of good news, Bagnulo's engineering representative announced a "work-around" schedule for the mobile launcher's swing arms. Hayes International of Birmingham would proceed with the late delivery of the first set of service arms to Huntsville. When checkout there had been completed, the arms would be moved to Florida for 500-F. Where there was insufficient time for testing, Hayes would ship the corresponding arms from the second set to KSC. After the latter had satisfied 500-F needs, KSC would exchange them with Marshall for the first set, which would have been tested by then.[33]

Although KSC managed to occupy portions of the first high bay (floors 1–7) on 1 October, the status of other facilities continued unsatisfactory. The construction firm of Morrison-Knudson-Hardeman expected to complete structural steel work on the mobile service structure in late November. This would leave eight months for the installation of ground support equipment, spacecraft piping, and instrumentation and communication cables. At the board meeting on 28 October, John Potate reported 40 weeks of negative slack in the service structure. KSC could eliminate 75% of the lag by performing some necessary modifications during the installation phase. The remainder involved the installation and testing of spacecraft checkout equipment. Potate asked the North American representative to determine

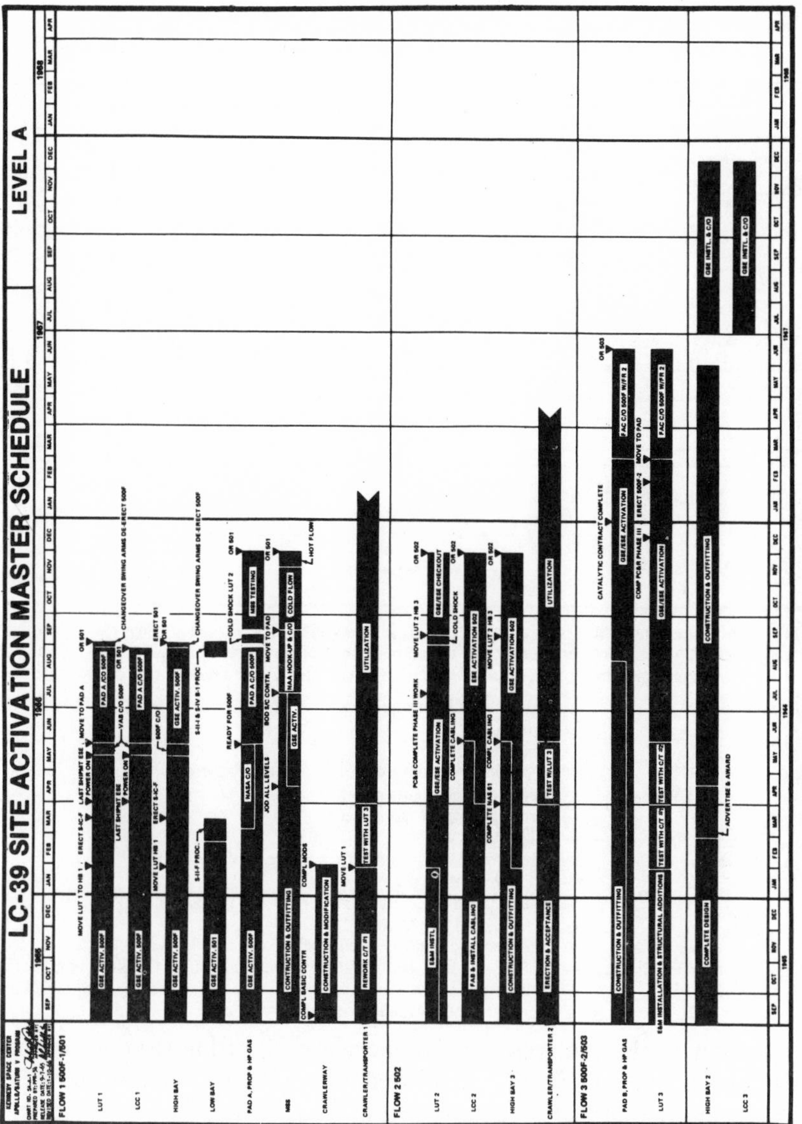

Fig. 109. LC-39 site activation master schedule, level A, 10 January 1966. The level B chart for approximately the same date is shown in fig. 105.

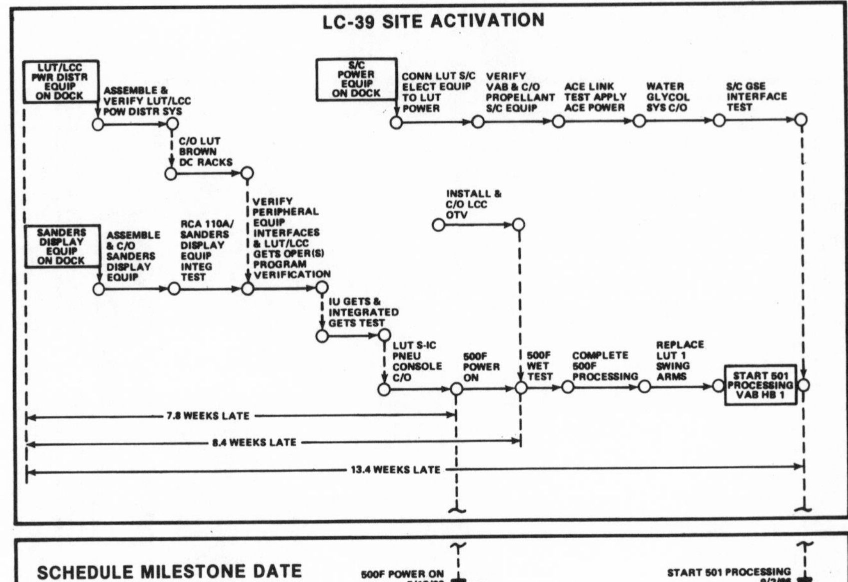

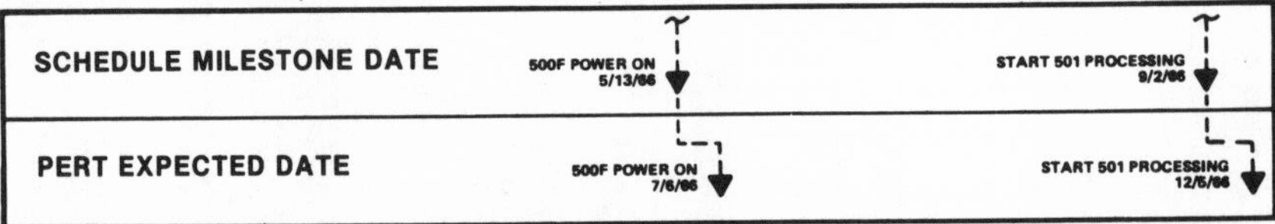

Fig. 110. Critical path summary for site activation, 20 January 1966. The sequence of events in the upper part of the chart is distilled to answer the key question on the bottom line: When will LC-39 be ready to use? Table 6 summarizes the slippage in a number of events related to site activation as of the same date.

Legend:
ACE — Apollo checkout equipment
C/O — Checkout
DC — Display console
500F — Facilities checkout vehicle, a dummy Saturn V
GETS — Ground equipment test set
GSE — Ground support equipment
HB — High bay
IU — Instrument unit, a major component of the Saturn
LCC — Launch control center
LUT — Launcher umbilical tower, part of the mobile launcher
OTV — Operational television
RCA 110A — The major computer used in checkout and launch
Sanders display — A firing room system that displayed data from the LCC computers
S/C — Spacecraft
VAB — Vertical (later, vehicle) assembly building

what portion of his requirements could be accomplished at the erection site rather than on the pad. Scheller appointed a NASA group to consider ways of shortening the cold flow and hot flow tests. In the former, engineers validated the spacecraft hypergolic systems with nontoxic freon, testing pumps, umbilical lines, and pressure valves; the latter test employed the toxic hypergolics used in flight.[34]

Huntsville's electronic support equipment continued to lag behind schedule, and in early November Marshall announced that it would miss the 7 February 1966 deadline for final deliveries by three months. Petrone protested to General Phillips:

> This results in a completely erroneous representation of activation constraints and precludes accurate status assessment and realistic planning. In order to use the KSC LC-39 PERT system effectively, and obtain credibility in the eyes of the users, I must reflect true status of MSFC [electronic support equipment] into the networks.[35]

He asked to revise the PERT charts in line with MSFC's May date. Although 500-F erection did not require the electrical support equipment, subsequent tests would use it. Petrone sought a new erection date of 15 April 1966.

Phillips's response set a deadline of 15 April for the delivery of the electrical support equipment and the erection of the test vehicle. On 7 December Petrone explained to Phillips how the new stacking date would affect operations. Of the nine swing arms, only seven would be installed by 15 April. While Marshall would have qualified only three of the seven, KSC would use substitutes from the second set. The launch team would install the service module and command module arms after 500-F erection and before its transfer to pad A. Petrone noted that the mid-April date allowed sufficient time to accomplish modifications to the assembly building and resolve mobile launcher–vehicle assembly building interface problems. The new date also placed some constraints on KSC. Assuming arrival of the AS-501 vehicle stages in early September 1966, the 500-F tests would require six-day, two-shift operations to prevent overlap with launch preparations for the first Saturn V. Even with the accelerated schedule, KSC faced the unpleasant task of installing qualified swing arms concurrently with AS-501 erection. Phillips accepted the mid-April schedule with the understanding that KSC would try to advance the date.[36]

While Petrone's office shuffled dates, rumors of another problem disturbed KSC leaders. Workers on pad A believed the foundation was sinking. The charge was serious; excessive settling might damage pipes that serviced the pad from nearby facilities. At three successive meetings Colonel Scheller pressed Steven Harris, the Engineering Division's site activation chief, for a detailed status report. In mid-November 1965, Colonel Bagnulo responded with reports to Debus and Petrone. There was minor, nonuniform settlement at the pad, but this lay within tolerance. After the Gahagan Dredging Company had removed some 24 meters of surcharge from the pad area in mid-1963, the soil had risen about 10 centimeters. This rebound, which represented the soil's elastic action, was expected to be the limit of any resettlement after pad construction. Measurements in July 1964, after pouring the 3.4 meters of concrete mat, indicated a maximum settlement of 9 centimeters. Settlement at the sides of the pad varied as much as a half centimeter. As recent measurements by the Corps of Engineers showed little change, the Engineering Division concluded that the launch pad was attaining a stable condition.[37]

Petrone's office advanced the activation schedule early in the new year as Marshall accelerated the flow of electrical support equipment to LC-39. At the 6 January meeting of the Site Activation Board, Scheller asked all organizations to consider the feasibility of moving the mobile launcher into high bay 1 on 28 January and beginning 500-F erection on 15 March. The discussion pointed up some confusion on the date for mating the crawler and mobile service structure. While the service structure contract indicated mid-July, the Launch Vehicle Operations Division preferred to delay it until completion of the 500-F tests on 1 September. Scheller indicated that the board would resolve the matter after consulting with Launch Vehicle Operations and Spacecraft Operations.[38]

The crawler brought mobile launcher 1 into the assembly building on 28 January, meeting the first major milestone of flow 1 (table 6). In the four months since the occupation of high bay 1, the stage contractors had outfitted their respective service platforms and the adjacent rooms.[39]

The Apollo team had no time to celebrate its accomplishment; PERT analysis showed considerable negative slack toward the next major milestones. Potential delays existed in the delivery and installation of service arms and electronic equipment from Huntsville, the installation and checkout of the operational TV system in the launch control center, and the delivery of spacecraft checkout equipment for the mobile launcher and mobile service structure. At the Site Activation Board meeting on 3 February, Potate noted

TABLE 6. SLIPPAGES IN LC-39 SITE ACTIVATION, 20 JANUARY 1966

Event	Flow 1 milestones (Apollo Program Office)	PERT dates (PERT analysis report)
Move LUT 1 to VAB	28 Jan. 1966	28 Jan. 1966
Start 500-F erection	15 March	8 April
500-F power turned on	13 May	6 July
500-F roll out from VAB to pad A	26 May	20 July
Pad A wet test	24 June	23 Aug.
501 operationally ready	2 Sept.	5 Dec.
501 at pad A	1 Dec.	6 March 1967

that PERT dates for erecting the 500-F were three weeks behind the scheduled milestone; the PERT dates for power-on and the pad A wet test lagged eight weeks. The 500-F delays in turn set back the ready dates for the first Saturn V launch. Although North American Aviation expected to be at least 13 weeks late in delivering the first S-II stage, General Phillips's office continued to list the AS-501 mission in January 1967. If the site activation team did not want the blame for holding up that launch, the PERT dates would have to be improved.[40]

Potate's review did not include KSC's newest emergency. When a subcontractor declared bankruptcy in December, American Machine and Foundry Company found itself without cables for its tail service masts. The delivery date for the masts was only four months away and the company still faced difficulties with hood fabrication, tube bending and flaring, and cleaning facilities for the mast lines. Furthermore, contract losses were rumored to exceed $500 000. At a meeting on 1 February 1966, American Machine and Foundry refused to accept new delivery dates since it had outstanding time and cost claims against 24 NASA change orders. KSC responded quickly, removing the cleaning requirement from the contract and dispatching technicians to the York, Pennsylvania, plant. In mid-February Bendix accepted responsibility for completing the cables. A NASA report from the York plant, however, painted a bleak picture. Contract losses on the tail service masts were disheartening. A recent Navy bomb contract offered large profits and the company, understandably, was concentrating on this project. According to the NASA observers, the company was also hiding behind the cable problem "assuming that a subcontractor would require weeks to produce them so they did not proceed with the production, cleaning, and assembly of tubing and components."[41]

KSC's concern apparently impressed American Machine and Foundry management, because the company voluntarily changed to two 12-hour shifts and a seven-day work week. In return NASA reconsidered American Machine's claim of additional costs. Despite the settlement, it seemed unlikely that the York plant could deliver the masts in time for 500-F's power-on date in May. KSC improved the likelihood through several work-around agreements. The center installed certain equipment at Merritt Island rather than at Huntsville, postponed line cleaning until after the power-on exercise on 13 May, and deleted the installation of vehicle electrical cables from the American Machine and Foundry contract. Gruene's group accomplished the latter task at KSC after the tail service masts had been mated to the mobile launcher. The shortcuts allowed the launch team to install the three masts by mid-April.[42]

Six service arms for the Saturn arrived by mid-March. In the transfer aisle of the assembly building, Pacific Crane and Rigging crews mounted two hinges to each swing arm. The hinges, 1.2 meters high and 2.1 meters wide, required careful alignment so that the arms would hang and retract properly on the mobile launcher. When this task had been completed, the 250-ton crane lifted each arm into high bay 1 where an eight-man rigging team secured the appendage to the launcher. The operation was expected to take 16 hours, but late deliveries forced the riggers to speed up their work. On 12 March, three days before the first stage was erected, they hung arms 1, 2, and 5. Mounting the sixth arm, 61 meters above the mobile launcher base, was a particularly impressive job, with riggers leaning from the top of the hinge to secure bolts in the tower. After the arms were hung, Pacific Crane ironworkers (craft unionists) routed the umbilical lines from the tower consoles to the swing-arm interface plate. Pipefitters and electricians (industrial unionists), employed by the stage contractors, then took over, routing the umbilical lines along the service arm to the launch vehicle.[43]

The rigging teams, supervised by KSC's Richard Hahn, faced another difficult task when the last swing arm for the Saturn arrived from Birmingham on 15 April. Four days later KSC mounted the arm, carefully working it into the narrow space between the tower, the 500-F vehicle, and the two

Fig. 111. Service arm 9 (the top one) on the floor of the assembly building, being prepared for mounting, May 1966. The hinge end is in the foreground. Service arms 1 through 7 supported the Saturn launch vehicle; arm 8 serviced the spacecraft while arm 9 provided access to the command module.

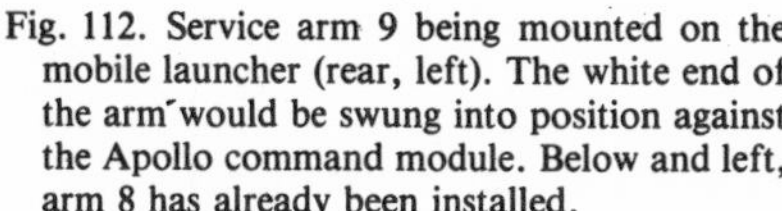

Fig. 112. Service arm 9 being mounted on the mobile launcher (rear, left). The white end of the arm would be swung into position against the Apollo command module. Below and left, arm 8 has already been installed.

adjacent arms. The late delivery threatened to delay the start of the power-on exercise, a little more than three weeks away. Potate solved the problem by rescheduling Boeing and North American activities in conjunction with the swing arm work of Pacific Crane and Rigging. In a rare spirit of cooperation, industrial union and craft union members labored alongside nonunion workers and civil servants to eliminate two weeks of negative slack. The last swing arm was ready on 8 May.[44]

500-F Up and Out

KSC passed its second major hurdle in March, erecting the 500-F launch vehicle in high bay 1. Crane operators began practice runs in February, using a 9.5-meter spherical water container. Stanley Smith, Bendix senior engineer for the crane and hoist group, simulated the different weights of the Saturn stages by varying the amount of water. On 15 March the 250-ton crane lifted the 500-F first stage from the transfer aisle to a vertical attitude and up 59 meters. After moving the S-IC-F stage through the opening in the bay trusswork, the crane operators lowered it gently to the platform of the mobile launcher. The second stage, S-II-F, followed the same route on the 25th, when it was mated with the first stage. The third stage and instrumentation unit joined the stack before the end of the month. Concurrent exercises out on pad A tested the interfaces between mobile launcher 3 and the pad. The final test pumped 1 135 000 liters of water through the deluge system on the launcher.[45]

The site activation schedule allowed three weeks in April for GETS (Ground Equipment Test Set) tests. The exercises, a long standing Marshall policy, verified Saturn V ground support equipment before its initial hookup with the launch vehicle, in this case the 500-F test vehicle. The checkout also acquainted stage-contractor crews with the equipment. Problems with the Brown Discrete Control System and the Sanders Saturn V Operational Display System threatened to delay the start of the GETS tests on 8 April. The Brown equipment, located in the launch control center, controlled the flow of signals into and out of the RCA 110 computer in the launch control center. The Sanders system in the firing room ran a series of consoles that displayed data from the control center computers for operational use. KSC technicians had the display systems working by 1 April.

The first week of GETS tests featured power and network checks of all electrical support equipment. The following week's tests of ground equipment included measuring and radio frequency, pneumatic systems, propellants, and the emergency detection system. The third week KSC conducted tests of the digital data acquisition system, camera control, and leak detection and purge panels. Although the contractors ran two-shift operations, shortcomings, particularly with the RCA 110 computer, forced a fourth week of individual stage tests. Integrated GETS tests followed during the week of 2 May.[46]

In mid-April the Site Activation Board added another to its list of major problems: the installation of electrical and pneumatic lines on mobile launcher 1. Late deliveries and last-minute modifications from Huntsville and Houston threw the Pacific Crane and Rigging crews behind schedule. Despite 14-hour shifts, seven days a week, the crews made little headway. A "tiger team," composed of members from KSC's Engineering Division, the stage contractors, and Pacific Crane and Rigging, supervised the rush work. As the power-on date of 13 May approached, pipefitters abandoned their practice of turning over complete pneumatic systems to operations personnel. They began working line-by-line to meet specific 500-F milestones. As late as 3 August, 39 electrical cables and 232 pneumatic lines remained to be installed. Temporary fittings and work-arounds, however, prevented any major delay in test dates.[47]

500-F rolled out from the assembly building on 25 May 1966, five years to the day after President Kennedy's challenge. Despite the attendance of many Apollo program dignitaries, the Saturn vehicle stole the show. Like the Trojan horse of old, the first glimpse of the emerging Saturn vehicle was awesome. The crawler experienced little difficulty carrying its 5700-metric-ton load to pad A. During portions of the trip, the transporter operated at full speed—1.6 kilometers per hour. The Saturn vehicle reached the top of the pad at dusk and was secured two hours later. Understandably, the success was a joyful occasion for KSC and the entire Apollo team. The operation proved the soundness of the mobile concept.[48]

Less than two weeks later, that concept received an unscheduled test under emergency conditions. In early June hurricane Alma skirted Florida's east coast. Debus put the mobile concept through its paces. At 1:00 p.m. on 8 June, he ordered 500-F back to the assembly building. Within three hours, the launch team had disconnected the mobile launcher from its moorings. Wind gusts over 80 kilometers per hour spurred the efforts. The crawler began the return trip at 5:33, taking one hour to descend the 392-meter sloping ramp. Sheets of rain and 96-kilometer-per-hour gusts accompanied the crawler team on the straightaway as they urged their ponderous vehicle to its top speed. The crew reached the assembly building at 11:43 p.m. and had the mobile launcher secure on its mounts one hour later.[49]

Lack of Oxygen Slows Apollo

For the first time KSC missed a major milestone in June when the Site Activation Board postponed the start of the wet test for 500-F. A series of events including hurricane Alma had slowed the Wyle Company's cleaning of LOX lines. When Quality Division inspected Wyle's cleaning of the cross-country LOX lines, a powdery residue was found in the pipes. The cleaning compound, when mixed with the local water supply, had formed a precipitate. KSC prepared new specifications for the job, directing Wyle to flush the residue from the lines with an acid solution. LOX lines on the mobile launcher, contaminated during welding operations, also required recleaning. At a board meeting on 23 June, Scheller asked Roger Enlow to expedite contractual arrangements with Spellman Engineering, the company responsible for the LOX lines on the mobile launcher. Despite Enlow's efforts, the cleaning lagged further behind schedule. On 1 July, Gruene and Scheller agreed that the wet tests could not start for another four weeks.[50]

KSC's program office may have underestimated the problem of keeping the cryogenic lines clean. In laying miles of pipe, workmen inevitably left debris behind. On one inspection half of a grinding wheel, broken pliers, and

AS-500F, the facilities checkout dummy vehicle, 25 May 1966

Fig. 113. AS-500F emerging from the assembly building, as seen from the ground.

Figure 114

Figure 115

Fig. 114. As seen from the air. Fig. 115. Starting up the incline to the pad.

a glove were found in a LOX line. The use of invar for the inner pipe of the vacuum-jacketed lines further complicated matters. Invar, a steel alloy containing 36% nickel, had a very low coefficient of expansion, making it ideal for a cryogenic line; but it also rusted easily. During fabrication and installation, NASA inspectors had to watch for minute particles of dirt or moisture that might cause corrosion. When contamination was suspected, inspectors employed bore-sighting equipment to evaluate the potential corrosion. A decision did not come easily; inspectors spent a couple of weeks trying to determine if the LOX lines on the mobile launcher were rusting or were simply discolored.[51]

At the 7 July meeting of the Site Activation Board, Scheller announced the Apollo Program Office's plan to delay the first Saturn V mission one month. Problems with the S-II test stage at the Mississippi Test Facility had prompted General Phillips's decision. Although the directive appeared to lessen the urgency of the activation schedule, Scheller insisted that the board strive to meet the old 501 erection date. Marshall was constructing a dummy spacer as a temporary substitute for the S-II stage, and KSC would probably erect the AS-501 with the spacer to check the instrument unit.[52]

During the delay caused in cleaning the LOX lines, the board scheduled the crawler's first lift of the mobile service structure for 20 July. Mobile

Fig. 116. The crawler carrying the service structure to the pad, 21 October 1966. The assembly building is in the background, right.

launcher 1, with 500-F aboard, was back at pad A where the crawler had transported it after hurricane Alma's departure. The Bendix crawler crew spent two days in preliminary runs on the crawlerway and pad ramp and then carried the mobile service structure to the top of pad A for compatibility tests. Interest centered on the fit of the structure's five clamshells around the Saturn. Launch officials also tested the service structure's water deluge system.[53]

The RP-1 fuel system was tested in mid-July by pumping 760 000 liters of kerosene from storage tanks to 500-F. As LOX line cleaning problems persisted, the Site Activation Board reduced the number of wet tests from twelve to eight and finally to five. Spellman Corporation completed its work on the mobile launcher LOX lines in early August, and KSC rescheduled the S-IC-F LOX loading for the 15th, when failure of both LOX replenishment pumps forced a cancellation. When the launch team tried again on the 19th, it ran into much bigger trouble.[54]

Technicians in the launch control center began pressurizing the LOX storage vessel at 1:15 p.m. Simultaneously they opened the pneumatically operated pump suction valve, located in front of the 90° elbow on the 46-centimeter suction line. This allowed a flow of LOX to cool down the 37 854-liter-per-minute pumps. What happened during the next two seconds kept several investigation boards busy for days. The gas caught in the 4.6 meters of piping between the storage vessel trap and the valve flowed out as

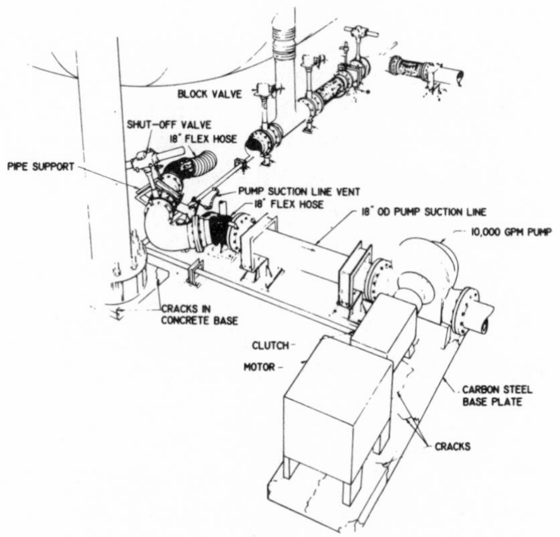

Fig. 117. Diagram of the big LOX spill, 19 August 1966. A 46-cm (18-in) flexible hose beneath the storage sphere ruptured between the block valve and the shut-off valve.

the valve began to open (the valve was timed to open fully in ten seconds and had an eccentric pivot to aid in closing against upstream pressure). The rapid evacuation of the gas increased the velocity of the LOX flowing down from the sphere. The butterfly valve was only about 20% open when the LOX hit it with a water-hammer effect. As the liquid column backed up in the restricted passage, the pressure closed the valve disc back on its eccentric pivot. The corrugated bellows in front of the valve ruptured. A Boeing console operator in the control center secured the LOX system shortly after the accident, but he could only shut off the valves downstream from the rupture. A Boeing team at the storage vessel attempted to shut the block valve above the break manually, but the men were soon driven back by freezing LOX vapors billowing over the area. Within an hour, more than 2 700 000 liters of LOX poured out.*

The sudden decompression caused the tank's inner sphere to buckle. The outer shell retained its shape, but the collapse of a corrugated bellows, connected to the tank's relief valve, indicated that a part of the inner sphere had caved inward. When technicians removed the perlite insulation between the two tanks, they found a depression in one quadrant of the inner tank. Some of the 9.5-millimeter stainless steel plates were bent 90° from their normal curvature. While KSC officials were not sure how long repairs would take, NASA headquarters announced that the accident would delay the AS-501 launch by 45 days.[55]

KSC faced two problems: repair of the LOX storage vessel and redesign of the system to prevent a recurrence. After draining the remaining 303 000 liters of LOX, the launch team used fans to circulate warm air through the inner tank. Following conferences with the manufacturer, KSC filled the tank with water. The stainless steel popped back into place at 0.4 kilograms per square centimeter (6 psi) and held its shape at higher pressures. KSC officials watched the operation over closed-circuit TV. While the water was draining, two engineers boarded a rubber raft to take a closer look from inside. Several days later, when engineers conducted a dye penetration test on the inner tank, no cracks were discovered. Technicians then replaced the perlite insulation between the inner and outer tanks. All damaged equipment had been replaced by 14 September.

Meanwhile, KSC and Boeing engineers, with advice from consultant Peter C. Vander Arend, modified the system to fill the LOX line from the

*The closed-circuit TV proved its worth during this accident. When Debus heard about the rupture, he rushed to the pad area, only to find the view obscured by the LOX vapor. When he returned to the headquarters building, the conference room screen had good pictures. A more detailed account of the accident and subsequent repairs appears in W. I. Moore and R. J. Arnold, "Failure of Apollo Saturn V Liquid Oxygen Loading System," *Advances in Cryogenic Engineering*, vol. 13.

storage vessel to the pneumatically operated valve gradually, prior to chilling the pump. KSC replaced the flex hose where the break had occurred with hard pipe and substituted a pneumatic valve for the manual block valve.[56] The new circulation and precooling system was installed by mid-September, and the second loading of LOX into S-IC-F went off without incident on 20 September. The remaining tests followed in rapid succession. With all of the AS-501 stages at the Cape (the S-II spacer in place of the live second stage), KSC officials were anxious to get on with the real show. 500-F came down in mid-October, ending seven months of valuable service.[57]

Management by Embarrassment

Although the Site Activation Board continued operations for another 20 months, it had made its major contribution to the Apollo program. The KSC team had successfully met its most difficult schedule, the activation of facilities for 500-F. Although some hectic days lay ahead, they involved the spacecraft rather than ground facilities. In July 1966, Rocco Petrone moved over from program management to launch operations. With him went much of the responsibility for site activation. From the beginning, program office representatives and launch operations personnel had argued over who should direct site activation. By the end of 500-F, the responsibility was shifting to Launch Operations and the Engineering Division.

Scheller, backed by Petrone, had won the opening rounds. At the time, the knowledge that the Site Activation Board had done its job was limited to KSC. Many outside observers did not believe that an American would be first on the moon. The Soviet Union had won the race to launch a multimanned spacecraft, sending Voskhod 1 aloft with three cosmonauts on 12 October 1964. The Russians conducted a second successful flight four days before the launch of America's first manned Gemini. Gus Grissom and John Young's three-orbit flight on 23 March 1965 went well, but earlier that month an Atlas-Centaur had exploded on the pad, causing over $2 million in damages. *Aviation Week and Space Technology* editor Robert Hotz commented after the successful Voskhod 2 flight: "Each Soviet manned space flight makes it clearer that the Russians are widening their lead over the U.S. in this vital area. It also makes it clear that the many billions the American people have poured willingly into our national space program for the purpose of wresting this leadership from the Soviets are not going to achieve that goal under present management."[58]

The activities described in this chapter helped render that judgment premature. One aspect of KSC management remains to be noted: the Site

Activation Board developed a keen sense of competition. A "hit parade board," prominently installed in the display room of the Activation Control Center, listed the ten most critical problems and the organization responsible for each activity. Unlike television's Lucky Strike Hit Parade Board of older days, no one wanted this recognition. Civil servants, as well as contractors, were frequently embarrassed at the biweekly meetings. Hard feelings were inevitable, but the program's goal helped pull organizations together. North American, a company that took as much criticism as anyone, reflected the spirit of fellowship in a going-away present to Rocco Petrone several years later. A common practice at board meetings (and also at later Apollo Launch Operations Committee meetings) had been to ask if anyone had constraints—situations that would hold up a schedule. North American officials presented Petrone a model of a Saturn V with the second stage missing. The sign on the space vehicle read, "Rocco, S-II is ready for roll except for one constraint." The constraint: no S-II.[59]

16

Automating Launch Operations

A participant in the U.S. space program likened the Apollo-Saturn to the ancient Tower of Babel. The moon rocket might have duplicated the chaos that marked that earlier dream. In a manual checkout of the Apollo-Saturn's many systems, hundreds of technicians would have swarmed on and around the space vehicle. Their reports flowing into a central control room would have indeed been a babel. Automated checkout equipment avoided this confusion. The Saturn ground computer checkout system tested 2700 Saturn functions and its computers monitored 150 000 signals per minute. Acceptance checkout equipment accomplished a similar task for the Apollo spacecraft. Wernher von Braun, after the first successful flight of an Apollo-Saturn V, credited success "to our automatic checkout procedure."[1] The story of Apollo is a study in automation.

Origins of Saturn Automated Checkout

Although automation has no precise meaning that is generally accepted in technical circles, there was considerable automation—however defined—in the missile programs of the 1950s. Engineers employed pressure gauges, temperature gauges, frequency detectors, and other devices to passively sequence a series of events. Using relay logic, if event A occurred, then event B took place, and so on. Interlocking circuitry and relay logic allowed ground support equipment to control portions of a countdown (for the SA-1 countdown, see pp. 60–62). In this chapter, the term *automation* will imply the use of digital computers and associated equipment.

In checking out the first Saturns, hundreds of control room switches sent signals over electrical lines to test points on the rocket. The launch vehicle responses, returning over the same wires or radio telemetry links, registered on strip charts and meters. The launch team then evaluated the test data. Automation began to change this procedure when, in 1960, Marshall engineers decided to design a test capability for Saturn's digital guidance computer (its maiden flight would come on SA-5). The first, tentative steps were to parallel, not replace, manual checkout. Quality Division representatives sought a flexible program that could be expanded to include other tests as automation proved itself.[2]

In early September Debus asked to have the Launch Operations Directorate participate in automated checkout discussions. Before the end of the year two computer systems were under study at Marshall. The Reliability Assurance Laboratory was testing a Packard Bell 250 for factory acceptance checkout while the Guidance and Control Division and the Quality Division investigated the use of an RCA 110 for launch tests. The RCA 110 was among the first priority-interrupt computers on the market. This feature provided for the division of the computer program into several sections, each one having an assigned priority level. The priority interrupt allowed an engineer to switch immediately from one test to another during operations. On the RCA 110 there were eight available levels and the computer could switch to a higher priority test in 100 microseconds. As the early use of the 110 indicated trouble-free operations, it became the workhorse of the Saturn checkout at the Cape.[3]

Marshall's first automation plan, published in September 1961, asked the question: "Why automation?" The advantages were speed and accuracy. The author, Ludie Richard, noted that man is a poor test conductor. He cannot run thousands of tests with uniform precision, and he frequently fails to observe the results. Machines ensure standardized testing and an accurate recording of the responses. Further, an automated operation required only a fraction of the time used in a manual procedure. The time savings would permit more testing, an important factor to operations personnel, particularly just prior to launch. With automation Marshall could duplicate the exact conditions under which a failure had occurred. Data would be available at the point of failure to aid in trouble-shooting and fault isolation. Richard also listed some disadvantages. Automated checkout procedures would complicate the Saturn, although the problem could be minimized by designing the automated test system into the vehicle. Another drawback was a lack of user confidence in the system. Richard attributed this to poor planning, either in training the users or in faulty machine language. A long-range problem involved the operators' possible loss of familiarity with the launch vehicle. Automation might work so well that its users would lose their "feel" of the rocket, with a corresponding drop in their ability to meet a crisis.[4]

Richard's automation plan proposed to phase the RCA 110 into Cape operations with the SA-5 launch from LC-37. The blockhouse computer would parallel the launch complex circuitry so that operations could proceed manually if necessary. At first the RCA 110 would check the digital flight computer and monitor other electrical systems. It was hoped that by SA-111 (the first Saturn I flight after the ten R&D launches) equipment reliability and user confidence would permit a fully automated launch.[5]

Planning for automation accelerated during the following months as Marshall moved from the C-2 to the C-5 version of the advanced Saturn. The Saturn V's size and LC-39's greater distance—4.8 kilometers—from launch control center to pad precluded a manual checkout. On 1 October 1961, von Braun established an Automation Board at Marshall to automate the Saturn V checkout. Thereafter, design of a computer checkout system paralleled launch vehicle development.[6]

Saturn I-IB Computer Complex

The development of the RCA 110 hardware for Saturn I tests proceeded at Huntsville under the direction of the Astrionics Laboratory with the collaboration of the electrical network group. Richard Jenke's automation team at KSC furnished operational requirements and participated in a number of design decisions. Some agreements between developer and future user came slowly. The two groups discussed the matter of control panel switches for months; the uncertainty centered on the desired status of the various test devices when the operations switched from a computer program to manual control. Eventually the two groups decided to treat each test device separately and modify the RCA 110 programs when experience dictated a change. About the same time the automation group approved a three-position switch for the Saturn IB control panel. During IB operations all signals, manual as well as automatic, would process through the computer. In the three-position switch, OFF manually terminated a function, such as opening a valve; ON manually initiated a function; and AUTO placed the computer in control for automatic testing. Experience with SA-5 demonstrated the need to prevent the computer from sending any command signals while it monitored the operation. The automation group added a discrete inhibit switch on SA-6 for the remaining Saturn I launches.[7]

The Cape played a larger role in the development of computer programs, called *software*. Marshall recognized the launch team's need to prepare its own tests and allowed KSC to manipulate "Boss," the 110's executive control program. Jenke's group, assisted by RCA, IBM, and Chrysler personnel, combined the 20 manual routines of the guidance computer checkout into four test sequences. While the performance of the RCA 110 on SA-5 left a few skeptics unconvinced, launch officials labeled the computer a success. On the following launch (SA-6) Jenke added a test sequence for automatic azimuth laying. The launch team added a cathode ray tube console (a television screen that displayed alpha-numeric characters received from the computer) for SA-7. The RCA 110 increased its monitoring role during the last three Saturn I launches, the Pegasus series.[8]

During the Saturn I program, automation moved forward at a slow, deliberate pace; at any time the launch team could have reverted to a manual operation. By the time of the first Saturn IB launch in February 1966, however, KSC was firmly committed to automated testing. While a completely automated checkout was still a long way off, the RCA 110A computer (a 110 with increased memory) was "on line" for the first IB operation. All test transmissions then went through the computer; if it failed, the entire checkout would stop. On-line status represented the decision to use the Saturn IB missions as a testbed for Saturn V automation.

Two RCA 110As—a "master" computer in the LC-34 blockhouse and a "slave" computer in the automatic ground control station—provided the brains for the Saturn checkout system. A high-speed data link, a coaxial cable running through the LC-34 cableway, connected the two computers. Engineers could initiate launch vehicle tests from display consoles or from programs stored within the master RCA 110A. In either case the computer in the launch control center digitized commands for transmission to the ground control station. The slave computer, housed beneath the umbilical tower, interfaced with the launch vehicle, issuing commands and receiving responses. Both computers were also tied into the spacecraft checkout system.[9]

In its design, the digital data acquisition system was typical of the digital systems employed at KSC. Sections, or modules, within each computer performed specific functions. The stage module had three distinct elements, each dedicated to one of the launch vehicle elements. Aboard the Saturn IB, three digital data transmitters (for the two stages and the instrument unit) multiplexed the pulse code modulated data, giving each signal a specific time slot on its channel. This data, reflecting the condition of the launch vehicle, was transmitted to ground receiving stations over coaxial cable or radio. The receiver decommutated, i.e., divided the data into its constituent parts, and then conditioned the signals for transmission. Another part of the digital data system, a high-speed memory core, stored the data for use by the 110A computers.[10]

A digital events evaluator determined whether or not a return signal indicated a change in the launch vehicle's status. After receiving data from the slave computer, the evaluator compared this signal with pre-programmed information in its memory or with a previous scan of the same function. The event was then time-tagged for identification and the results either printed for display or stored for retrieval by the master RCA 110A at a later date. The evaluator logged every event within approximately two milliseconds, providing real-time—virtually immediate—printouts of event changes.[11]

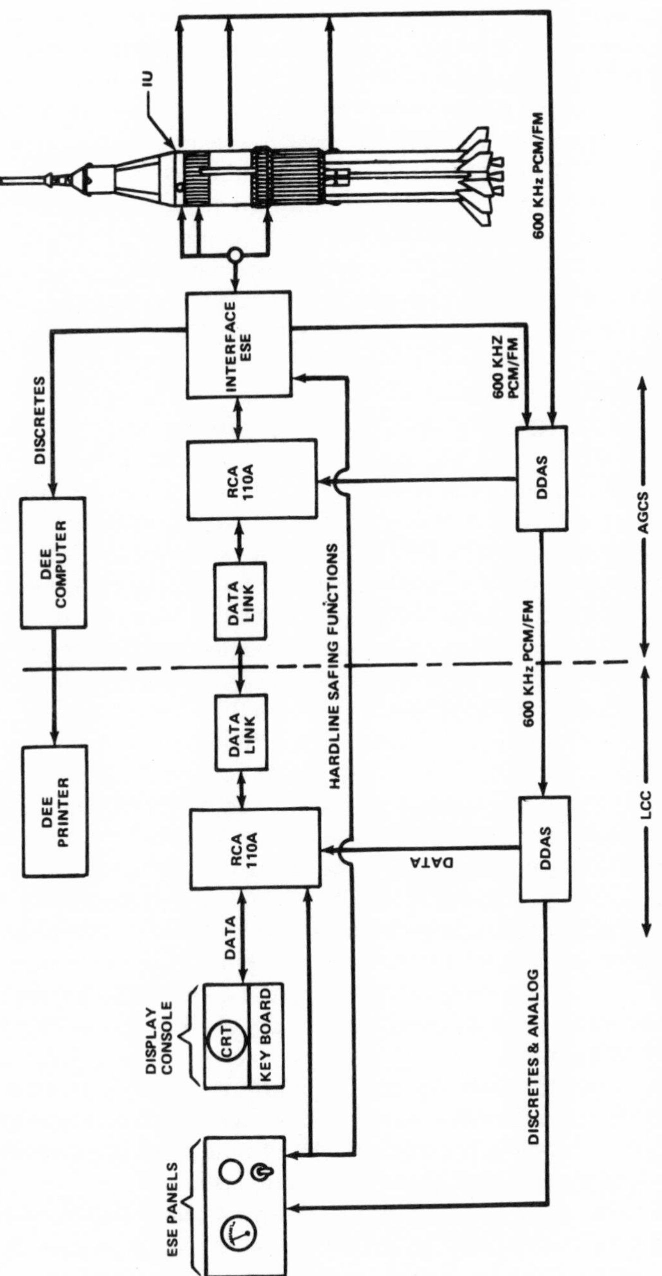

Fig. 118. LC-34/37 computer system schematic. From B. E. Duran, "Saturn I/IB Launch Vehicle Operational Status and Experience," pamphlet 680739, Aeronautic and Space Engineering and Manufacturing Convention, Los Angeles, 7–11 October 1968. Courtesy of the Society of Automotive Engineers, Warrendale, PA.

Although the Saturn checkout complex was usually identified by the RCA 110A computers, the related electrical support equipment was the larger part of the system. Included was test equipment and a complex distribution system. The RCA 110As relied on input-output address lines to place test equipment in a receive or transmit condition, set sequence control relays, select analog or digital signal lines, drive digital-to-analog converters, and issue warnings. Input-output sense lines informed the computers of test equipment status. The computers employed input-output buffer registers to handle the heavy data flow with the Saturn guidance computer and telemetry system. Hardwire connections tied the 110As in with LC-34's timing system: the countdown clock, a Greenwich Mean Time clock, and interval timers.[12]

Saturn V Computer Complex

Dual computers on LC-34 were largely redundant since the distance from launch vehicle to blockhouse was not great enough to attenuate analog signals. On LC-39, however, the dual arrangement was essential. The computer could not accurately process analog signals from the launch vehicle five kilometers away. In this regard, the earlier installation was intended as a testbed for the Saturn V complex. There were also important differences between the two installations. The slave computer in LC-39 was housed on the mobile launcher. Dual data links, running beneath the crawlerway, tied it to the 110A computer in the launch control center. Saturn IB operations were delayed occasionally because test engineers could not reach the slave computer's peripheral equipment during hazardous pad activities. On LC-39, magnetic tapes and other related equipment were moved back from the mobile launcher to the control center, to be available at all times. Another change involved the display systems. At LC-34, the master computer in the blockhouse processed display signals to the various consoles. LC-39 included a Sanders display system with its own digital computer. The greater size of the Saturn V dictated other changes; for example, LC-39's checkout system could issue and receive twice as many discrete signals. Perhaps the biggest difference was the increased capacity for testing. In effect there were three of everything: computers for each of the equipped firing rooms, computers in the mobile launchers for interfacing with the launch vehicles, digital data systems, and all the peripheral equipment. As a result, KSC could conduct automated tests on three Saturn V vehicles simultaneously.

As on LC-34, the 110A computers were central elements. The mobile launcher's computer tied in with the Saturn V vehicle, its peripheral equipment (line printer, card reader–punch, magnetic tapes), the launch control center computer via the data link, and the range clock system. Commands to the launch vehicle went through a discrete output system which employed triple-modular redundancy to minimize errors. Approximately 2000 test responses returned from the Saturn V. A remote control capability in the control center allowed engineers to continue Saturn tests in the event of a computer breakdown on the mobile launcher. The RCA 110As in the control center had even more interfaces; their data channels controlled signals to and from peripheral equipment, control consoles, the dual data links to the pad, digital data systems of the launch vehicle, computer display systems, the spacecraft computer system, and the countdown clock system.[13]

The control center was the focal point of Saturn checkout. While the original design included four firing rooms, only three were fully equipped. The various control consoles and display devices were physically grouped by stages or function. Management officials, including the launch director, test supervisor, and test conductors, occupied the first three rows. Within each stage area, test personnel were organized according to functional subsystems. Thus, in the instrument unit area were consoles for the emergency detection system, networks, guidance, stabilization, flight control, and measuring and telemetry. During prelaunch testing, approximately 400 people occupied stations in the firing room. A lesser number manned a backup control room. The test supervisor and conductors directed the operation by means of test procedures, countdown clock readouts, and an intercommunication system.[14]

Although equipment in a firing room varied with launch vehicle requirements, about 400 consoles were employed. Of these, approximately 100 were cathode ray tube displays. Four large overhead screens provided an additional means of displaying information. Presentations on the 2 × 3-meter translucent screens normally paralleled the test procedures, but the sequence could be altered to display data on a particular launch problem. Information from various sources appeared on the screen—telemetered data, closed-circuit and commercial television, slides, and viewgraphs. The data kept NASA managers and test conductors abreast of the vehicle status.[15]

An important addition to firing room equipment came after the start of Saturn V operations. While systems engineers could monitor vehicle outputs, the RCA 110A lacked the means to provide full coverage of launch vehicle measurements. Saturn engineers added an alert monitor capability to the launch vehicle display system. The equipment was first tried out in the backup firing room. In 1970 it became an integral part of the control center's display equipment. The alert monitor system—ten dual sets of cathode ray tube displays tied in with the digital data acquisition system—automatically indicated when certain measurements were out of tolerance.[16]

Fig. 119. An operation under way in the launch control center, firing room 1, March 1967.

The Transition to Automation

With the RCA 110A computers on line for the first Saturn IB mission, Hans Gruene's Launch Vehicle Operations team found itself totally dependent on computers and computer programs. They were not dependent, however, on automation. Having approached the idea of automated testing with reservations, they insisted that the Saturn design provide for manual testing as well as automation. Hence, LC-34's complex provided a dual capability. Test engineers could proceed in the traditional manner—initiating commands from switchboards and checking the results on meters, strip recorders, and cathode ray tube displays. In this mode, testing would remain essentially manual, with the computer complex serving as an expensive data link. Or the launch team could convert the manual test procedures into computer programs for interpretation and execution by the 110As. During the early IB missions, manual procedures predominated. Besides the inevitable resistance to change, systems engineers had trouble converting their test procedures into machine language. Development of a special computer test language alleviated the latter problem, while Gruene's leadership prevailed over personal inertia. By the end of Apollo, most launch vehicle tests were fully automated programs.[17]

Saturn ground computers employed two types of programs: operating system and test. As the name implied, the operating system program was the computer's basic software. It operated continuously, seeking alternate paths when a failure was detected. Manual testing of the Saturn vehicle was accomplished through this program. Test and monitor programs provided the means to automate the checkout process; they were the system engineer's primary software tool. Test programs could be prepared in machine language, using the full capability of the computer's logic, or in Atoll (Acceptance Test Or Launch Language), a form of engineer's shorthand. Once prepared, the test programs performed a number of important tasks: sequencing required events during a test, evaluating system responses, monitoring, displaying anomalies, and tying together a series of programs. A test program could preselect the operation sequence, change the limits of tested values, and intervene during the operation. On the later Apollo missions, test programs accomplished much of the routine checkout. Engineers would initiate programs through console keyboards and react when problems arose. Many of the tests would start on their own—the launch team having programmed the computer to call up test programs at a certain time in the countdown or when another test was successfully completed. While test programs were the key to automated checkout, maintenance and post-test processing programs played an important role. Maintenance programs tested the interfaces between RCA 110As and related equipment. With the post-test processing programs, engineers converted raw data into usable printouts.[18]

Changing test procedures into computer programs may well have been the highest hurdle in Saturn automation. Early Saturn I automation rested largely with IBM, Huntsville's contractor for computer software. From KSC requirements, IBM programmers and system analysts prepared machine-language programs. The Automation Office provided the coordination between Saturn test engineers and IBM computer experts. Unfortunately, it was not simply a matter of converting a few lines from English into a computer program. KSC's manual procedures did not spell out every detail; many contingency actions depended upon an engineer's intimate knowledge of a system. Inevitably, misunderstandings arose. Some KSC systems engineers viewed the process as a one-way street: IBM programmers were gaining a knowledge of Saturn hardware while KSC engineers learned little about automation. Furthermore, as Saturn automation grew, requirements for IBM support increased. Clearly, KSC needed some way to simplify the conversion from test procedure to computer program—a route that would bypass machine language.[19]

The solution was Atoll, a computer language under development at Huntsville's Quality Laboratory. By 1965 the Astrionics Laboratory and

IBM were incorporating Atoll into IB plans. KSC's automation team helped define launch site requirements for the new software system. The AS-201 mission in February 1966 used only a half-dozen Atoll procedures, but subsequent launches showed increasing numbers as KSC engineers converted from manual procedures to computer programs. There were 21 Atoll programs on AS-501, 43 on AS-506 (Apollo 11), and 105 by the Apollo 14 launch in early 1971. Atoll proved particularly valuable for Saturn systems that changed from one mission to the next. While modifications to machine language procedures required approval from Marshall, KSC's Automation Office controlled Atoll. Changes could therefore be approved quickly at the launch center. More importantly, Atoll involved the test engineers directly and its use was instrumental in their acceptance of computerized checkout.[20]

Automating Telemetry Operations

The move from the Redstone to the Saturn era brought a pressing need to automate data reduction.* The Saturn I's telemetry, the primary source of postflight test data, represented a three-fold increase over previous rockets and the Saturn V would be an even greater jump (see table 1). The increase posed a problem of space. Continued reliance on analog strip charts would have forced Saturn V engineers to review thousands of feet of chart after each launch. Time was a second consideration. In September 1961 Fridtjof Speer, chairman of Marshall's Saturn System Evaluation Working Group, expressed concern about possible delays in the delivery of postflight data. Although LOD had agreed to submit telemetry data within 12 hours, Speer wanted backup blockhouse records (strip charts and event recorders) within 24 hours. He asked Debus to "exploit every possible course of action to satisfy this requirement."[21]

A month after Speer's letter, Dr. Rudolf Bruns's data reduction team pioneered the use of computers in Saturn I launch operations. Digital computers offered two advantages. During a launch the computer could record incoming telemetry data on a magnetic tape and subsequently process compact printouts in a relatively short time. The computer, supported by peripheral equipment such as cathode ray tube consoles, could display critical information in real time.[22]

**Data reduction* means the transformation of observed values into useful, ordered, or simplified information. With telemetry it could involve transferring an analog electrical signal onto a brush recording or eliminating unnecessary portions of a message (e.g., the address), restructuring the data, and directing it to various users.

The Flight Instrumentation Planning and Analysis Group set up a Burroughs 205 computer alongside the telemetry station in hangar D prior to the SA-1 launch. A tarpaper shack housed the computer—a far cry from facilities the team would enjoy in a few years. Despite its primitive surroundings, the 205 provided guidance data and some measurement reduction in real time. A General Electric computer replaced the Burroughs machine on SA-3; the GE computer's solid state circuitry and core memory provided a faster sampling of Saturn telemetry.[23]

During the last hours of the SA-3 countdown, the launch team periodically tested telemetry transmitters and the data reduction computer. The team "dumped" the telemetry data from the GE computer at several predetermined times, taking a quick look at the printout to see if the measurements were within calibration. At T – 30 minutes the computer began processing data from the Saturn's ten commutators. The GE 225 took approximately ten seconds to complete one sample of the Saturn's ten telemetry links; most of that time was taken up by the telemetry station switching device, switching from one commutator to the next. As the 225 printer could not match the computer's calculating speed, the real-time printout listed only the out-of-tolerance values. The GE 225 processed telemetry data until the signal faded a few minutes after launch. During postlaunch activities, the telemetry station rewound its analog tapes and played them back for the computer. The digital magnetic tape data, produced by the playback, were then converted in another computer process to specific engineering units (e.g., degrees Celsius), which were displayed on a printout or a Stromberg-Carlson 4020 plotter.[24]

Efficient data reduction depended on parallel advances in the telemetry station's digitizing system, the equipment that converted the Saturn's analog telemetry into a digital message for the GE computer. The earliest digitizer employed one analog-to-digital converter and one synchronizer. Each of the ten commutators on SA-3 fed its measurements (27 each for the low-speed commutators and 216 for the high-speed commutator) into a subcarrier oscillator. The oscillators sent the signals out over the launch vehicle's six telemetry transmitters as a pulse-amplitude-modulated wave. The receiver in the ground telemetry station removed the subcarrier and directed it to one of three discriminators where it was demodulated. The analog-to-digital converter changed each signal voltage from a magnitude to a pulse. A 0.6-second delay in synchronizing the converter's switch from one commutator to another limited the digitizer's output to the GE 225. Due to the switching delay, the computer received approximately one measurement of every 120 that came from the Saturn's high-speed commutator.[25]

Work on a faster digitizing system began in mid-1962. An analog-to-digital converter was added for each vehicle commutator; a parallel programmer-addresser generated a 12-bit address word to identify the data. A digital scanner, essentially an electronic 16-position switch, scanned the outputs of the data channels (convertor and addressor), transferring new data to a core memory. The memory stored each measurement according to its digital address and provided the computer with random access to any data. Although the telemetry team experienced problems interfacing the scanner and core memory, the specifications were ready by November.[26]

The telemetry scanning and digitizing system was added to a GE 235 computer for the block II series of the Saturn I program. The GE computer, given immediate access to all data through the digital scanner and core memory, recorded data in real time on a magnetic tape. Since the 235's printing lagged behind its computations, real-time display was still limited. The 235's other functions included: separating data by program (S-I, S-IV, instrumentation unit, and spacecraft) and measurement, converting measurements to engineering units, arranging data for use on the 4020 plotter, and comparing engineering units versus time on printouts and 4020 plots (microfilm or hard-copy graphs). Within two hours of the SA-7 launch, all the 4020 plots had been processed. The total data reduction program was completed eight hours later.[27]

Measuring techniques needed to be improved to keep up with the advances in digitizing and telemetry reduction. Automation of measurements during checkout of the Saturn vehicle was begun in March 1962, using an IBM card system. Punched cards, placed in a card reader, selected the appropriate channel (and relay if calibration was necessary). The card reader compared the signal returning from the launch vehicle's measuring device with data prepunched on the card and gave a "go, no-go" evaluation.* Following experimentation during a checkout on LC-34, the system was installed in LC-37's measuring station for the block II launches.[28]

As the automation plans gained momentum, Debus expressed concern about their impact on LOC's relations with the Air Force. On 21 February 1962 Debus penned a brief note to Gruene's weekly report:

> Hans: One day we have to start an analysis of what this entire automatic checkout with computers will mean in our countdown-and-test interfaces with the range! For instance: timing, on-off commanded to the Range, TM [telemetry] receiving . . . and a

---

*A *go, no-go* indication told the operator whether a device was functioning properly, without indicating how far out of tolerance it might be or what was wrong.

> host of other interactions. . . . Does it still make sense to plan a "joint" TM station. . . ?[29]

After Debus and General Davis discussed some of LOC's scheduling problems in May, Air Force and NASA officials held a series of meetings that summer, some as the Joint Instrumentation Planning Group, others in informal sessions. Their work led to the development of a new telemetry station and the central instrumentation facility.[30]

Automatic Checkout for the Spacecraft

Automated checkout of the Apollo spacecraft had its origins at Cape Canaveral in 1961. Preflight Operations Division engineers, members of the Space Task Group, realized that Mercury launch methods would not satisfy Apollo requirements. The Mercury preflight tests resembled an aircraft checkout. One test team worked from a command post near the spacecraft while a second group monitored the test results at a remote station. During the checkout hundreds of wires ran through the open hatch into the cockpit, leaving barely enough room for an astronaut or a test engineer. There were other limitations. During the prelaunch operations, the spacecraft would likely move several times, and each move required disconnecting and reconnecting the various test lines. As the checkout grew more complicated, the test conductor found it increasingly difficult to coordinate activities at the spacecraft and monitoring station. Less than 100 telemetered measurements in Mercury had occupied the Instrumentation Branch. The 2000 measurements projected by Apollo feasibility studies made some form of automated checkout inevitable.[31]

Following a September 1961 briefing on Apollo, G. Merritt Preston, Preflight Operations Division's chief, asked his staff to consider the proposed spacecraft's impact on launch operations. Jacob Moser and his Flight and Ground Instrumentation chiefs, Walter Parsons and Harold Johnson, responded with an automation proposal, and Preston gave the project a green light. Mercury operations limited progress during the next two months, but with further urging from Preston, the instrumentation team formalized a presentation in December. Two young engineers, Thomas Walton and Gary Woods, joined in this early conceptual work. For their efforts the five subsequently won a patent on the checkout system. The group's pre-Christmas briefing favorably impressed the staff. A Marshall delegation displayed less enthusiasm but failed to halt the project.[32]

The automation team began the new year with a search for available equipment. Since money was scarce, only off-the-shelf hardware could be used. Walton and Woods scoured American factories, finding all the necessary components except a digital command system. At the same time Preston secured the support of Robert Gilruth and Walter Williams, the Director and Associate Director for NASA's Manned Spaceflight Center. In February the team conducted a series of formal briefings for NASA's manned spaceflight organizations and for supporting contractors. The road show, complete with a projector and more than 500 slides, drew a mixed response. Headquarters officials questioned some of the team's technical assumptions (e.g., James Sloan, OMSF's Deputy Director for Integration and Checkout, doubted that the software planned for the system could be perfected). The principal user, North American, perhaps hoping to develop a checkout system itself, was particularly critical of the concept. Despite the numerous objections, acceptance checkout equipment (ACE)* was approved by mid-1962.[33]

While gaining support within NASA was, perhaps, the most difficult hurdle, the design also involved some challenges. The spacecraft had not yet been clearly defined when the group began work on a report in February 1962. Woods concentrated on the system's uplink. As the name implies, the uplink carried commands from operator consoles to the spacecraft via coaxial cable or radio. Woods demonstrated the feasibility of his uplink in June, using 32 kilometers of cable stretched from a Patrick Air Force Base command post to the Cape. Meanwhile Walton pursued the problems of the downlink, the portion of the checkout system that brought encoded signals from the spacecraft, through a decommutator and computer, to display devices. Johnson focused on another part of the downlink, the analog display recorders.[34]

In July the acceptance checkout equipment team began procuring equipment for an experimental station at the Cape. Gemini officials helped fund the laboratory in hopes that the system might benefit their program. The Instrumentation Branch activated the station in September; its original equipment consisted of a small computer, an alphanumeric display device, a decommutation system, and the manual uplink prototype. A downlink prototype was put in operation the following month. By April 1963 the team was working two digital computers in a non-synchronized mode, exchanging data through a shared memory base. Gordon Cooper's 22 revolutions around the world in May 1963 marked another milestone for the station. The experimental equipment provided real-time support of preflight checkout and inflight operations for the last Mercury mission. The station's computers displayed

*ACE was initially SPACE, Spacecraft Prelaunch Automatic Checkout Equipment. Cape officials changed the title to Prelaunch Automatic Checkout Equipment for Spacecraft, PACE-S/C only to find that PACE was already a legal name. They then dropped the Prelaunch and changed the Automatic to Acceptance.

Faith 7's telemetry data on screens and high-speed line printers. The laboratory was fast becoming one of the tourist attractions at Cape Canaveral; during their visits to the Cape, new astronauts spent a half-day in the station.[35]

The General Electric Company entered the ACE story in November 1962. GE's Apollo roles, as delineated by NASA management, included the development of "overall system checkout equipment" (pp. 173–76). Since ACE would test North American's command and service modules and Grumman's lunar module, the checkout system fell within GE's area of responsibility. At first GE provided engineering support. Within three months Leroy Foster had 20 engineers working on equipment specifications. The decision at NASA Headquarters to have GE produce the Apollo checkout stations (as a modification to its existing contract) touched off ten months of proposals and counterproposals. The main dispute between GE and Cape officials centered on the issue of government-furnished equipment. The Preflight Operations Division intended to provide GE most of the components, buying parts already developed by other companies. GE, understandably, thought it could improve on some of the equipment. At a stormy July session in Daytona, Jack Records, GE's number two man at the Apollo plant, and Dr. Lyndell Saline questioned the suitability of Control Data Corporation's 160G computer. When Preston asked for proof of the computer's inadequacy, however, the GE executives withdrew their charge.[36]

Negotiations with General Electric were complicated by officials at NASA Headquarters; Joseph Shea, OMSF's Deputy Director for Systems, supported GE. In September 1963, he called the ACE team to Washington for a showdown on the spacecraft checkout. Shea and his Bellcomm* advisors attacked ACE on several grounds, including insufficient memory and interrupt capability. Cape officials refuted the criticisms point by point. Before the end of the day Shea had given up his opposition to ACE.[37]

After settling the issue of government-furnished equipment, GE and the Florida Operations group (the new name for Houston's launch team at the Cape) moved swiftly to meet the September 1964 deadline for the first operational ACE station. At the Cape, Douglas Black's team conducted a series of critical interface tests at the experimental station in the first half of 1964. By June the first computer programs had been verified. GE shipped components for the first station to Downey, California, in July. Within 60 days North American was using the station to check out Apollo 009, the spacecraft that would fly on AS-201. GE installed 13 more ACE stations: 2

*Bellcomm, Inc., was a subsidiary corporation of AT&T, organized to assist OMSF's Systems Office in the overall integration of Apollo. The work resembled that being done by GE, but was at a higher level and on a much smaller scale.

at Downey; 3 at Grumman's Bethpage, New York, plant; 2 in Houston; and 6 at the Cape. KSC's first station became operational in March 1965.[38]

Spacecraft Checkout

ACE's first major test at KSC came with the checkout of Apollo 009 (AS-201 mission) in late 1965. The spacecraft team directed the checkout from a control room in the operations and checkout building. Engineers from Spacecraft Operations and North American, working in pairs, tested the nine functional systems: communications, instrumentation, service propulsion and reaction control, stabilization and control, guidance and navigation, power and sequence, fuel cell and cryogenics, aeromedical and astronaut communications, and environmental control. Commands were initiated at the test consoles, e.g., an engineer might test the freon level in the command module's environmental control system. His signal went to the command computer for conversion to a binary instruction. The digital message traveled a complicated electrical path to the spacecraft, where it triggered a sensor in the command module. The sensor noted the condition of the freon and transmitted an appropriate response. Data acquisition equipment routed the signal back to a display computer, which processed the message for presentation on the same test console whence the command had come seconds earlier. Command and display computers and much of the data acquisition and recording equipment were located in an ACE computer room.[39]

Three different groups of sensors obtained data concerning the Apollo spacecraft: ground service equipment, carry-on equipment that was removed prior to flight, and sensors built into the spacecraft. Coaxial cable and radio connected the various sensors to the control rooms in the operations and checkout building. There, data traveled through one of three different paths. The most important, from the standpoint of real-time display, was the display computer. Its functions included: comparing machine words to determine whether data fell within predetermined limits, converting data into engineering units (such as heat rise in degrees per second), and generating signals that would produce alphanumeric displays on consoles. The display system was impressive but not foolproof. An engineer recalls that on its first day of operation, the console welcomed them: "GOOD MORING." ACE had failed its first spelling test.[40]

During lunar missions, four control rooms would be used for spacecraft checkout: primary and backup rooms for the command and service modules and another pair for the lunar module. Each room had 20 master consoles and additional slave consoles. The latter displayed the same data

Fig. 120. An operation under way in the automatic checkout equipment (ACE) control room, February 1967.

Fig. 121. Automatic checkout equipment room, February 1967.

shown on a master, but did not provide the means to select information. Nine TV monitors carried pictures from portable cameras located around the spacecraft. The overhead monitors were part of an operational TV network that carried spacecraft and launch vehicle pictures to the launch control center and central instrumentation facility, as well as the operations and checkout building. Although the equipment had a similar appearance, configurations differed, depending on the requirements of particular systems. During checkout, between 40 and 50 men occupied each of the primary control rooms. In the backup rooms, the consoles were kept in operation but usually were not manned. Each control room was supported by a computer room with its uplink and downlink equipment.[41]

17

Launching the Saturn IB

The Apollo program made another major advance toward its goal in 1966 with three successful launches of the Saturn IB. The IB had been added to the program in 1962 as a means of conducting early manned Apollo missions in earth orbit. The IB launch vehicle was a hybrid, combining the Saturn I's booster with the S-IVB stage that would fly as the third stage on the moon rocket. Three research and development flights were scheduled for 1966; two would check out the Apollo–Saturn IB configuration while a third tested the liquid-hydrogen propellant system in the S-IVB stage. A fourth Saturn IB launch, scheduled toward the end of 1966, would put the first Apollo crew into space. The launches posed a challenge for KSC. In the midst of a major site activation—LC-39—the launch team faced a new operation. There was a new launch vehicle stage and, with the RCA 110A computers, a new checkout system. Before completing the missions, the launch team would experience some of the most frustrating moments in the entire Apollo program.

Remodeling LC-34 for Bigger Things

First sign of the Saturn IB series at the Cape was NASA's rebuilding of the LC-34 facilities. The complex had last been used to launch SA-4 in March 1963. During the rest of the year, LC-34 was earmarked for back-up service during the Saturn I, block II series. Contractors had completed a gas storage building and begun work on liquid-hydrogen facilities. Mueller's revised launch schedule of 1 November 1963 had prompted Debus to recommend cancellation of further Saturn I work at the complex. NASA then began the task of readying LC-34 for the launching of AS-201, first of the Saturn IBs.[1]

The old LC-34 service structure was almost completely rebuilt. Previously open to the winds, it was now equipped with hurricane gates and four weather-tight silo enclosures. Anchor piers were strengthened to hold the service structure in place over the pad. The modifications also included eight vertically adjustable service platforms and new traveling hoist machinery. On the umbilical tower, the swing arms were rebuilt to meet the new rocket's dimensions; testing was completed in June 1965. Astronauts would board the command module through a new arm at the 67-meter level. The addition included a white room to control the temperature and cleanliness inside the module. While AS-201 would be an unmanned flight, the launch complex would be man-rated in almost every particular.[2]

The change from the Saturn I to the IB meant larger fuel requirements, for the upper stage a 130% increase. Major alterations were made in LC-34's propellant facilities. The RP-1 main storage tanks were reinsulated and the liquid-hydrogen system was enlarged. A new tanking control system loaded propellants to prescribed levels and maintained those levels until liftoff. Pneumatic requirements involved modification of the high-pressure gaseous nitrogen and helium installations and construction of a gaseous hydrogen system.[3]

Colonel Bagnulo reported on 5 August 1965 that, "after a full measure of blood, sweat, and tears," the basic modifications to the service structure were essentially complete. The initial contract cost had risen from $3.5 million to $5.3 million, partly because of changes to the design, but more from the additional overtime required to keep the work near the original schedule. Minor work continued almost up to launch time; the last change requirements were released on 4 January 1966.[4]

LC-34 Wet Tests

The erection of the S-IB stage and the dummy stages for the S-IVB and instrument unit marked the start of LC-34 facility tests on 18 August 1965. Although the mating went well, the launch team soon fell behind schedule. Hans Gruene reported a four-day lag the following week, attributing most of the delay to faulty electrical support equipment from Huntsville. He listed among the shortcomings missing connectors, cables improperly marked, and schematics that did not reflect engineering changes already accomplished.[5] Similar problems threatened in early September to postpone the start of tests on the ground equipment test sets. More than 250 power cables had not arrived. About 100 GE cables were of the wrong length. Gruene also singled out computer problems, an area that would plague Launch Vehicle Operations throughout the 201 mission. The shortage of spares was also critical. A power supply failure on the 26th had necessitated the air delivery of a new component from California. Computer breakdowns during the test of the ground equipment test sets could cause a day-for-day slip in the schedule.

Delays in Marshall's breadboard* testing of the RCA 110A operating program could also impact the checkout.[6]

The wet test in September disclosed some problems in LC-34's new propellants system. Hydrogen did not flow from its storage tank during the first H_2 "cold shock" test. When no mechanical block could be found in the valves, lines, or filter, the obstruction was blamed on frozen nitrogen. The gas had leaked into the hydrogen system through a hand valve during the nitrogen pressurization test. The launch team also had trouble loading the S-IVB auxiliary propulsion system.† The surprisingly slow flow rate of the hypergolic oxidizer, coupled with a thunderstorm, left no time for the flow test of the fuel. As Launch Vehicle Operations planned to remove the dummy stage the following day, the second half of the hypergolic loading was postponed until after the erection of the live stage.[7]

Another highlight of the facilities test was the replacement of an S-IB fuel tank. The tank had been damaged during a load test, and repressurization left numerous wrinkles in its skin. Although a Chrysler crew subjected the tank to above-normal pressures without mishap, Marshall representatives wanted a replacement. The new fuel tank arrived from Michoud, Louisiana, on 24 September and was installed in eight hours on the 29th. This delayed erection of the S-IVB stage by two days, but numerous breakdowns in the RCA 110A computer had already thrown the tests 12 days behind schedule.[8]

The Apollo-Saturn IB Space Vehicle

The LC-34 modifications were designed to accommodate a 68.2-meter Apollo-Saturn (AS-201), which could count as many "firsts" as any of the Saturns. Its upper stage (S-IVB) would be the first to use a hydrogen-burning J-2 engine (900 000 newtons or 200 000 pounds of thrust); it had a new instrument unit, nerve center for guidance and control; it was the first to carry a live (though unmanned) Apollo command module, powered by a service module, the engine of which was intended to start and restart in space. Perhaps most important, and certainly most troublesome, was the first installation of an on-line, automated checkout system. These innovations were

---

**Breadboard* means an assembly of circuits or parts used to prove the feasibility of a device or system. Huntsville used breadboards as a design tool (those for Saturn circuitry occupied a half-dozen large rooms). The breadboards were kept in the same configuration as the vehicle and ground support equipment. When a problem arose, Huntsville engineers verified any proposed solution on the breadboard before KSC applied it to the flight equipment.

†The auxiliary propulsion system provided attitude control for the S-IVB stage and payload during the coast phases of flight.

the cause of many delays in the launch program—and justification for the delays, as well: what was worked out successfully for AS-201 would be available for Saturn V.[9]

The first piece of AS-201 to arrive at the Cape was Chrysler Corporation's S-IB stage. It arrived from the Michoud Assembly Facility aboard the barge *Promise* 14 August 1965. It was the first Saturn to enter the Banana River and KSC through the Canaveral locks. The new S-IB was basically the S-I stage, redesigned to reduce weight and increase thrust. The empty weight was 42 048 kilograms, some 11% lighter than the S-I. North American Aviation had improved the operation of the eight H-1 engines so that the stage produced 7 200 000 newtons (1 600 000 pounds of thrust), some 6% greater than the S-I. The stage would reach an altitude of 60 kilometers in 2.5 minutes of flight.[10]

The S-IVB second stage went through its acceptance test at the Douglas Aircraft Company's Sacramento Test Center on 8 August and made its first appearance at the Cape on 1 October. While the Cape had welcomed an old friend back in the S-IB, the S-IVB was a newcomer. And an important newcomer: not only would it serve as second stage in AS-201, but it would also be the third stage in the all-important Saturn V. Its single J-2 engine (by

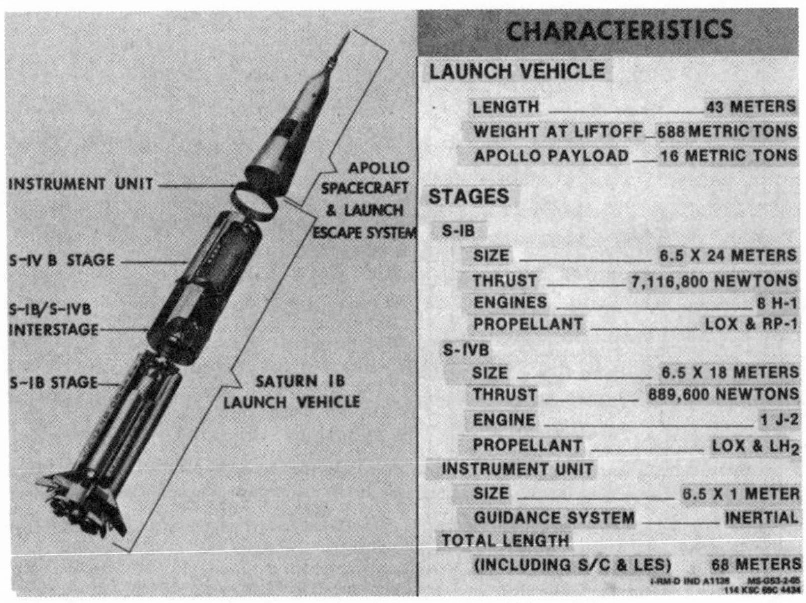

Fig. 122. Schematic of Saturn IB, with characteristics.

North American Rocketdyne Division), burning 7.5 minutes, could put it into earth orbit (though not on AS-201); as the Saturn V third stage, it would put Apollo into translunar trajectory.

Spacecraft components arrived at the Cape in late October, the command module on 25 October and the service module two days later. The base of the service module housed the spacecraft's main propulsion unit, a single engine that used a half-and-half mixture of unsymmetrical-dimethylhydrazine and nitrogen tetroxide to achieve 97 400 newtons (21 900 pounds of thrust). It would be ignited twice on the AS-201 flight, once for three minutes and again for ten seconds. Beneath it was the vehicle's only big piece of boilerplate, the lunar module adapter joining the service module and the S-IVB instrument unit. On AS-201 it consisted of aluminum alloy bracing; in future flights it would house the lunar excursion module.[11]

The Troubled Checkout of AS-201

By late September Merritt Preston's Launch Operations staff and Rocco Petrone's Programs Division had set the 201 launch for late January. That schedule assumed the spacecraft would arrive at the Cape on 9 October. Mueller refused, at first, to approve KSC's recommended date. He hoped for an earlier launch, perhaps in late December. Although Petrone promised to continue looking for possible shortcuts, delays in the spacecraft delivery precluded a 1965 launch.[12]

The Douglas crew erected the S-IVB stage on 1 October and completed the initial checkout of the ground equipment test sets shortly thereafter. KSC received some bad news on the 7th: the RCA 110A computer in the breadboard at Huntsville would not be operational for another ten days. Since it would take two weeks beyond that to check out the computer's program, the 110A executive routine would not reach the Cape before 1 November. Without the executive routine to test the computer's internal systems, the 110A could not apply power to the launch vehicle. On the 15th John Twigg, chief test conductor for the Saturn IB, reported that pad operations were virtually at a standstill. Representatives from Huntsville helped devise a temporary computer program, and the launch team finally applied power to the S-IB stage on 22 October. By the end of the month, KSC had begun limited testing with an uncertified program tape. Meanwhile the instrument unit arrived and underwent inspection at hangar AF. IBM engineers corrected several deficiencies, but an environmental control system coolant pump continued to give the launch team trouble after the instrument unit was stacked above the S-IVB stage on 25 October.[13]

S-IVB stage for AS-201

Figure 123

Figure 124

Figure 125

Figs. 123–125. The second stage for AS-201 arriving at pad 34, 1 October 1965; the stage hoisted; and eased into place.

After arrival of the major modules of spacecraft 009 in late October, the command module went to the hypergolic building for environmental control system servicing and electrical power checks. North American technicians moved the service module out to pad 16 for an electrical systems check. At the operations and checkout building, other workmen installed measuring instruments in the boilerplate lunar module adapter. Although the service module passed leak and functional tests, the 9 November static firing was postponed ten days. A dirty filter in a ground oxidizer system caused much of the delay. On the 18th Preston notified Debus that the late static firing and an accumulation of spacecraft modifications might cause a two-week slip in the launch schedule.[14]

The Saturn ground computer system continued to cause grief. On 10 November John Twigg reported:

> The RCA 110A computer developed problems increasing in number with time of operations. It was detected that some

Fig. 126. The service module for AS-201 in the operations and checkout building, November 1965.

> capacitors which normally breathe in open air developed problems when under protective coating. Most of these cards [electrical printed-circuit boards] in the blockhouse computer which developed these problems first were exchanged. At the same time, it was decided to exchange identical cards in the pad computer as soon as they become available.[15]

Two weeks later Twigg announced the failure of the first computer-run tests, a switch selector functional test and the emergency detection system test.* Isom Rigell's report of 3 December said of the computer:

> We have experienced a number of high-speed memory parity errors in the last few days. No solution has been found to date. December 2d, we have experienced some problems with random discretes to the S-IV-B stage and also apparently random outputs of the computer operating the switch selector. Investigation of this problem is under way at this time. I would like to discuss with you the feasibility to obtain the services of an outside expert (preferably some University instructor) to assess the criticality and problems of our computer system. (Army had good success with this approach, I understand.)[16]

Despite the RCA 110A computer problems, the launch vehicle checkout was nearly on schedule in early December while spacecraft and spacecraft facility tests lagged two weeks behind. On the 13th technicians in protective suits started a week of hypergol tests on the spacecraft feed lines at LC-34. The "hot tests" (using toxic propellants) indicated a need for additional facility modifications. For one thing, the tanks of the command module's reaction control system were not filling properly. On the 15th engineers reported a 48-hour delay in a combined command module–service module systems test. A circuit interrupter malfunction had allowed an electrical signal to interfere with the stabilization and control system's attitude and rate control. Spacecraft Operations completed the test on the 17th after waiting another day for a spare part.[17]

With the spacecraft hot tests tying up pad 34 during daylight hours, the Saturn team continued its checkout from 7:00 p.m. to 3:00 a.m. The Saturn ground computer complex did not work any better in the dark. In five successive switch selector functional tests, the checkout team registered only

---

*The emergency detection system alerted the astronauts to a space vehicle failure and initiated escape procedures. The system sensed hundreds of space vehicle functions and provided triple redundancy, i.e., three sensors checked a function to guard against sensor malfunction. While a manual test of this system took 12–14 hours, an automated checkout ran about 20 minutes.

a partial success. On the 15th Gruene's team traced a bad interface between the digital data acquisition system and the 110A computer to a defective printed-circuit board. After further investigations disclosed five more faulty circuit boards, Launch Vehicle Operations began a survey of all boards.* By Christmas the Saturn IB checkout had fallen 16 days behind schedule. Matters were even worse with Apollo; the spacecraft showed a 20-day slip. A January launch for AS-201 appeared highly unlikely.[18]

North American technicians enjoyed a one-day Christmas holiday. Sunday, the 26th, found a spacecraft team at pad 34 erecting the Apollo on top of the launch vehicle. After mounting the launch escape system, workmen linked the spacecraft with the pad facilities, checked the service module's umbilical arm fit and the white room's interface with the command module. On 5 January the spacecraft began a series of electrical tests with a launch vehicle simulator. Launch vehicle operations, during the same period, included tests of the electrical bridge wire and emergency detection system, a sequence malfunction test, and a LOX simulate and malfunction test (checking the electrical portion of the LOX system).[19]

A new 201 schedule, published 12 January, moved the launch date back to the week of 6 February.[20] The delay was not sufficient, however, as problems continued to plague the operation. Weekly reports on computer failures ran five to six pages long. Launch officials began to wonder if they would ever get through the mission. Norman Carlson, Saturn IB Test Conductor, recalled:

> It [the computer] and we had many hours of grief. You know I really predicted that we would never launch AS-201 by using the computer. It was that bad. We would power up in the morning, and sometimes we were lucky if we got two hours of testing in the whole day. It was up and down all the time.[21]

Some of the 110A idiosyncrasies are amusing in retrospect. On one occasion, an engineer noticed that the computer was repeating a program it had run several hours before. A memory drum had reversed itself and was feeding information back into the computer, which accepted the memory data as new commands. Early in the launch operations, the computers kept going out of action at midnight (Greenwich time). The computers, unable to make the transition from 2400 to 0001, "turned into a pumpkin."[22]

---

*KSC engineers had coated the circuit boards for protection from the Cape's salt air. During computer operations, heat built up under the coating and cracked some solder joints. Thereafter a wire would open intermittently during operations. The boards were sent back to RCA's California plant after the 201 mission for modification; the "fix" was to place a stress relief eyelet around the wire.

Computer problems with ACE and the 110A caused 13 hours of hold time during the plugs-in (umbilicals connected) test on 24 January. After similar difficulties on the plugs-out overall test, Launch Operations Manager Paul Donnelly scheduled another run for 1 February. The repeat was a success and NASA announced a launch date of 22 February. The countdown demonstration test lasted four days (3–6 February). The launch vehicle team followed a script prepared back in October:

Phase 1. Ordnance installation and S-IVB propulsion checks.
Phase 2. Battery installation, power transfer test, and guidance and control checkout.
Phase 3. Command checks and propellant loading.

Workmen hurriedly corrected a number of deficiencies found during the propellant loading and rescheduled the last 22 hours for 8–9 February. Following the flight readiness test on the 12th, North American technicians began loading hypergols aboard the Apollo for a 23 February launch.[23]

In mid-February General Phillips, Apollo Program Director, asked KSC to review the shortcomings of the automated checkout. Phillips hoped to use Marshall's breadboard more effectively in the next Saturn IB checkout.[24] KSC's response listed 15 general recommendations (and more than 50 specific corrections). One problem area involved configuration differences between the breadboard and LC-34, e.g., the Cape's telemetry interrupter for the S-IVB stage was apparently quite different from its counterpart at Huntsville. This illustrated a bigger problem: engineering orders accomplished at one site were overlooked at the other. Since the breadboard had not duplicated all of 34's automation, Marshall was unable to assist with certain problems. KSC recommended that Huntsville have more duplication in the breadboard and a capability to respond faster for emergency tests. The launch team accepted much of the blame for the long delays in initializing and reinitializing the computer—which meant the practice of loading the executive routine (in the form of magnetic tapes) into the computer at the start of a work day. Frequently, a failure in the computer hardware would scramble the operating program and force a technician to reinitialize the 110As. The launch team believed the problem would diminish as operators gained confidence in the computers. The report scored the lack of reliable up-to-date documentation (e.g., unit schematics) and concluded:

> Possibly the most significant single problem area during AS-201, from complex activation through launch, was the never-ending struggle to obtain Engineering Orders to work changes in Electrical Support Equipment. Very few of the changes in Complex

> 34 ESE were subject to any level of technical arbitration. Delays and difficulties were primarily simple matters of overcoming the system inertia.[25]

Launch veterans, in retrospect, have singled out another factor as the biggest challenge on AS-201: the psychological problems of persuading engineers to accept automation. Paul Donnelly recalled that electrical systems personnel were generally receptive since "to check out a computer, the easiest thing to do is use another computer."[26] Convincing the mechanical engineers was another matter. Saturn engineers had little faith that a computer program tape would actuate a hydraulic valve at the proper moment; balky computers compounded the problem. In Hans Gruene's words:

> It was the hardest thing to do to convince engineers who are used to manual operations that the black box out there, which he cannot fully understand, does a job for him and he will not see the little green lights any longer but the box will do the checking for him. I think this convincing of the engineers was the most complicated task in automation.[27]

The 201 Launch

AS-201 finally went down in Cape annals as the "scrub and de-scrub" launch. After their many weeks of problem after problem, delay after delay, the launch team began the countdown at midnight, 20 February. Bad weather imposed three holds, two for 24 hours each; and terminal countdown did not get underway until 5:15 p.m., 25 February. It was held at T – 266 minutes in the early morning hours of 26 February on account of an Apollo access arm problem. A faulty helium regulator took up the remaining 30 minutes of scheduled hold time. At 35 seconds before liftoff, a nitrogen regulator commenced a high flow purge of the S-IB stage's LOX dome and thrust chamber fuel injector.* The reading on the stage's high-pressure nitrogen spheres, normally at 211 kilograms/square centimeter, fell rapidly. At T – 4 seconds the pressure dropped below 199 and the automatic sensor stopped the count.[28]

After some discussion, the launch team decided that the purge of the booster's LOX dome and thrust chamber fuel injector was using most of the nitrogen flow from the ground supply, in effect starving the high-pressure spheres. A technician increased the flow by resetting the pressure on the equipment supplying the nitrogen. At T – 5 minutes, although the nitrogen sphere on the S-IB stage read a satisfactory 203.2 kg/sq cm, Marshall and Chrysler stage engineers requested another hold. Their calculations indicated that, if the low readings on the nitrogen spheres were caused by excessive purge flow or leakage, the existing pressure might not prove sufficient to maintain the minimum needed to pressurize engine gear boxes, actuate LOX and fuel lines, and purge the LOX seal area of the engine turbopumps through stage burnout. The stage engineers recommended eliminating the calorimeter purges. These instruments on the base heat shield measured heat radiation from the stage engines. No serious problems in this area were anticipated and the measurement had no influence on the flight, so the purge was expendable. The operation, however, would take longer than the launch window allowed, and the mission was scrubbed.[29] But a few members of the launch team refused to quit. Gruene reported later:

> A few of my people, including [A. J.] Pickett and [L. E.] Fannin, had an idea that if they could just run one test and convince the [Marshall] people this test was valid, . . . we could still launch the vehicle. We ran the test, de-scrubbed and launched—all in the same day.[30]

The test involved a simulated liftoff and 150-second flight. The simulation demonstrated that 203.2 kg/sq cm of nitrogen in the high-pressure spheres at liftoff would provide adequate pressure in the spheres at burnout.* Hurried calculations by stage engineers supported KSC's findings, and Marshall engineers then agreed to resume the count at 10:57 a.m.[31]

The trouble-plagued AS-201 lifted its 585 metric tons off the pad 15 minutes later. During the 39-minute trip down the Eastern Test Range, the S-IVB stage and the main propulsion engine in the service module increased the Apollo's velocity to nearly 29000 kilometers per hour, a speed greater than manned Apollos would face at reentry. The command module splashed down east of Ascension Island where Navy forces recovered it.[32] With the flight a success, KSC released a general sigh of relief. Carlson said later: "We had struggled so long and so hard. . . . We were all glad to see it go."[33]

The pad suffered substantial damage from flame and vibration at launch. Three seconds after liftoff, high voltage fuses in the pad area substation vibrated loose from their holders and blew a 300-ampere fuse in the industrial power feeder. LC-34 and other Cape facilities were powerless for an hour. One casualty was the launcher water deluge system. Its failure accounted for much of the fire damage on the pad and nearby structures. The power failure also short-circuited the Eastern Test Range's impact computer B,

*Telemetered flight data confirmed that the residual pressure at S-IB cutoff exactly equalled the prediction.

used by Houston to make an abort decision. Computer B tried to transfer to the alternate power system and failed; the back-up computer came on for six seconds and then quit. As a result, Range Safety could not determine vehicle abort impact points during the first five minutes of flight and Mission Control (Houston) operated without trajectory data.[34]

A Reorganization

During the latter launches of the Saturn I program, contractors began to assume responsibility for mission operations—responsibility that civil servants had previously exercised. The transition, completed during the Saturn IB launches, proved a difficult one for many government employees. Many did not want to manage other men, preferring instead to apply their engineering skills directly to the hardware. Veterans of the Debus team recall the change in their status as one of the significant events in the Apollo launch program. Aside from the personal impact, the molding together of the various contractor teams under government management ranks as one of the great accomplishments at KSC.

The problems brought on by the changing role of contractor and civil servant gave impetus to a center reorganization in early 1966. On 17 January Debus told his senior staff that the Office of Manned Space Flight, while voicing the highest praise for KSC's launch operations to date, was concerned about its readiness to handle the upcoming Apollo-Saturn launch preparations. The ensuing study of the management structure was conducted by a KSC task force headed by Deputy Director Albert Siepert, assisted by John Young from NASA Headquarters. General Medaris, former commander of the Army Ballistic Missile Agency, contributed an independent study for the launch center. The study groups concentrated on two problem areas that affected Apollo: the need to clarify and separate the duties of Apollo program management from other center-wide activities, and the liaison of the center with its contractors.[35]

Following the review and evaluation, Debus sent to Headquarters formal proposals to realign KSC's administrative organization. A major change involved the creation of two deputy director posts. The Deputy Director, Operations, would be responsible for engineering matters and technical operations. The Deputy Director, Management, would handle relations with contractors, other government agencies, and the community, and direct the development of management concepts and policies. Two new departments were added. Most of KSC's design functions were centralized under a Director of Design Engineering. He would be responsible for monitoring and issuing technical directions to design support contractors, hardware contractors, and the Corps of Engineers. The other new department, Installation Support, would take over housekeeping services: plant maintenance, supply transportation, documentation security, safety, and quality surveillance. In both cases, the new departments concentrated functions that had previously been scattered among several elements of the launch center.[36]

Debus proposed an important change in the launch operations organization to provide strong and clear direction during the performance of preflight and launch operations. Test management, as a discrete function, was set up at the top Launch Operations level, with counterparts at the Launch Vehicle and Spacecraft Operations directorates. These offices would plan and direct launch operations, with a specific individual in charge of each mission. The test manager would be just that—a *manager*, not merely a *coordinator* as had generally been the case in the past. In this capacity, he would be responsible for the mission hardware from the time of its arrival at the center to the launch. Engineers in various operational areas would be assigned to assist the test manager when required. These specialists, however, would not have authority to give formal instructions to the contractors performing the work; they were to provide only informal technical guidance. Formal instructions could come only from the test manager.

The reorganization altered the civil servant–contractor relationship in several important ways. The Director of Design Engineering assumed responsibility for all KSC hardware development contracts, construction and modification contracts, as well as the design engineering support contracts. Lines for reporting were streamlined so that other major contractors reported to a single KSC element. The changes established a specific chain of command for each launch and helped the government provide the contractors with formal direction, informal instruction, and a better evaluation of performance. Administrator Webb signed the new KSC organizational chart on 27 April and the changes were phased in through the remainder of the year.[37]

More Launches of the Saturn IB

When spacecraft problems in the spring of 1966 delayed the preparation of Apollo module 011, AS-203 became the second Saturn IB flight. The AS-203 carried no spacecraft; its primary purpose was to test the dynamics of liquid hydrogen in the weightlessness of space. On a lunar mission, the S-IVB stage would orbit the earth one and one-half times and then restart its J-2 engine to propel Apollo toward the moon. Marshall engineers wondered whether the ten tons of liquid hydrogen would settle to one part of the fuel tanks or slosh violently about. The S-IVB stage of the AS-203 was equipped with 83 special measuring devices and two television cameras to study the chilldown of the J-2 engine (the preliminary cooling of the propellant systems with small amounts of cryogenic hydrogen). The mission also tested IBM's new instrument unit. AS-203 was launched from pad 37B, which had been modified extensively since the SA-10 launch the previous summer.[38]

Chrysler technicians erected the S-IB booster on 19 April. On subsequent days the S-IVB stage, the instrument unit, and the nose cone joined the stack. The checkout soon bogged down in another epidemic of computer ills. Most of the blame was laid to cracked solder joints in the printed-circuit boards, the same defect that had troubled AS-201. By 24 May technicians had exchanged 2000 printed boards and planned to remove 6000 more. Other portions of the Saturn checkout proceeded on schedule. On 27 May Albert Joralan reported that S-IB measuring calibration was 70% complete; the calibration of the S-IVB and instrument unit stood at 60 and 87%.[39]

The month of June saw an unusual spectacle at the Cape—three Saturns looking skyward, and menaced briefly by a hurricane. Saturn 500F stood on LC-39, AS-202 on LC-34, and AS-203 on LC-37. The simultaneous operations taxed KSC's propellant reserves, but essential needs were met.[40]

The AS-203 launch, originally scheduled for 30 June, was delayed by an Explorer launch and minor problems. It was almost scrubbed when one of the television cameras failed, but on 5 July the rocket achieved a virtually perfect orbital insertion. The remaining television camera operated perfectly, and apparently answered any questions about S-IVB's readiness to serve as the Saturn V third stage. In September Douglas Aircraft announced that the S-IVB stage had no serious unsolved technical problems.[41]

Computer problems also characterized the AS-202 operations at LC-34. Printed-circuit boards continued to frustrate Gruene's Launch Vehicle Operations Division and, after the AS-203 launch, KSC transferred all of LC-37's printed-circuit boards to LC-34. The change reduced the downtime of the RCA 110As considerably. Despite the launch vehicle team's misfortunes, NASA spokesmen cited spacecraft delays in postponing the AS-202 launch until after the 203 mission. Late deliveries of equipment and engineering orders plagued spacecraft operations. The patching of the ACE system (rerouting the electrical lines to various pieces of test equipment) was particularly troublesome. The spacecraft team found, to their sorrow, that Apollo 011 did not duplicate the 009 modules. The spacecraft team corrected most of the problems in three months and erected Apollo 011 on 2 July 1966. The countdown demonstration test began on the 29th and ran for one week. During that period, KSC also conducted two spacecraft emergency egress tests. The launch team completed the flight readiness test on 16 August.[42] Alfred O'Hara, chief of the Saturn I-IB Operations Office, reported that all

Saturn tests had been completed satisfactorily. Richard Proffitt, spacecraft test conductor, described the Apollo checkout as "a good clean test and we feel that we are 100 per cent ready."[43]

AS-202 lifted off on 25 August. A communications problem between Mission Control in Houston and a tracking ship in the Atlantic had caused the only significant delay in the countdown. In the final minutes, however, the launch team barely outraced hurricane Faith; the tropical storm shut down the Antigua tracking station 45 minutes after launch. AS-202's 93-minute suborbital flight covered 33 000 kilometers. Although the spacecraft splashed down 370 kilometers short of its target in the Pacific Ocean, the mission was judged a success. A design certification review board, meeting in September, declared that Apollo–Saturn IB could now be used for manned flight.[44]

The Apollo–Saturn IB launches of 1966 represented important gains for NASA's launch team. LC-34 and LC-37, testbeds for automated checkout, were found wanting. In the 20 months between AS-201 and AS-501, KSC corrected the major automation problems. Without these trial and error advances, AS-501, the toughest launch in Apollo's history, would have been far more difficult.

18

The Fire That Seared the Spaceport

The thirteenth Saturn flight (the third Saturn IB) on 25 August 1966 was the thirteenth success. It fulfilled all major mission objectives. For the first manned mission NASA had selected two veterans and one rookie. Command Pilot Virgil Ivan Grissom had flown Mercury's Liberty Bell 7, America's second suborbital flight, in July 1961, and Molly Brown, the first manned Gemini, in March 1965. Edward White had become the first American to walk in space while on the fourth Gemini flight, three months later. Flying with these two would be the youngest American ever chosen to go into space, Roger B. Chaffee, 31 years of age.

NASA gave Grissom the option of an open-ended mission. The astronauts could stay in orbit up to 14 days, depending on how well things went. The purpose of their flight was to check out the launch operations, ground tracking and control facilities, and the performance of the Apollo-Saturn. Grissom was determined to keep 204 up the full 14 days if at all possible.

North American Aviation constructed the Apollo command and service modules. The spacecraft, 11 meters long and weighing about 27 metric tons when fully fueled, was considerably larger and more sophisticated than earlier space vehicles, with a maze of controls, gauges, dials, switches, lights, and toggles above the couches. Unlike the outward-opening hatches of the McDonnell-built spacecraft for Mercury and Gemini flights, the Apollo

Fig. 127. The crew of AS-204: Grissom, White, and Chaffee.

hatches opened inward. They required a minimum of ninety seconds for opening under routine conditions.[1]

Predictions of Trouble

Many men, including Grissom, had presumed that serious accidents would occur in the testing of new spacecraft. A variety of things could go wrong. But most who admitted in the back of their minds that accidents might occur, expected them somewhere off in space.

Some individuals had misgivings about particular aspects of the spacecraft. Dr. Emmanuel Roth of the Lovelace Foundation for Medical Education and Research, for instance, prepared for NASA in 1964 a four-part series on "The Selection of Space-Cabin Atmospheres." He surveyed and summarized all the literature available at the time. He warned that combustible items, including natural fabrics and most synthetics, would burn violently in the pure oxygen atmosphere of the command module. Even allegedly flame-proof materials would burn. He warned against the use of combustibles in the vehicle.[2]

In 1964 Dr. Frank J. Hendel, a staff scientist with Apollo Space Sciences and Systems at North American and the author of numerous articles and a textbook, contributed an article on "Gaseous Environment during Space Missions" to the *Journal of Spacecraft and Rockets*, a publication of the American Institute of Aeronautics and Astronautics. "Pure oxygen at five pounds per square inch of pressure," he wrote, "presents a fire hazard which is especially great on the launching pad. . . . Even a small fire creates toxic products of combustion; no fire-fighting methods have yet been developed that can cope with a fire in pure oxygen."[3]

Further, oxygen fires had occurred often enough to give safety experts cause for extra-careful procedures: at Brooks Air Force Base and at the Navy's Equipment Diving Unit at Washington, D.C., in 1965; and at the Airesearch Facility in Torrance, California, in 1964, 1965, and 1966.[4]

One man saw danger on earth, from hazards other than fire. In November 1965, the American Society for Testing and Materials held a symposium in Seattle on the operation of manned space chambers. The papers gave great attention to the length of time spent in the chambers, to decompression problems, and to safety programs. The Society published the proceedings under the title of *Factors in the Operation of Manned Space Chambers* (Philadelphia, 1966). In reviewing this publication, Ronald G. Newswald concluded: "With reliability figures and flight schedules as they are, the odds are that the first casualty in space will occur on the ground."[5]

Since Newswald was a contributing editor of *Space/Aeronautics*, it may well be that he contributed the section entitled "Men in Space Chambers: Guidelines Are Missing" in the "Aerospace Perspective" section of that magazine during the same month that his review appeared in *Science Journal*. The editorial reflects the ideas and the wording of his review. The "Guidelines" writer began: "The odds are that the first spaceflight casualty due to environmental exposure will occur not in space, but on the ground." He saw no real formulation of scientific procedures involving safety—such as automatic termination of a chamber run in the event of abnormal conditions. "By now," he stated, "NASA and other involved agencies are well aware that a regularly updated, progressive set of recommended practices—engineering, medical and procedural—for repressurization schedules and atmospheres, medical monitoring, safety rescue and so on, would be welcome in the community."[6]

Gen. Samuel Phillips, Apollo Program Director, had misgivings about the performance of North American Aviation, the builder of the spacecraft, as early as the fall of 1965. He had taken a task force to Downey, California, to go over the management of the Saturn-II stage and command-service module programs. The task force included Marshall's Eberhard Rees and the Apollo Spacecraft Program Manager, Joseph Shea; they had many discussions with the officials of North American. On 19 December 1965, Phillips wrote to John Leland Atwood, the President of North American Aviation, enclosing a "NASA Review Team Report," which later came to be called the "Phillips Report."[7] The visit of the task force was not an unusual NASA procedure, but the analysis was more intensive than earlier ones.

In the introduction, the purpose was clearly stated: "The Review was conducted as a result of the continual failure of NAA to achieve the progress required to support the objective of the Apollo program."[8] The review included an examination of the corporate organization and its relationship to the Space Division, which was responsible for both the S-II stage and the command-service module, and an examination of North American Aviation's activities at Kennedy Space Center and the Mississippi Test Facility. The former area belongs more properly to the relations of North American Aviation with NASA Headquarters, but the latter directly affected activities at Kennedy Space Center.

Despite the elimination of some troublesome components and escalations in costs, both the S-II stage and the spacecraft were behind schedule. The team found serious technical difficulties remaining with the insulation and welding on stage II and in stress corrosion and failure of oxidizer tanks on the command-service module. The "Report" pointed out that NAA's inability to meet deadlines had caused rescheduling of the total Apollo program and, with reference to the command-service module, "there is little confidence that NAA will meet its schedule and performance commitments."[9]

Phillips and his task force returned to Downey for a follow-up week in mid-April 1966. He did not amend the original conclusions, but he told President Atwood that North American was moving in the right direction.[10]

The astronauts themselves suggested many changes in the block I spacecraft design. In April 1967, Donald K. Slayton was to tell the Subcommittee on NASA Oversight of the House Committee on Science and Astronautics that the astronauts had recommended 45 improvements, including a new hatch. North American had acted on 39 of these recommendations. They were introducing the other six into later spacecraft. "Most of these," Slayton testified, "were of a relatively minor nature."[11] The only major change for later spacecraft was to have been a new hatch. And the astronauts had recommended this not so much for safety as for ease in getting out for space-walks and at the end of flights.[12]

The Spacecraft Comes to KSC

In July and August 1966, NASA officials conducted a customer acceptance readiness review at North American Aviation's Downey plant, issued a certificate of worthiness, and authorized spacecraft 012 to be shipped to the Kennedy Space Center. The certificate listed incomplete work: North American Aviation had not finished 113 significant engineering orders at the time of delivery.[13]

The command module arrived at KSC on 26 August and went to the pyrotechnic installation building for a weight and balance demonstration.[14] With the completion of the thrust vector alignment on 29 August, the test team moved the command module to the altitude chamber in the operations and checkout building and began mating the command and service modules. Minor problems with the service module had already showed up, and considerable difficulties with the new mating hardware caused delays.

On 7 September NASA released a checkout schedule. By 14 September, while the Saturn launch vehicle moved on schedule, the Apollo spacecraft already lagged four days behind. On the same day, a combined systems test was begun. Discrepancy reports numbered 80 on 16 September and had risen to 152 within six days. One of the major problems was a short in the radio command system. In the meantime, the test team had installed all but one of the flight panels.

At Headquarters during this time, a board chaired by the Associate Administrator for Manned Space Flight, Dr. George Mueller, and made up of OMSF center directors, conducted a detailed review of the spacecraft. On 7 October this board certified the design as "flightworthy, pending satisfactory resolution of listed open items."[15]

The simulated altitude run, originally scheduled for 26 September, had gradually slipped back in schedule. It was run on 11 October, but plans for an unmanned altitude run on 12 October, a flight crew altitude run on 14 October, and a backup crew run on 15 October also slipped. So did the projected dates of mechanical mating of the spacecraft with the launch vehicle and the launch itself.

The unmanned altitude chamber run finished satisfactorily on 15 October. The first manned run in the altitude test chamber, on 18 October, experienced trouble after reaching a simulated altitude of 4000 meters because of the failure of a transistor in one of the inverters. With the replacement of the inverter, the system functioned satisfactorily. The prime crew of Grissom, White, and Chaffee repeated the 16-hour run the next day with only one major problem developing in the oxygen cabin supply regulator. This problem caused a delay of the second manned run with the backup crew scheduled for 21 October. Continued trouble with the new oxygen regulator caused the indefinite suspension of the second manned test before the end of October. By this time it had become clear that the spacecraft needed a new environmental control unit. Technicians removed the old unit on 1 November.

Meanwhile, at North American Aviation's Downey plant a propellant tank had ruptured in the service module of spacecraft 017. This provoked a special test of the propellant tanks on the 012 service module at KSC. In order to conduct this testing in parallel with further checking of the command module, the test team removed the command module from the altitude chamber.[16] Later they removed the fuel tanks from the service module in the chamber. After pressure-integrity tests, they replaced the tanks and returned the command module to the chamber.[17] The test team installed and fit-checked the new environmental control unit on 8 November and hooked up the interface lines two days later. But this did not completely solve the difficulties. Problems in the glycol cooling system surfaced toward the end of November and on 5 December forced a removal of the second environmental control unit.

The Apollo Review Board was to say of this glycol leakage several months later,

> water/glycol coming into contact with electrical connectors can cause corrosion of these connectors. Dried water/glycol on wiring insulation leaves a residue which is electrically conductive and combustible. Of the six recorded instances where

> water/glycol spillage or leakage occurred (a total of 90 ounces leaked or spilled is noted in the records) the records indicate that this resulted in wetting of conductors and wiring on only one occasion. There is no evidence which indicates that damage resulted to the conductors or that faults were produced on connectors due to water/glycol.[18]

The difficulties in the materials that already had arrived at KSC and the endless changes that came in from North American Aviation—623 distinct engineering orders—presented major problems for the NASA-NAA test teams. As many workmen as could possibly function inside the command module continually swarmed into it to replace defective equipment or make the changes that NAA suggested and Houston approved. The astronauts came and went, sometimes concerned with major and sometimes with minor matters on the spacecraft.[19]

These difficulties at KSC and concurrent problems at Mission Control Center, Houston, forced two revisions to the schedule, one on 17 November, the next on 9 December. The test team kept up with or moved ahead of the latter schedule during the ensuing weeks. The third environmental control unit arrived for installation on 16 December.

The test teams had been working on a 24-hour basis since the arrival of the spacecraft at Kennedy, taking off only on Christmas and New Year's Day. On 28 December, while conducting an unmanned altitude run, the test team located a radio frequency communications problem and referred it to ground support technicians for correction. On 30 December a new backup crew of Schirra, Eisele, and Cunningham (McDivitt's original backup crew had received a new assignment) successfully completed a manned altitude run.[20] Six major problems on the spacecraft surfaced, one in very-high-frequency radio communications; but a review board was to give a favorable appraisal not long afterward: "This final manned test in the altitude chamber was very successful with all spacecraft systems functioning normally. At the post-test debriefing the backup flight crew expressed their satisfaction with the condition and performance of the spacecraft."[21]

By 5 January the mating of the spacecraft to the lunar module adapter and the ordnance installation were proceeding six days ahead of schedule. The following day the spacecraft was moved from the operations and checkout building to LC-34. KSC advanced the electrical mating and the emergency detection system tests to 18 January, and these were completed that day. The daily status report for 20 January 1967 reported that no significant problems occurred during the plugs-in overall test. A repeat of the test on 25 January took 24 hours. A problem in the automatic checkout equipment link-up caused the delay. Further, the instrument unit did not record

Fig. 128. Mating the AS-204 spacecraft modules in the operations and checkout building, 4 January 1967.

simulated liftoff—a duplication of an earlier deficiency. The schedule called for a plugs-out test at 3:00 p.m. on 26 January, a test in which the vehicle would rely on internal power. NASA did not rate the plugs-out test as "hazardous," reserving that label for tests involving fueled vehicles, hypergolic propellants, cryogenic systems, high-pressure tanks, live pyrotechnics, or altitude chamber tests.[22]

The Hunches of Tom Baron

All the tests and modifications in the spacecraft did not go far enough or fast enough in the view of one North American employee, Thomas R. Baron of Mims, Florida. Baron's story has significance for two reasons. His attitude reflected the unidentified worries of many who did not express them until too late. Also, the reaction of KSC managers indicated a determination to check every lead that might uncover an unsafe condition. The local press at the time gave ample but one-sided coverage of the Baron story.

Baron had a premonition of disaster. He believed his company would not respond to his warnings and wanted to get his message to the top command at KSC. While a patient at Jeff Parrish Hospital in Titusville, Florida, during December 1966, and later at Holiday Hospital in Orlando, Baron expressed his fears to a number of people. His roommate at Jeff Parrish happened to be a KSC technical writer, Michael Mogilevsky.[23] After Baron claimed to have in his possession documentary evidence of deficiencies in the heat shield, cabling, and life support systems, Mogilevsky went to see Frank Childers in NASA Quality Control on 16 December. Childers called in an engineer of the Office of the Director of Quality Assurance, and Mogilevsky related Baron's complaints and fears again.[24]

That evening Rocco Petrone asked John M. Brooks, the Chief of NASA's Regional Inspections Office, to locate and interview Baron. Brooks interviewed Baron twice and briefed Debus, Albert Siepert, and Petrone on Baron's complaints: poor workmanship, failure to maintain cleanliness, faulty installation of equipment, improper testing, unauthorized deviations from specifications and instructions, disregard for rules and regulations, lack of communication between Quality Control and engineering organizations and personnel, and poor personnel practices.

Baron claimed to possess notebooks that would substantiate his charges. He promised to cooperate with KSC and with North American Aviation if someone above his immediate supervisor would listen to what he had to say. He did not believe his previous complaints had ever gone beyond that supervisor. He asked to be allowed to talk to John Hansel, Chief of Quality Control for North American. Baron's complaints were against North American, not KSC. He believed that the center needed additional personnel to enforce compliance with procedures in the Apollo program. Brooks later reported: "Baron was assured that an appropriate level of NAA management would be in touch with him in the next day or two."[25]

On 22 December 1966, Petrone and Wiley E. Williams, Test and Operations Management Office, Directorate for Spacecraft Operations, received a briefing on Baron's complaints. The two men recognized that these were primarily North American Aviation in-house problems and that the company should inquire into Baron's complaints and advise KSC officials of the results. NAA officials W. S. Ford, James L. Pearce, and John L. Hansel met with Petrone that same day. They arranged to talk with Baron the following day.[26]

Since Baron had confidence in Hansel, who was an expert in Quality Control, Hansel's testimony is especially valuable. Baron had lots of complaints but, Hansel insisted, no real proof of major deficiencies, either in the papers Baron had in his possession or in the report that Baron wrote (and

Hansel was to read) a short time later. Lastly, Hansel stated, Baron was not working in a critical area at that time.[27]

North American informed Petrone of the interview by 4 January, but sent no written report to Petrone's office.[28] On 5 January a North American spokesman told newsmen that the company was terminating Baron's services.[29] Since his clearance at the space center had been withdrawn, Baron phoned John Brooks, the NASA inspector, on 24 January and invited him to his home. Brooks accepted the invitation, and Baron gave him a 57-page report for duplication and use. Brooks duplicated it and returned the original to Baron on 25 January.[30] Brooks assured Baron that KSC and NAA had looked into his allegations and taken corrective action where necessary.

Petrone received a mimeographed copy of Baron's report on 26 January. John Wasik of the Titusville *Star Advocate* telephoned Brooks to ask about KSC's interest in Baron's information. Wasik indicated that he was going to seek an interview with Petrone. On the following morning, Gordon Harris, head of the Public Affairs Office at KSC, heard that Wasik had spent approximately one and one-half hours with Zack Strickland, of the North American Aviation Public Relations Information Office, going over the Baron report.[31]

That same day Hansel, North American's head of Quality Control—the man Baron had hoped his report would reach—told Wasik that Baron was one of the most conscientious quality control men he ever had working for him and that his work was always good. "If anything," Hansel related in the presence of Strickland, "Baron was too much of a perfectionist. He couldn't bend and allow deviations from test procedures—and anyone knows that when you're working in a field like this, there is constant change and improvement. The test procedures written in an office often don't fit when they are actually applied. Baron couldn't understand this." Wasik also stated: "Hansel readily agreed that Baron's alleged discrepancies were, for the most part, true."[32] What Wasik did not say was that none of the discrepancies, true though they were, was serious enough to cause a disaster.

Hansel was not alone in his misgivings about Baron. Hansel did not know of Frank Childers's report nor had he ever talked to Childers about Baron. Childers, too, had doubts about the man's reliability. Even though he had sympathetically reported to NASA officials the fears of the North American employee, Childers admitted that Baron, who signed himself T. R. Baron, had the nickname "D. R. (Discrepancy Report) Baron."[33] R. E. Reyes, an engineer in KSC's Preflight Operations Branch, said Baron filed so many negative charges that, had KSC heeded them all, NASA would not have had a man on the moon until the year 2069.[34] To confirm the opinions of these men, Baron himself admitted before a congressional investigating committee a short time later that he had turned in so many negative reports that his department ran out of the proper forms. Further—in confirmation of Hansel's view of Baron's report—Baron based his testimony on hearsay, not on any personal records in his possession.[35] Baron's forebodings were to prove correct, but not for any reason he could document.*

Both NASA and North American Aviation, a historian must conclude, gave far more serious consideration to Baron's complaints than a casual perusal of newspapers during the succeeding weeks, or even close reading of such books as *Mission to the Moon*, would indicate.[36]

Disaster at Pad 34

While top administrators were checking out the fears of Tom Baron, two NASA men, Clarence Chauvin and R. E. Reyes, and two North American Project Engineers, Bruce Haight and Chuck Hannon, met on the morning of 26 January at launch complex 34 to review the general spacecraft readiness and configuration for one of the last major previews, the plugs-out test. The craft looked ready.[37]

That same night the prime and backup crews studied mission plans. The next day a simulated countdown would start shortly before liftoff and then the test would carry through several hours of flight time. There would be no fuel in the Saturn. Grissom, White, and Chaffee would don their full spacesuits and enter the Apollo, breathing pure oxygen to approximate orbital conditions as closely as possible. After simulated liftoff, the spacecraft center in Houston would monitor the performance of the astronauts. The plugs-out test did not rate a hazardous classification; the spacecraft had successfully operated in the test chamber for a greater period of time than it would on the pad.[38]

The astronauts entered the Apollo at 1:00 p.m., Friday, 27 January 1967. Problems immediately arose. NASA Spacecraft Test Conductor Clarence Chauvin later described them: "The first problem that we encountered was when Gus Grissom ingressed into the spacecraft and hooked up to his oxygen supply from the spacecraft. Essentially, his first words were that there was a strange odor in the suit loop. He described it as a 'sour smell'

*The Chairman of the House Subcommittee on NASA Oversight, Congressman Olin Teague of Texas, said in thanking Baron for his testimony: "What you have done has caused North American to search their procedures." House Subcommittee on NASA Oversight, *Investigation into Apollo 204 Accident*, 1: 499.

somewhat like buttermilk." The crew stopped to take a sample of the suit loop, and after discussion with Grissom decided to continue the test.

The next problem was a high oxygen flow indication which periodically triggered the master alarm. The men discussed this matter with environmental control systems personnel, who believed the high flow resulted from movements of the crew. The matter was not really resolved.

A third serious problem arose in communications. At first, faulty communications seemed to exist solely between Command Pilot Grissom and the control room. The crew made adjustments. Later, the difficulty extended to include communications between the operations and checkout building and the blockhouse at complex 34. "The overall communications problem was so bad at times," Chauvin testified, "that we could not even understand what the crew was saying."[39] William H. Schick, Assistant Test Supervisor in the blockhouse at complex 34, reported in at 4:30 p.m. and monitored the spacecraft checkout procedure for the Deputy of Launch Operations. He sat at the test supervisor's console and logged the events, including various problems in communications.[40] To complicate matters further, no one person controlled the trouble-shooting of the communications problem.[41] This failure in communication forced a hold of the countdown at 5:40 p.m. By 6:31 the test conductors were about ready to pick up the count when ground instruments showed an unexplained rise in the oxygen flow into the spacesuits. One of the crew, presumably Grissom, moved slightly.

Four seconds later, an astronaut, probably Chaffee, announced almost casually over the intercom: "Fire. I smell fire." Two seconds later, Astronaut White's voice was more insistent: "Fire in the cockpit."

In the blockhouse, engineers and technicians looked up from their consoles to the television monitors trained at the spacecraft. To their horror, they saw flames licking furiously inside Apollo, and smoke blurred their pictures. Men who had gone through Mercury and Gemini tests and launches without a major hitch stood momentarily stunned at the turn of events. Their eyes saw what was happening, but their minds refused to believe. Finally a near hysterical shout filled the air: "There's a fire in the spacecraft!"

Procedures for emergency escape called for a minimum of 90 seconds. But in practice the crew had never accomplished the routines in the minimum time. Grissom had to lower White's headrest so White could reach above and behind his left shoulder to actuate a ratchet-type device that would release the first of a series of latches. According to one source, White had actually made part of a full turn with the ratchet before he was overcome by smoke. In the meantime, Chaffee had carried out his duties by switching the power and then turning up the cabin lights as an aid to vision. Outside the white room that totally surrounded the spacecraft, Donald O. Babbitt of North

American Aviation ordered emergency procedures to rescue the astronauts. Technicians started toward the white room. Then the command module ruptured.[42]

Witnesses differed as to how fast everything happened. Gary W. Propst, an RCA technician at the communication control racks in area D on the first floor at launch complex 34, testified four days later that three minutes elapsed between the first shout of "Fire" and the filling of the white room with smoke. Other observers had gathered around his monitor and discussed why the astronauts did not blow the hatch and why no one entered the white room. One of these men, A. R. Caswell, testified on 2 February, two days after Propst. In answer to a question about the time between the first sign of fire and activity outside the spacecraft in the white room, he said: "It appeared to be quite a long period of time, perhaps three or four minutes. . . ."[43]

The men on the launch tower told a different story. Bruce W. Davis, a systems technician with North American Aviation who was on level A8 of the service structure at the time of the fire, reported an almost instantaneous spread of the fire from the moment of first warning. "I heard someone say, 'There is a fire in the cockpit.' I turned around and after about one second I saw flames within the two open access panels in the command module near the umbilical." Jessie L. Owens, North American Systems Engineer, stood near the pad leader's desk when someone shouted: "Fire." He heard what sounded like the cabin relief valve opening and high velocity gas escaping. "Immediately this gas burst into flames somewhat like lighting an acetylene torch," he said. "I turned to go to the white room at the above-noted instant, but was met by a flame wall."[44]

Spacecraft technicians ran toward the sealed Apollo, but before they could reach it, the command module ruptured. Flame and thick black clouds of smoke billowed out, filling the room. Now a new danger arose. Many feared that the fire might set off the launch escape system atop Apollo. This, in turn, could ignite the entire service structure. Instinct told the men to get out while they could. Many did so, but others tried to rescue the astronauts.

Approximately 90 seconds after the first report of fire, pad leader Donald Babbitt reported over a headset from the swing arm that his men had begun attempts to open the hatch. Thus the panel that investigated the fire concluded that only one minute elapsed between the first warning of the fire and the rescue attempt. Babbitt's personal recollection of his reporting over the headset did not make it clear that he had already been in the white room, as the panel seemed to conclude.[45] Be that as it may, for more than five minutes, Babbitt and his North American Aviation crew of James D. Gleaves,

Figure 129

Figure 130

Fig. 129. The interior of the AS-204 spacecraft after the fire: Dale Carothers, Spacecraft Operations Directorate, in the white room, looking through the open hatch. Fig. 130. Exterior of AS-204, with the white room to the left.

Jerry W. Hawkins, Steven B. Clemmons, and L. D. Reece, and NASA's Henry H. Rodgers, Jr., struggled to open the hatch. The intense heat and dense smoke drove one after another back, but finally they succeeded. Unfortunately, it was too late. The astronauts were dead. Firemen arrived within three minutes of the hatch opening, doctors soon thereafter. A medical board was to determine that the astronauts died of carbon monoxide asphyxsia, with thermal burns as contributing causes. The board could not say how much of the burns came after the three had died. Fire had destroyed 70% of Grissom's spacesuit, 25% of White's, and 15% of Chaffee's.[46] Doctors treated 27 men for smoke inhalation. Two were hospitalized.

Rumors of disaster spread in driblets through the area. Men who had worked on the day shift returned to see if they could be of help. Crewmen removed the three charred bodies well after midnight.[47]

The sudden deaths of the three astronauts caused international grief and widespread questioning of the space program. Momentarily the whole manned lunar program stood in suspense. Writing in *Newsweek*, Walter Lippman immediately deplored what he called the pride-spurred rush of the program.[48] The Washington *Sunday Star* spoke of soaring costs and claimed that "know-who" had more to do than "know-how" in the choice of North American over Martin Marietta as prime contractor for the spacecraft.[49] A long-time critic of the space program, Senator William J. Fulbright of Arkansas, Chairman of the Senate Foreign Relations Committee, placed the "root cause of the tragedy" in "the inflexible, but meaningless, goal of putting an American on the moon by 1970" and called for a "full reappraisal of the space program." The distinguished scientist Dr. James A. Van Allen, discoverer of radiation belts in space, charged that NASA was "losing its soul." It had become "a huge engineering, technological and operational agency with less and less devotion to the true spirit of exploration and to the advancement of basic knowledge."[50] A lead editorial in the *New York Times* spoke of the incompetence and negligence that became apparent as the full story of disaster came to light, but put the central blame on "the technically senseless" and "highly dangerous" dedication to the meaningless timetable of putting a man on the moon by 1970.[51] An article in the American Institute of Chemical Engineers *Journal* had the long-anticipated title: "NASA's in the Cold, Cold Ground."[52] But President Johnson held firm to the predetermined goal and communicated his confidence to NASA.[53]

The Review Board

After removal of the bodies, NASA impounded everything at launch complex 34. On 3 February, NASA Administrator Webb set up a review board to investigate the matter thoroughly. Except for one Air Force officer and an explosives expert from the Bureau of Mines, both specialists in safety, all the members of the board came from NASA.* North American Aviation had a man on the board for one day. At least George Jeffs, NAA's chief Apollo engineer, thought he was on the board. After consultation with Shea and Gilruth of the Manned Space Flight Center, North American officials recommended him as one who could contribute more than any other NAA officer. Jeffs flew to the Cape and sat in on several meetings until, as Jeffs was to report later to the House Subcommittee on NASA Oversight, "I was told that I was no longer a member of the Board." The representative of the review board who dismissed Jeffs gave no reason for the dismissal.[54] Thus all members of the board were government employees, a fact that was to cause NASA considerable criticism from Congress.

Debus asked all KSC and contractor employees for complete cooperation with the review board. He called their attention to the Apollo Mission Failure Contingency Plan of 13 May 1966 that prohibited all government and contractor employees from discussing technical aspects of the accident with anyone other than a member of the board. All press information would go through the Public Affairs Office. In scheduled public addresses, speakers might discuss other aspects of the space program but "should courteously but absolutely refuse to speculate at this time on anything connected with the Apollo 204 investigation or with factors that might be related, directly or indirectly, to the accident."[55] Debus's action muted at KSC the wild rumors that had prevailed in east Florida and spread throughout the country after the fire.[56]

Under authorization from the review board, ground crews carefully removed the debris on the crew couches inside the command module on 3 February. They recorded the type and location of the material removed. Then they laid a plywood shelf across the three interlocked seats so that combustion specialists could enter the command module and examine the cabin more thoroughly. On the following day they removed the plywood and the three seats. Two days after that, they suspended a plastic false floor inside the command module so that investigators could continue to examine the

---

*The NASA members were: the chairman, Dr. Floyd L. Thompson, Dir., Langley Research Center; Astronaut Frank Borman, Manned Spacecraft Center; Dr. Maxime A. Faget, Dir., Engineering and Development, MSC; E. Barton Geer, Assoc. Chief, Flight Vehicles and Systems Div., Langley; George C. White, Jr., Dir. of Reliability and Quality, Apollo Program Off.; and John J. Williams, Dir., Spacecraft Operations, Kennedy. The non-NASA members were Dr. Robert W. Van Dolah, Research Dir., Explosive Research Center, Bureau of Mines, Dept. of the Interior; and Col. Charles F. Strange, Chief of Missiles and Space Safety Div., Off. of the Air Force Inspector General, Norton AFB, CA. *Report of Apollo 204 Review Board*, p. 5. The only non-government person on the original board, Dr. Frank Long of Cornell Univ., a member of the President's Scientific Advisory Committee, soon resigned because of the press of other activities and was replaced by Van Dolah. *Aviation Week and Space Technology*, 13 Feb. 1967, p. 33.

command module interior without aggravating the condition of the lower part of the cabin.[57]

Engineers at the Manned Spacecraft Center duplicated conditions of Apollo 204 without crewmen in the capsule. They reconstructed events as studies at KSC brought them to light. The investigation on pad 34 showed that the fire started in or near one of the wire bundles to the left and just in front of Grissom's seat on the left side of the cabin—a spot visible to Chaffee. The fire was probably invisible for about five or six seconds until Chaffee sounded the alarm. "From then on," a *Time* writer stated, "the pattern and the intensity of the test fire followed, almost to a second, the pattern and intensity of the fire aboard Apollo 204."[58]

The members of the review board sifted every ash in the command module, photographed every angle, checked every wire, and questioned in exhausting detail almost everyone who had the remotest knowledge of events related to the fire. They carefully dismantled and inspected every component in the cockpit.[59]

In submitting its formal report to Administrator Webb on 5 April 1967, the board summarized its findings: "The fire in Apollo 204 was most probably brought about by some minor malfunction or failure of equipment or wire insulation. This failure, which most likely will never be positively identified, initiated a sequence of events that culminated in the conflagration."*[60]

To the KSC Safety Office, the next finding of the Review Board seemed to be the key to the entire report: "Those organizations responsible for the planning, conduct and safety of this test failed to identify it as being hazardous."[61] Since NASA had not considered the test hazardous, KSC had not instituted those procedures that normally would have accompanied such a test.[62]

The Review Board had other severe criticism:

> Deficiencies existed in Command Module design, workmanship and quality control. . . .
>
> The Command Module contained many types and classes of combustible material in areas contiguous to possible ignition sources. . . . The rapid spread of fire caused an increase in

---

*The review board ignored and a congressional committee later vehemently rejected the hypothesis of Dr. John McCarthy, NAA Division Director of Research, Engineering, and Test, that Grissom accidentally scuffed the insulation of a wire in moving about the spacecraft. (*Investigation into Apollo 204 Accident*, 1: 202, 263.) In the same congressional investigation, Col. Frank Borman, the first astronaut to enter the burnt-out spacecraft, testified: "We found no evidence to support the thesis that Gus, or any of the crew members kicked the wire that ignited the flammables." This theory that a scuffed wire caused the spark that led to the fire still has wide currency at Kennedy Space Center. Men differ, however, on the cause of the scuff.

> pressure and temperature which resulted in rupture of the Command Module and creation of a toxic atmosphere. . . . Due to internal pressure, the Command Module inner hatch could not be opened prior to rupture of Command Module. . . . The overall communications system was unsatisfactory. . . . Problems of program management and relationships between Centers and with the contractor have led in some cases to insufficient response to changing program requirements. . . . Emergency fire, rescue and medical teams were not in attendance. . . . The Command Module Environmental Control System design provides a pure oxygen atmosphere. . . . This atmosphere presents severe fire hazards.[63]

A last recommendation went beyond hazards: "Every effort must be made to insure the maximum clarification and understanding of the responsibilities of all the organizations involved, the objective being a fully coordinated and efficient program."[64]

The review board recommended that NASA continue its program and get to the moon and back before the end of 1969. Safety, however, was to be a prime consideration, outranking the target date. The board urged, finally, that NASA keep the appropriate congressional committees informed on significant problems arising in its programs.

Astronaut Frank Borman, a member of the board, summed up the fact that everyone had taken safety in ground testing for granted. The crewmen, he stated, had the right not to enter the spacecraft if they thought it was unsafe. However, "none of us," Borman insisted, "gave any serious consideration to a fire in the spacecraft."[65]

The board members sharply criticized the fact that the astronauts had no quick means of escape and recommended a redesigned hatch that could be opened in two to three seconds instead of a minute and a half. They proposed a number of other changes in the design of both the spacecraft and the pad and recommended revised practices and procedures for emergencies. Many of these, incidentally, KSC already had in its plans for "hazardous" operations.[66]

One of the most amazing facts to come out in the testimony of so many at KSC was the complicated process of communications. A contractor employee would confer with his NASA counterpart, who would in turn get in touch with his supervisor, who would in turn report to someone else in the chain of command. It must have seemed to the review board easier for a man on the pad to get through to the White House than to reach a local authority in time of an emergency.[67]

Congress Investigates

When the review board began its investigation in February, the Senate Committee on Aeronautical and Space Sciences held a few hearings but confined its queries to major NASA officials.[68] When the Apollo 204 Review Board turned in its report to Administrator Webb, the Senate Committee enlarged the scope of its survey; and the House Committee on Science and Astronautics, more particularly the Subcommittee on NASA Oversight, went into action.

Congress had wider concerns, however, than the mechanics of the fire that had occupied so much of the review board's time. Both houses, and especially two legislators from Illinois, freshman Senator Charles Percy and Representative Donald Rumsfeld, showed great interest in the composition of the review board, especially its lack of non-government investigators.[69] Members of Congress questioned the board's omission of any analysis of the possibility of weakness in the managerial structure that might have allowed conditions to approach the point of disaster. Senator Edward Brooke of Massachusetts wondered about the extensive involvement of North American Aviation and its capacity to handle such a huge percentage of the Apollo contracts.[70] To the surprise of both NASA and NAA officials, members of both the Senate and House committees were to take a growing interest in the report of the Phillips review team of December 1965. This probing was to lead to some embarrassing moments for Mueller of NASA and Atwood of North American Aviation.[71] But these aspects of the hearings belong more properly to the NASA Headquarters history.

Questioning of Debus by two members of the House Committee on Science and Astronautics at a hearing in Washington on the evening of 12 April bears directly on the KSC story. Congressman John Wydler of New York asked Debus to clarify his secrecy directive, which Wydler believed had caused some misunderstanding. Debus read his initial directive of 3 February, which asked for total cooperation with the board and squelched other discussion of the disaster; and then his second announcement of 11 April, after the review board had submitted its report, which removed all restraints.[72] Wydler seemed satisfied.

When Congressman James Fulton of Pennsylvania asked Debus a few minutes later if he would like to make a short statement for the record, Debus came out candidly:

> As director of the installation I share the responsibility for this tragic accident and I have given it much thought. It is for

> me very difficult to find out why we did not think deeply enough or were not inventive enough to identify this as a very hazardous test.
>
> I have searched in my past for safety criteria that we developed in the early days of guided missile work and I must say that there are some that are subject to intuitive thinking and forward assessment. Some are made by practical experience and involved not only astronauts but the hundreds of people on the pads. . . .
>
> It is very deplorable but it was the known condition which started from Commander Shepard's flight . . . from then on we developed a tradition that . . . considered the possibility of a fire but we had no concept of the possible viciousness of this fire and its speed.
>
> We never knew that the conflagration would go that fast through the spacecraft so that no rescue would essentially help. This was not known. This is the essential cause of the tragedy. Had we known, we would have prepared with as adequate support as humanly possible for egress.[73]

Congressman Fulton congratulated Debus on his statement. "This is why we have confidence in NASA. We have been with you on many successes. We have been with you on previous failures, not so tragic. . . . The Air Force had five consecutive failures and this committee still backed them and said go ahead." By looking at matters openly and seeking better procedures, Fulton felt that NASA was making progress.[74]

The House Subcommittee on NASA Oversight, under the chairmanship of Olin Teague of Texas, held hearings at the Kennedy Space Center on 21 April. When the investigation opened, it soon became clear—as the review board had already learned—that any emergency procedures at the space center would be extremely complicated matters involving conferences between NASA and contractor counterparts, and even in certain instances with representatives of the Air Force safety section. Beyond this the most noteworthy event of the hearing was the recommendation of Congressman Daddario that the members commend the brave men on the pad* who had tried to save the astronauts.[75]

---

*Six spacecraft technicians who had risked their lives to save the astronauts received the National Medal for Exceptional Bravery on 24 October 1967. They were Henry H. Rodgers, Jr., of NASA, and Donald O. Babbitt, James D. Gleaves, Jerry W. Hawkins, Steven B. Clemmons, and L. D. Reece, all of North American Aviation. Taylor, *Liftoff*, p. 267.

While the Senate committee in Washington spent a great deal of time on the Phillips report, and embarrassed NASA and NAA officials with questions about the document, the committee finally had to agree with the testimony that "the findings of the Phillips task force had no effect on the accident, did not lead to the accident, and were not related to the accident."[76] On the positive side, the committee learned from President Atwood that North American Aviation had made substantial changes in its management. The firm had placed William B. Bergen, former president of Martin-Marietta, in charge of its Space Division; obtained the full-time services of Bastian Hello and hired as consultant G. T. Wiley, both former Martin officials; and transferred one of its own officers, P. R. Vogt, from the Rocketdyne Division to the Space Division. Atwood testified that North American would probably make other changes.[77] In the end, the Senate committee recommended that NASA move forward to achieve its goal within the prescribed time, but reaffirmed the review board's insistence that safety take precedence over target dates, and reminded NASA to keep appropriate congressional committees informed of any significant problems that might arise in its program.[78]

Reaction at KSC

During the ensuing months, NASA took many steps to prevent future disasters. It gave top priority to a redesigned hatch, a single-hinged door that swung outward with only one-half pound of force. An astronaut could unlatch the door in three seconds. The hatch had a push-pull unlatching handle, a window for visibility in flight, a plunger handle inside the command module to unlatch a segment of the protective cover, a pull loop that permitted someone outside to unlatch the protective cover, and a counterbalance that would hold the door open.[79] NASA revised flight schedules. An unmanned Saturn V would go up in late 1967, but the manned flight of the backup crew for the Grissom team—Schirra, Eisele, and Cunningham—would not be ready before the following May or June.[80] In the choice of materials for space suits, NASA settled on a new flame-proof material called "Beta Cloth" instead of nylon. Within the spacecraft, technicians covered exposed wires and plumbing to preclude inadvertent contact, redesigned wire bundles and harness routings, and increased fire protection.

Initially, NASA administrators said they would stay with oxygen as the atmosphere in the spacecraft. But after a year and a half of testing, NASA was to settle on a formula of 60% oxygen and 40% nitrogen. NASA provided a spacecraft mockup at KSC for training the rescue and the operational teams. At complex 34 technicians put a fan in the white room to ventilate any possible smoke. They added water hoses and fire extinguishers and an escape slide wire. Astronauts and workers could ride down this wire during emergencies, reaching the ground from a height of over 60 meters in seconds.[81]

NASA safety officers were instructed to report directly to the center director. At Kennedy this procedure had been the practice for some time. A Headquarters decision also extended the responsibilities of the Flight Safety Office at Kennedy. Test conductors and all others intimately involved with the development of the spacecraft and its performance sent every change in procedure to the Flight Safety Office for approval.[82]

The fire had a significant impact on KSC's relations with the spacecraft contractors. When KSC had absorbed Houston's Florida Operations team in December 1964, the launch center was supposed to have assumed direction of the spacecraft contractors at the Cape. The North American and Grumman teams at KSC, however, had continued to look to their home offices, and indirectly to Houston, for guidance. This ended in the aftermath of LC-34's tragedy. With the support of NASA Headquarters, KSC took firm control of all spacecraft activities at the launch center.

The Boeing-TIE Contract

To strengthen program management further, NASA entered into a contract with the Boeing Company to assist and support the NASA Apollo organization in the performance of specific technical integration and evaluation functions. NASA retained responsibility for final technical decisions.[83] This Boeing-TIE contract, as it came to be called at KSC, proved the most controversial of all post-fire precautions. Many in middle or lower echelons at KSC criticized it. They looked upon it as a public relations scheme to convince Congress of NASA's sincere effort to promote safety.

Even NASA Headquarters found it difficult to explain to a congressional subcommittee either the expenditure of $73 million in one year on the contract, or that it had hired a firm to inspect work which that firm itself performed. As a matter of fact one segment of the Boeing firm—that working under the TIE contract—had to check on another, the one that worked on the first stage of Saturn V. Mueller explained to the committee that "the Boeing selection for the TIE contract . . . was based upon the fact that this was an extension of the work [Boeing personnel] were already doing in terms of integrating the Saturn V launch vehicle."[84]

When a member of the committee staff called Mueller's attention to the fact that Boeing had problems with its own specific share of the total effort, Mueller's defense of the contract rested on the old adage that "nothing succeeds like success." He felt that if the total program succeeded, the nation would no longer question specific aspects and expenditures.[85]

Boeing sent 771 people to KSC, one-sixth of the total it brought onto NASA installations under the TIE contract. In such a speedy expansion, the quality of performance was spotty. The "TIE-ers" were to find it difficult to get data from other contractors, as well as from NASA personnel. The men at KSC felt they had the personnel to do themselves what the TIE-ers were attempting to do.

The TIE statement of work at KSC carried a technical description of twelve distinct task areas: program integration, engineering evaluation, program control, interface and configuration management, safety, test, design certification reviews, flight readiness reviews, logistics, mission analysis, Apollo Space System Engineering Team, and program assurance.[86]

Many KSC personnel felt that the TIE contract was too much like the General Electric contract they had fought a few years before. In this they forgot that the earlier contract had been a permanent one, which would have given GE access to its competitors' files, and thus involved a conflict of interest. The Boeing–TIE contract had a specific purpose and a time limit. NASA made the arrangement on an annual basis. Further, those who criticized the number of Boeing personnel forgot that one could not assess the size of the problem until he investigated it.

The TIE personnel located and defined delays in the progress of equipment to the Kennedy Space Center. They spotted deficiencies in equipment. They discovered erroneous color coding of lines, for instance, that might have caused a disaster. The insulation of pipes had obscured the color and men had improperly tagged the sources of propellants and gases. When tests at KSC proved changes of equipment necessary, the TIE personnel expedited these changes. They set down time schedules for necessary adjustments. They eliminated extraneous material from the interface control documents. But it remains difficult to assess the exact contribution of the TIE contract.[87]

Far more important than the efforts of the 771 Boeing–TIE personnel, or any specific recommendation of the review board (except perhaps that calling for a new hatch design), the most significant difference at Kennedy Space Center was a larger awareness of how easily things could go wrong. For a long time no test or launch would be thought of as a foregone success.

Most important of all, in spite of the disaster, the President, the Congress, the nation, and NASA itself determined that the moon landing program would go on with the hope of coming as close to President Kennedy's target date as possible.

19

Apollo 4: The Trial Run

The Significance of AS-501

The problems of the spacecraft threatened, but did not extinguish, the hopes of reaching the moon within the decade. Much depended on the outcome of the first Saturn V mission. If the largest launch vehicle and launch complex yet built both performed satisfactorily, the Apollo program could still meet its schedule.

A successful mission would achieve several significant goals. It would mark: the first launch from launch complex 39, the first flight of the integrated Apollo–Saturn V space vehicle, the initial trials of the first (S-IC) and second (S-II) stages of the Saturn V launch vehicle, the first shutdown and restart in space of the third stage (S-IVB) engine, and the first demonstration of the Apollo spacecraft's ability to reenter the earth's atmosphere at the speeds and temperatures it would reach on return from a mission to the moon. Many other benefits would accrue if the unmanned earth-orbital mission succeeded. The adequacy of ground tracking, telemetry, and communications operations at stations around the world could be evaluated. The launch vehicle stages and spacecraft modules would carry additional research and development instrumentation to measure the performance of their internal components. A total of 4098 in-flight measurements—about ⅔ of them for the launch vehicle, ⅓ for the spacecraft—were scheduled.[1]

The results of this mission would confirm or deny the validity of a major management decision made in the fall of 1963—the use of all-up flight testing. Designed to result in an overall time saving, all-up testing meant that all launch vehicle stages and spacecraft modules (essentially in their final configuration) would be tested together on each flight. Previous practice had favored a gradual buildup of subsystems, systems, stages, and modules in successive flight tests.[2] Based in part on the unqualified successes of the first four Saturn I missions, but made before any Apollo spacecraft had flown, the eggs-in-one-basket decision involved a calculated risk. Success in all-up testing was the quickest way to accomplish a manned lunar landing. On the other hand, failure of the first Saturn V mission would be a major catastrophe.

For KSC the first flight of the Apollo–Saturn V had a narrower, but more important, objective than that of the total mission. For the first time the facilities, equipment, procedures, and checkout crews would be put to the test. The 500-F facility checkout tests had instilled a certain degree of confidence (while revealing much that remained to be done), but this would be "the real thing." This time, every action would lead toward those moments when the first-stage engines would ignite, the hold-down arms on the launcher platform would retract, and the Apollo–Saturn V vehicle would be committed to flight. In the process of receiving, assembling, testing, and launching this first Apollo–Saturn V, KSC civil service managers and the launch vehicle, spacecraft, and launch support contractor crews would be learning to work together as a unit. It would prove a difficult task for all concerned—and not without its rough moments—but, in the end, a well-functioning launch team would be the reward.

The Parts of AS-501

The first of the Apollo–Saturn V space vehicles had received its official designation in April 1965 when Maj. Gen. Samuel C. Phillips, Apollo Program Director, announced: "Apollo flight missions to be flown on Saturn IB and Saturn V will be designated as Apollo/Saturn followed by the number of the launch vehicle assigned to the flight mission (i.e., Apollo/Saturn 201, Apollo/Saturn 202, etc., and Apollo/Saturn 501, Apollo/Saturn 502, etc.)."[3] The AS-501 space vehicle consisted of Saturn V launch vehicle number 501 and Apollo spacecraft number 017. The launch vehicle had three stages and an instrument unit. The spacecraft included a spacecraft lunar module adapter, a lunar module, a service module, a command module, and a launch escape system.

Components of the AS-501's first stage (S-IC) were constructed by the Boeing Company at Michoud, Louisiana, and assembled at the Marshall Space Flight Center in Huntsville, Alabama. The S-IC stage consisted of a structural framework to which the engines were attached, an RP-1 (kerosene) fuel tank, a LOX (liquid oxygen) tank, an intertank structure separating the fuel and LOX tanks, and a forward skirt that connected to the second stage. The five Rocketdyne F-1 engines would develop a total of 33.4 million newtons (7 500 000 pounds of thrust) at liftoff. The center engine was fixed in position, but the others were mounted on gimbals to provide attitude control and steering for the vehicle. Two hydraulic actuators swiveled each engine in response to signals from the flight control computer located in the instrument unit. In less than 3 minutes of powered flight, the first stage engines would consume almost 2000 metric tons of propellants. Eight small solid-propellant retrorockets were attached to the framework to slow the first stage after engine shutdown, guaranteeing separation of the first and second stages.

The second stage (S-II), built by North American Aviation, Inc., Canoga Park, California, consisted of an aft interstage, an aft skirt and framework to which the engines were attached, integral LOX and liquid hydrogen (LH_2) tanks with a single common bulkhead, and a forward skirt. The five Rocketdyne J-2 engines were arranged similarly to those of the first stages, with the center engine fixed and the four outer engines gimbaled by hydraulic actuators in response to signals from the instrument unit. The aft interstage, which surrounded the rocket engines, was the means of attaching the second stage to the first stage; it also supported the weight of the second and third stages and the spacecraft. In flight when the first and second stages separated and the second stage engines ignited, the aft interstage was jettisoned. During the second stage's 6 minutes of powered flight, the five J-2 engines would consume about 425 metric tons of propellants while developing nearly 4.5 million newtons (one million pounds of thrust).

The third stage (S-IVB) of the launch vehicle was built by Douglas Aircraft Company. It consisted of the aft interstage, an aft skirt, a thrust structure to which the single J-2 engine was attached, a LOX tank and an LH_2 tank, and a forward skirt. Because the third stage was smaller in diameter than the first and second stages, the aft interstage tapered from a diameter of 10 meters at its base to 6.6 meters where it joined the aft skirt. The single Rocketdyne J-2 engine would develop 889 600 newtons (200 000 pounds of thrust) and was capable of being shut down in space, and then reignited. Hydraulic actuators gimbaled the engine, in response to signals from the instrument unit, to provide pitch and yaw control during powered flight. Two self-contained auxiliary propulsion system modules, mounted 180 degrees apart on the aft skirt, would provide roll control during powered flight, and pitch, yaw, and roll control while the J-2 engine was shut down. During the approximately 7½ minutes of third-stage powered flight (including first and second burns), about 105 metric tons of propellants would be consumed.

IBM's instrument unit (S-IU), atop the third stage, was 6.6 meters in diameter, slightly less than one meter in height, and weighed about 10 metric tons. The unit consisted of segments of honeycomb material sandwiched between inner and outer skins and looked like a narrow collar or ring that had been slipped part way down the vehicle. Mounted on the inner skin were 16

The components of Saturn V

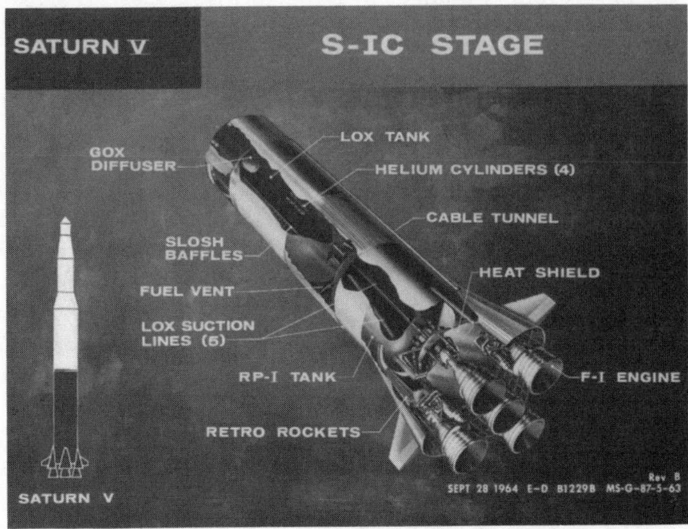

Figure 131

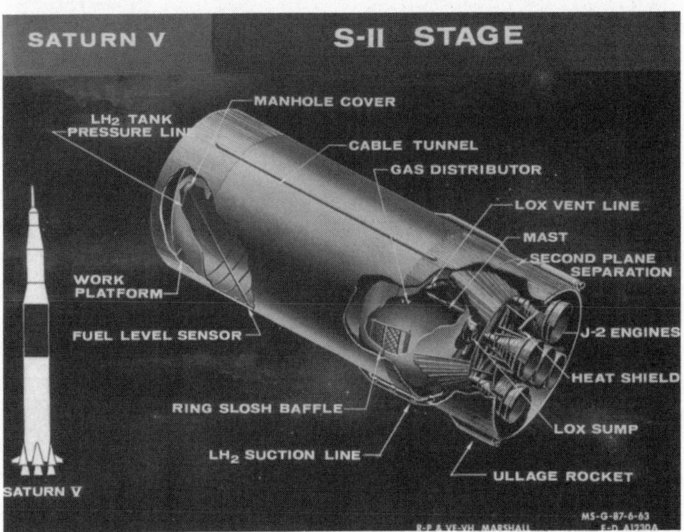

Figure 132

Fig. 131. The first stage, S-IC. GOX stands for gaseous oxygen, LOX for liquid oxygen; RP-1 was a rocket propellant similar to kerosene. Fig. 132. The second stage, S-II. LH_2 means liquid hydrogen.

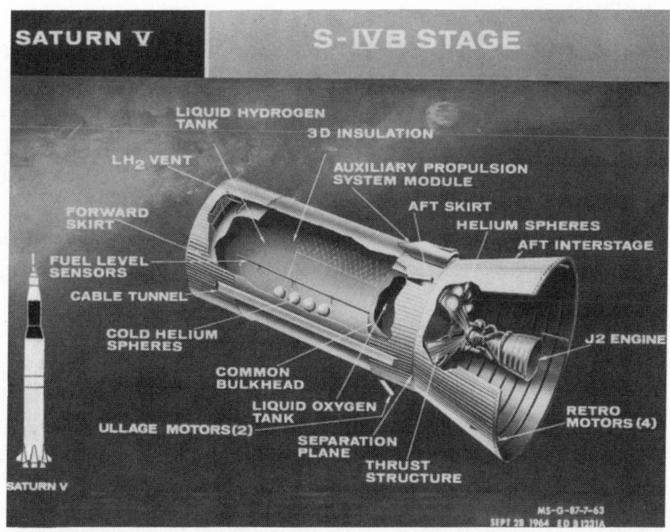

Figure 133

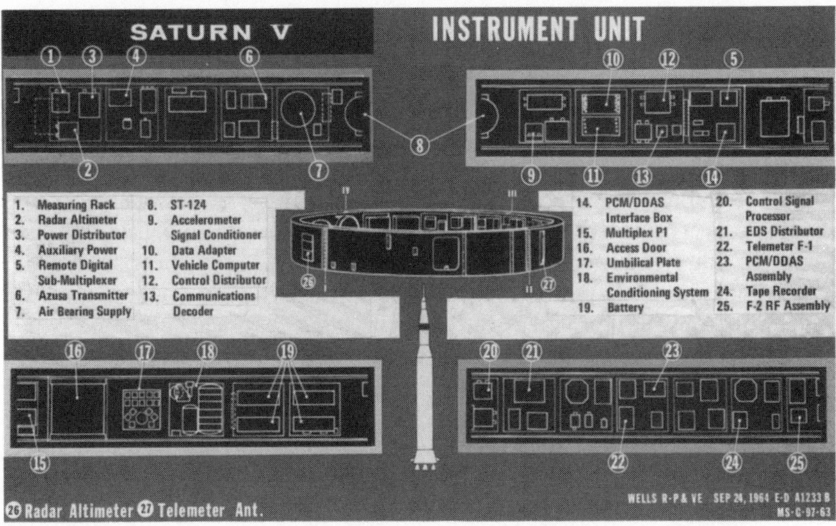

Figure 134

Fig. 133. The third stage, S-IVB. Fig. 134. Schematic of the instrument unit, which was shaped like a ring or collar and placed around the upper end of the propellant tankage in what would otherwise have been wasted space.

cold plates, each 76 centimeters square. Coolant fluid circulated through these plates to dissipate heat generated by the operation of the guidance and control, instrumentation, and electrical power and distribution equipment installed on them. By attaching the equipment to the skin, space was left in the center for the domed bulkhead of the third-stage liquid-hydrogen tank, which extended into the instrument unit, and for the landing gear of the lunar module to be included on later missions.[4]

The flight plan called for the Saturn V to place the spacecraft and third stage into a circular orbit. After completing two orbits, the third stage would ignite a second time. Separating from the third stage, the spacecraft would rise to an apogee of approximately 18 500 kilometers by firing its service propulsion system engine. A second firing during descent from apogee would boost the command module's reentry velocity to 11 075 meters per second or 40 234 kilometers per hour. Protected by its heat shield, the command module would reenter the atmosphere and return to earth northwest of Hawaii.

Delay after Delay after Delay

Apollo 4 was not ill-starred. In fact, it eventually went into space trailing a sizable cloud of glory. But no mission was so plagued by vexatious delay, due in part to the teething troubles of a new rocket and new stages, especially the S-II; in part, to the aftermath of the fire; and in part, to the all-up procedure which put a premium on prelaunch preparations. The delays were not unproductive. Many involved the learning of lessons that, once mastered, were needed in succeeding Saturn V launches. Some serious problems did not delay the launch. For example, early in the checkout LC-39's LOX line ruptured, threatening to hold up operations for several weeks (pp. 343–44). The line was repaired, and could have been repaired two or three times over, while other and more serious problems were being solved.

In mid-1966 General Phillips hoped to launch the first Saturn V early the following year. Few Apollo officials were very confident about the target date. The S-II second stage had become the pacing item in the program. Development problems had already delayed its delivery at KSC from July to October 1966. On 13 August the S-II reached the Mississippi Test Facility, only to be held up again when technicians found cracks. The discovery delayed the acceptance firings and forced Phillips to reschedule the arrival of the S-II stage at KSC for mid-November. That month the Apollo Program Office issued a revised schedule calling for delivery of the S-II stage at KSC on 9 January, with launch three months later. Meantime, checkout of the

501 vehicle proceeded without S-II. In its place the launch team employed a spacer, referred to as the "spool" because of its shape—a cylinder that flared out at both ends. With the spacer the launch team could stack the stages and begin checkout in the assembly building. The spool also gave KSC the opportunity to test handling equipment for the second stage.[5]

The third stage (S-IVB) was the first major component of Apollo 4 to be delivered at KSC. It arrived from Sacramento aboard the Guppy aircraft on 14 August 1966 and went immediately into a low bay of the assembly building for inspection and checkout. The following week the spacer and instrument unit arrived. On 12 September, as Peter Conrad and Richard Gordon prepared to blast off in Gemini 11, the barge *Poseidon* sailed into the Banana River with the first stage. Boeing gave it a lengthy checkout in the transfer aisle of the high bay before erecting the booster on 27 October. During the following week, technicians stacked the remaining launch vehicle stages, using the spool for the absent S-II. There were a few problems—the checkout of the swing arms took an extra two days and a cooling unit for the instrument unit sprang a leak—but the launch team, still counting on the mid-November delivery date for the S-II, hoped to roll the complete vehicle out to pad A by 13 January 1967.[6]

By late November the Apollo Program Office had moved the S-II's arrival back to January, and the launch back to April. Since spacecraft 017 would not arrive for another three weeks, KSC erected the facilities verification model of Apollo on 28 November. This allowed North American to check out some of its spacecraft support equipment. The first week in December the memory core in a digital events evaluator failed after intermittent troubles; cracked solder joints were blamed. A hurried repair put the computer back on line.[7]

The command-service module arrived at KSC on Christmas Eve and was mated to the launch vehicle on 12 January 1967. That tardy prima donna, the S-II stage, finally appeared on 21 January. Tank inspection, insulation, and engine work were in progress by the 23d. Test crews found damaged connectors on three recirculation pumps and set about investigating the extent of the rework that would be necessary. While inspecting the liquid hydrogen tank on the second stage, the North American team found 22 cracked gussets. These triangular metal braces, used to support the horizontal ribs of the stage framework, had to be replaced. Plans to move the second stage into a low bay checkout cell on the 29th were temporarily set aside because of a late shipment of the aft interstage (the cylindrical aluminum structure that formed the structural interface between the first and second stages). The interstage arrived on 31 January, and by the end of the next day the stage was in a low bay cell with work platforms around it.[8]

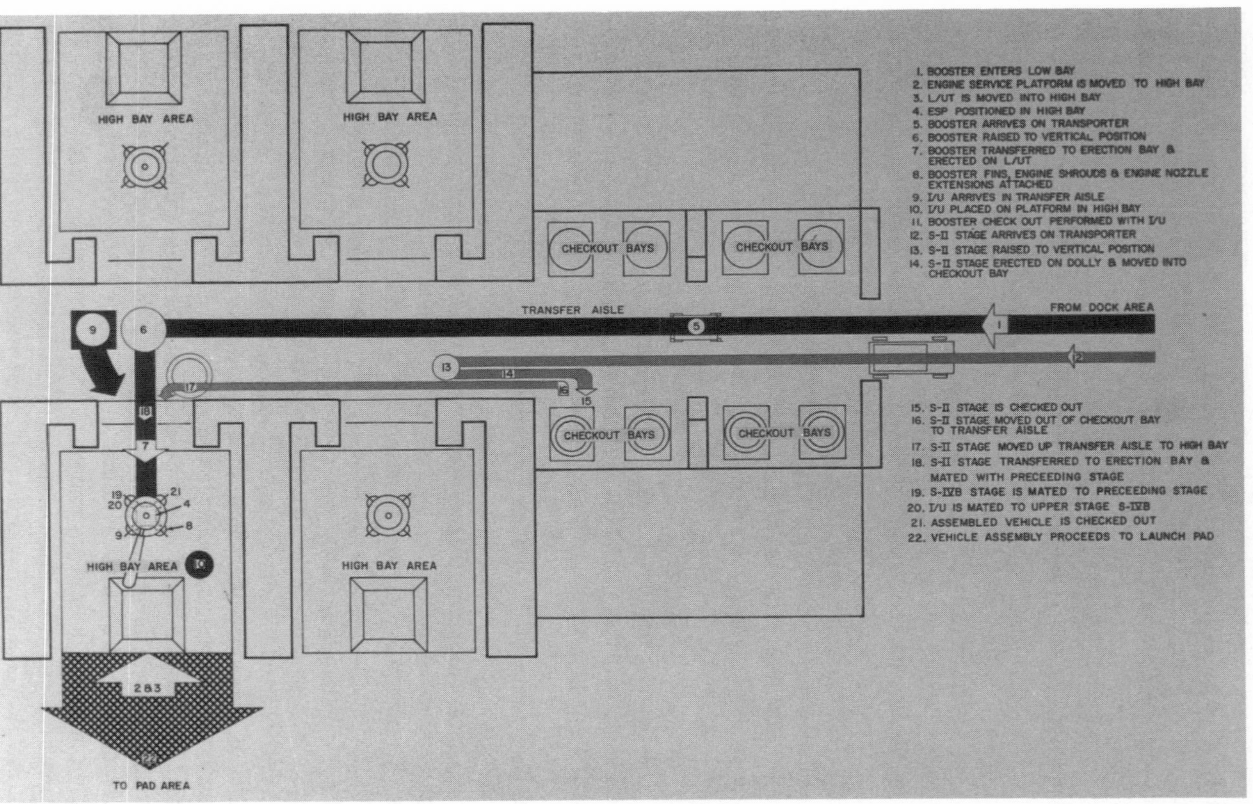

Fig. 135. Flow chart for assembling a Saturn V.

Fig. 136. An S-IC stage in the assembly building.

Despite the delay with the S-II stage, KSC officials expected to meet the new launch date in May. The fire on 27 January placed all schedules in question. Although Apollo 4 was an unmanned mission, NASA officials wanted to give command-module 017 a close examination. On 14 February, a week before the S-II could be inserted into a fully assembled vehicle, the spacecraft was removed from the stack and taken to the operations and checkout building. When inspection disclosed a number of wiring errors, KSC's Operations Office cancelled the restacking of the spacecraft. By 1 March electrical engineers had discovered so many wiring discrepancies that the test team stopped their repair work, pending a thorough investigation of all spacecraft wiring. Within two weeks the North American and NASA quality control teams recorded 1407 discrepancies. While North American repaired about half of these on the spot, modifications, repair work, and validations continued into June. During the break technicians performed pressure tests on service module systems at pad 16. It would be mid-June, with the wiring modifications for the command module finally completed, before North American could remate the spacecraft and take it back to the assembly building.[9]

As the extent of the wiring problems was not immediately recognized, the launch vehicle team forged ahead to recoup the time lost on the S-II stage. In mid-February Boeing's airframe handling and ordnance group removed the instrument unit and spacer from the 501 stack and on the 23d erected the S-II. The operation involved incredibly close tolerances. To qualify crane handlers, Stanley Smith, Bendix senior engineer of the crane and hoist group, stated, "We give them a technical examination and then check their reflexes and response to commands in training sessions." During a mating, an operator and an electrician boarded the crane and another man helped guide movements from the floor by communicating with the operator via a walkie-talkie. Smith set a high goal for his team: "We strive to train our men to the point where they could conceivably lower the crane hook on top of an egg without breaking the shell."[10]

After a stage was properly aligned on the Saturn stack, a crew of one engineer, two quality control inspectors, one chief mechanic, and eight assistants took eight hours to complete the mating. Three 30-centimeter pins on the second stage fitted into brackets located 120° apart on the periphery of the first stage. Then the mechanics inserted 216 one-centimeter, high-strength fasteners into matching holes around the perimeter where the two stages joined. The team torqued the fasteners in a staggered sequence to secure the bolts evenly and ensure a uniform distribution of stress. The mating of the second and third stages was conducted in much the same manner.[11] The 501 was now set up except for the missing CSM.

The lengthy delays with the flight hardware aided the Site Activation Board in its efforts to get LC-39 ready for its first launch. The board's first flow (pp. 318-19) included firing room 1, mobile launcher 1, high bay 1, and the other facilities required for the support of Apollo 4—1280 activities altogether. During the first quarter of 1967, PERT charts showed less than 1% of these activities behind schedule. The decision in mid-April to modify the LOX system on launcher 1 and pad A put five weeks of negative slack into the site activation schedule. The modifications were made necessary by excessive pressure in the LOX system. KSC engineers added an automatic bleed system, relief valve supports, and a block valve that prevented purging through the drain line. As continued vehicle problems further delayed the rollout, the five weeks of negative slack disappeared.[12]

On 24 May the S-II stage was in trouble again. NASA announced it would be dismantled for inspection, consequent on the discovery of hairline cracks in the propellant tank weld seams on another S-II at the factory in California. The additional checks were not expected to delay the flight of 501 "more than a week or so." By mid-June the inspection, which included extensive x-ray and dye penetrant tests, was completed and the stage returned to the stack. On 20 June, the command-service module was mechanically mated to the Saturn V, and 501 was—at last—a fully assembled space vehicle. A revised schedule on 21 July set rollout for mid-August. On 26 August 1967, the big rocket emerged from the high bay slightly more than a year after its first components had arrived at KSC, and a good six months after its originally scheduled launch date. It had been a year of delay and frustration, and the end was not yet.[13]

The Tests

While KSC officials were fighting the seemingly endless delays with the S-II stage and command module wiring, the launch team was putting Apollo 4 through the tests that would verify its flight readiness. The 456 tests in the Apollo 4 catalog fell into nine categories: electrical networks (90); measuring, fire detection, etc. (49); telemetry (27); RF and tracking (21); gyroscopes, navigation, control, and ground operations computers (86); mechanical and propulsion (146); combined systems (9); launch support equipment (13); and space vehicle (15).[14]

Saturn V tests, like those of the Saturn I and IB, progressed from component and subsystems tests, through systems, to combined systems or integrated tests. Hans Gruene's launch vehicle operations team began by checking out the various pieces of support equipment in the low and high bays. The "ESE qualification test, low bay" was a typical procedure. As the initial KSC checkout of the low bay's electrical support equipment, the test verified the performance of panels, consoles, cables, and the digital data acquisition system—all the electrical equipment that would be used to test the upper stages of the Saturn.[15]

After checkout of all the support equipment, the launch vehicle teams began testing components and subsystems within the separate stages. The checkout of the first stage was performed on the mobile launcher in a high bay, while the upper stages were tested in the low bay cells. Technicians tested valves, electrical networks, radio frequencies, measuring instruments—all the items that made up the various systems within the stages. For example, North American conducted a "pressure transducer, potentiometer type systems test" that verified the performance of the S-II's pressure transducers. (The Saturn's transducers converted such things as temperature and pressure to electrical signals.) Before conducting the test, North American checked out the second stage's digital data acquisition

Stacking the space vehicle

Figure 137A

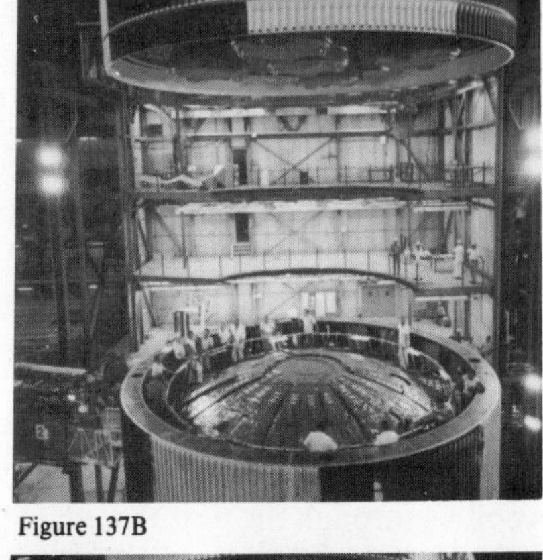

Figure 137B

Figure 137C

Figure 137D

Figs. 137–140. The piece-by-piece buildup of the space vehicle in the assembly building. Figs. 137A–D. The S-II is placed on top of the S-IC. Fig. 138. The S-IVB goes on the S-II. On following pages: Fig. 139. In the upper reaches of the assembly building, the instrument unit is added to the stack. In the foreground, a swing arm is in use. Fig. 140. The last major piece is the Apollo spacecraft, which rests on the instrument unit. An extensible work platform has been moved up to the vehicle at the S-IVB level.

Figure 138

Figure 139

Figure 140

Fig. 141. The view from the top of AS-501, with the work platforms retracted, May 1967. Access arms 8 and 9 are visible at the top. The top-most piece—the launch escape rocket—has not been installed.

system and the connections to the assembly building's measurement calibration station. Then, with stage instrumentation power on, readings were taken on each pressure transducer.[16]

The erection of the launch vehicle in the high bay marked the first major milestone in KSC's operations and prompted a series of tests such as the "S-IC-S-II electrical mate." Three men, working eight hours, checked out the electrical interface between the two stages. Another stage test in the high bay was the "umbilical interconnect verification test, S-IVB flight stage." Through a series of measurements, a Douglas crew verified the proper plug fit and electrical continuity between power sources on the swing arm and the S-IVB networks.[17]

The weeks after erection were spent in system and subsystem testing and in modifying the Saturn rocket. One day of Saturn activities illustrates the extent of the launch vehicle operation:

- Leak and functional test of the first stage nitrogen control pressure and purge system
- Checkout of the engine-bearing coolant valve
- Retest of the earlier engine-cutoff modification
- Engine leak checks
- Instrumentation system checkout
- Range safety receiver and decoder checks
- Guidance and control test
- S-IVB auxiliary propulsion system relay functional tests.[18]

Fig. 142. The test cells in the low bays of the assembly building.

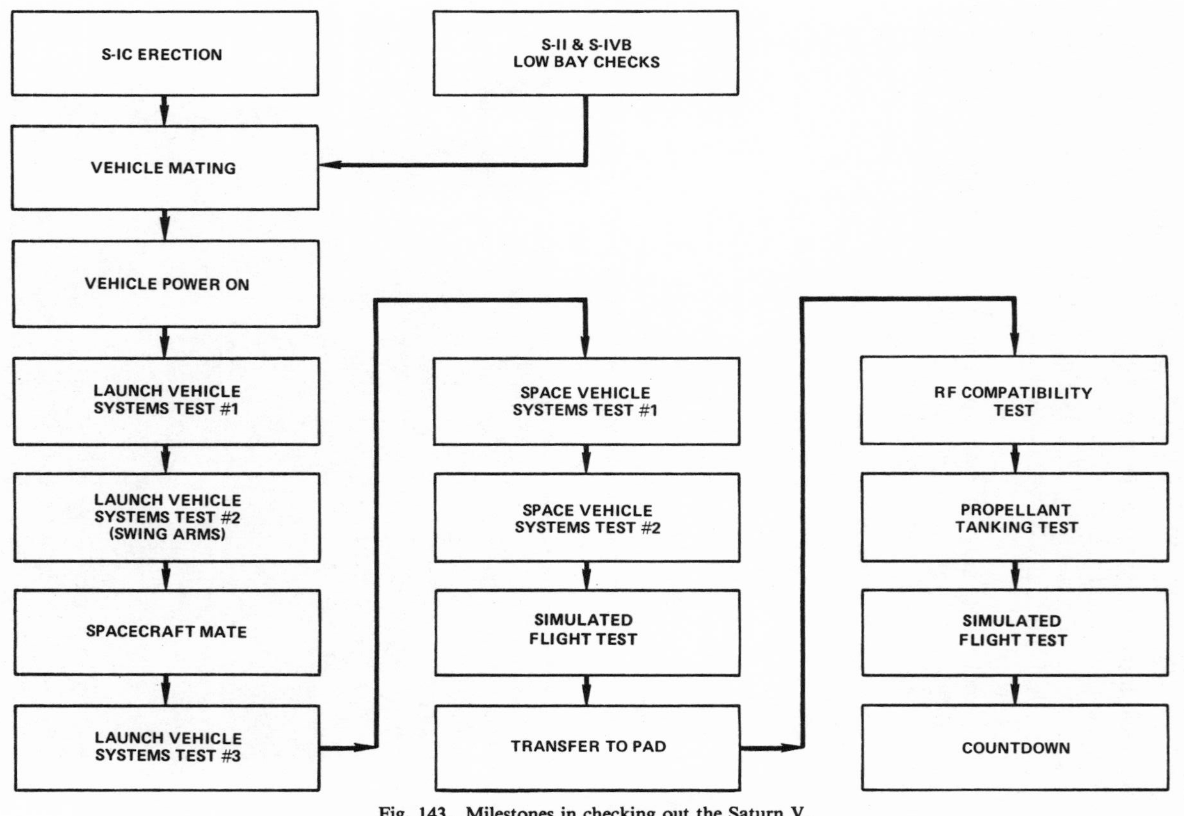

Fig. 143. Milestones in checking out the Saturn V.

When the various subsystems and systems procedures were completed satisfactorily, the launch team moved on to the Saturn's integrated system tests.

The combined or integrated systems procedures tested vehicle functions involving several systems in one or more stages. These included the operation of the range safety command receivers and the Saturn's destruct systems, the electrical interfaces of the combined vehicle, and the transfer from external to internal power. The flight sequence test took the launch vehicle past liftoff to exercise the switch selectors that keyed the flight systems. The emergency detection system test checked out the launch vehicle's response to an abort situation. Since this system was one of the most intricate in the space vehicle, its test was one of the few automated for the Apollo 4 operation. The test employed an Apollo simulator and consisted of five parts: engine out, excessive rate (attitude), rate gyro, verification of the command module indicator, and a test of the vehicle's abort logic plan.

The integrated tests on the launch vehicle culminated with the sequence malfunction procedures and the swing-arm tests. The former was actually a series of ten tests that ran a day or more. They verified the compatibility and operation of the launch vehicle and electrical support equipment in case of a malfunction and cutoff in the last seconds of the terminal count. For example, in test 5 the launch team would simulate a malfunction in the service arms just prior to their swinging clear of the vehicles. The test would determine whether the vehicle could shut down properly. The swing arm overall test verified the operation of all Saturn and ground support equipment systems during a normal firing sequence and on into flight. The test included the actual release of the hold-down arms, umbilical ejection, and the withdrawal of the swing arms and the tail service masts. Following the simulated liftoff, the flight computer directed the various switch selectors in the Saturn stages through the operation. The exercise terminated with the engine cutoff of the S-IVB stage and the issuance of propellant dispersion system commands.[19]

After the spacecraft joined the stack, integrated testing continued. Several tests, such as the emergency detection system procedure, covered familiar ground but now involved a complete space vehicle. The space vehicle overall tests 1 and 2 climaxed the test operations in the assembly building. Overall test 1, popularly known as "plugs in," tested the electrical systems and some of the mechanical systems of the Apollo-Saturn, along with pertinent ground support equipment and range facilities, during a simulated normal liftoff and flight. The Saturn went through an internal power check while the spacecraft's environmental control and navigation systems were

Fig. 144. Preparing to move to the pad: the crawler-transporter under the mobile launcher and AS-501, the base of which is largely hidden by the plumbing on the launcher.

Fig. 145. AS-501 en route from the assembly building, left, to the pad, off right, having passed the mobile service structure in its parking position, 26 August 1967.

Fig. 146. AS-501 being tested on launch pad 39A, August 1967. The mobile service structure is on the right; the access arms are extended from the mobile launcher, left.

Fig. 147. A firing room in the launch control center; testing of AS-501 is in progress.

checked out. After liftoff, the test simulated stage cutoffs as they would occur in normal flight. Overall test 2, the plugs-out test, came several days later. By actually releasing the hold-down arms and the umbilical plugs, this test verified that there was no electrical interference during the umbilical disconnect. In both tests the Saturn telemetering channels operated "closed loop" over lines back to the central instrumentation facility. The spacecraft operated its radio equipment "open-loop" to the Eastern Test Range and operation and checkout building. At the close of the plugs-out test, KSC and Marshall compared data on the Saturn's operation with similar data collected during the swing arm and plugs-in tests.[20]

After the rollout to the pad, integrated tests, such as the "space vehicle power-on check," verified the interface between the space vehicle and the pad facilities. The power-on test involved the ACE and RCA 110A automated checkout systems and the mobile launcher (pp. 359–62). A radio frequency compatibility test ensured that the pyrotechnic circuitry on the spacecraft would not be triggered by radio signals. This test was conducted in two stages, first with the mobile service structure around the Apollo, and then back at its parking site. The launch team ran another malfunction test at the pad and yet another check of the emergency detection system.[21]

The flight readiness test represented one of the last major milestones. The test verified the proper operation of the space vehicle and associated ground support equipment before and after liftoff in a normal countdown, following terminal procedures as closely as possible. For this test the launch team brought the Apollo-Saturn as near as possible to its flight configuration. A minimum amount of test equipment was employed. Electrical circuits that could inadvertently damage the space vehicle were by-passed. The test conductor first ran the space vehicle through a simulated terminal count ending in a pad abort. A second run put it through tower clearance and ended with a service propulsion system abort and an earth landing for the spacecraft. After a second recycling of the count, the space vehicle flew a successful mission.[22]

Propellant loading tests came next, followed by the countdown demonstration test. While the flight readiness test focused on the space vehicle and its systems, countdown demonstration was intended to test the performance of the launch team and the ground support equipment. The objectives were to:

- Demonstrate the time phasing of the normal sequences necessary to prepare the space vehicle for launch;
- Verify that the space vehicle and support equipment were in a satisfactory status for launch as if launch were imminent, thus demonstrating the countdown procedure adequacy; and
- Verify propellant system integrity by loading the cryogenics.[23]

Like an actual 6-day countdown, the test was divided into a precount and a countdown. Spacecraft operations during the precount included powering up and testing the Apollo systems; servicing the liquid (water), gas (helium and nitrogen), and cryogenic (LOX and LH_2) tanks; and installing pyrotechnic simulators. Two of the last actions were to stow spacecraft provisions and equipment and activate the service module's fuel cells. At the same time the launch vehicle team turned on the Saturn's power, tested electrical circuits, and installed batteries and ordnance items. Usually a hold would come at the end of the precount. These open periods provided the launch team a needed break, as well as time to make unscheduled corrections on the space vehicle.

At the beginning of the simulated countdown, the spacecraft team completed all work requiring access from the mobile service structure and removed it. Then the team continued with radio checks, the closeout of the command module, a hatch leak check, and a power transfer check. The launch vehicle team performed power-on and guidance system checks, loaded cryogenics, and tested the range safety command and other radio frequencies. Finally, with the ignition systems blocked to prevent an accidental launch, the terminal sequencer took the count to T – 8.9 seconds where the test ended.[24]

The time required to prepare for a launch varied considerably during the Apollo program. As the trial run of the Saturn V and LC-39, everyone expected troubles with AS-501. On subsequent missions, however, the checkout took longer than KSC officials expected. A major reason was the condition of the flight hardware. NASA started with the premise that stages would arrive at KSC in a nearly flight-ready condition. The prelaunch checkout would require no more than four to six weeks in the assembly building and one week on the pad. Events proved otherwise. The launch vehicle and spacecraft contractors, beset by problems, delivered stages and modules that required extensive modifications. These changes contributed in large measure to the extended launch operations.[25]

KSC did not achieve a standard or routine for launch operations until the Apollo 9–11 missions, each of which required over five months. The receiving, inspection, and preliminary checkout ran four to six weeks for the S-II and S-IVB stages, one to two weeks for the S-IC and the instrument unit. After the launch vehicle was erected, subsystems tests took another month. Seven to ten days of integrated tests on the Saturn were followed by the erection of the spacecraft. A month more of tests in the assembly building culminated with the space vehicle overall test 1 (plugs in). Out on the pad, three weeks of tests preceded the flight readiness test. Propellant loading tests followed two weeks later. Several days thereafter KSC began the week-long countdown demonstration test that immediately preceded launch.[26]

More Delays for AS-501

Despite all the trials KSC had gone through with AS-501 by September, more were ahead. On 31 August the launch operations office issued a new schedule with the countdown demonstration test to begin 20 September. In less than a week the schedule was broken. When Boeing had to replace the hydraulic engine actuators on the first stage, Petrone's office rescheduled the test for 25 September. A major milestone, the space vehicle malfunction overall test, was scrubbed on the 12th because of rain and high winds. The test team concluded the exercise the following day, but lightning slowed down operations on the 14th. For later flights the sequence would be flight readiness test, countdown demonstration test, countdown and launch. For the first flight of the Saturn V, however, the test directors wanted to have the flight readiness test as close to the launch date as possible and scheduled it after the demonstration test. KSC officials were not at all sure how well the new launch complex would perform. Events would justify their concern.[27]

The six-day countdown test started on the evening of 27 September. By 2 October the launch team was two days behind schedule. Following a hold, the test went smoothly from T – 18 to T – 13 hours, when computer problems forced another delay. The count reached T – 45 minutes on 4 October when a computer, monitoring the propellant loading operation, failed. As a result 1 900 000 liters of kerosene and liquid oxygen had to be removed from the S-IC stage. The count, set back to T – 13 hours, was resumed on 9 October. More computer problems and a faulty regulator on the helium gas system marred operations that day. By the time the count reached T – 5 hours, the launch team was exhausted. Petrone called a two-day recess. Shortly after the test resumed on 11 October, a problem appeared with a battery heater on the S-II stage. As the battery could not be repaired or replaced quickly, another day's work was cancelled. KSC finally completed the test on 13 October, after 17 frustrating days.[28]

As Paul Donnelly later noted: "In spite of the many problems encountered in the test, the crew had received an education that money couldn't buy."[29] The launch date for Apollo 4 was postponed, pending the outcome of the test. After it was completed, few at KSC seriously believed that 501 could be launched on the new date of 7 November. Phillips acknowledged that "this is a target date. We are in a very complex learning process and we are going to take all the time we need on this first launch."[30] The growing concern of higher NASA management expressed itself at the flight readiness review on 19 October. The purpose of this meeting was to assess the readiness of the overall mission in general and the S-II in particular. Because it was unmanned, Apollo 4 was cleared for launch assuming the satisfactory completion of the remaining tests and modifications.[31]

The Launch of Apollo 4

KSC began an abbreviated countdown (56½ hours) on 6 November 1967 pointing toward a 9 November launch. The propellant loading was a feat in itself; the propellant systems pumped 89 trailer-truck loads of LOX, 28 trailer loads of liquid hydrogen, and 27 rail cars of kerosene aboard Apollo 4. The day before the launch, representatives from the groups supporting the mission met at KSC for an informal review. The meeting gave Apollo 4 a "go" for launch, contingent on the resolution of a few minor problems.[32]

To recognize individuals who had performed in an exemplary manner on the manned spaceflight program, KSC invited Apollo contractors to select employees to visit the launch center for the liftoff. On 8 November, 43 of these "Manned Flight Awareness" honorees were guests of the center for a tour of facilities, a social evening that included a visit with six astronauts, and a view of the launch the next morning.

On the morning of 9 November, cars clogged the access roads as visitors filled every available spot. The countdown continued to its climax, when the five engines ignited. The small but astonishingly strong hold-down arms held back the giant ship for a few seconds. Suddenly the 36-story vehicle seemed to stand for an instant above the launch umbilical tower, and then it moved skyward with increasing speed. The bleachers at the press site shook, their light fixtures bounced, a flock of ducks changed course without breaking their V formation. Men shouted in triumph.

If distinguished guests in the stands to the northwest of the assembly building, the press corps, and the thousands of other visitors felt a sense of triumph, it paled before the feelings of the experts at their consoles in the launch control center. KSC's last official act was Launch Operations Manager Paul Donnelly's statement: "The vehicle has cleared the tower." At that moment, responsibility left KSC's hands. The Manned Space Flight team at Houston might refer back to Kennedy on specific problems for unmanned flights like Apollo 4, but in flights with men on board, corrections would come from the astronauts.

Wernher von Braun spoke of the mission as "an expert launching all the way through, from lift-off exactly on time to performance of every single stage." General Phillips said:

> I was tremendously impressed with the smooth teamwork that this combined government/multi-industry team put together. It was smooth, it was professional, it was confident. It was perfect in every respect. It was a powerful operation. You could almost feel the will with which it was being carried out. Apollo is on the way to the moon.[33]

During the course of the following week, George E. Mueller, NASA's Associate Administrator for Manned Space Flight, put the success of Apollo 4 in focus. Noting the space age was ten years old, he said that the voyage of Apollo 4 dramatically increased the confidence of people across the nation and showed the maturing of a management structure that could administer the largest single research and development program ever undertaken in the Western world. He discussed the crucial flights of the lunar module coming in the near future and predicted that it would be possible for astronauts to land on the moon about the middle of 1969.[34]

Press, VIPs, Tourists, Dependents

Elaborate plans for the reception of guests paid off both at the launch of Apollo 4 and at the 13 subsequent Apollo launches. Five days before the launch, the Office of Public Affairs had opened a news center on the 10th floor of the Cape Royal Building in Cocoa Beach. The news center issued badges to representatives of industry and the news media, including TV technicians, for access to the space center. Bus tours of the entire center were conducted twice daily for reporters and photographers. Starting twelve hours before launch time, three NASA buses operated a shuttle service every half hour between the Cape Royal and the LC-39 press site. The last bus departed one hour before launch time, but by then most media personnel were in their seats on a "first-come, first-served" basis. Southern Bell installed 360 telephones at the press site, with the news organizations paying individually for service. A mobile food service unit supplied hot snacks.[35]

The news center held status briefings on the mission twice daily preceding the launch. The day before the launch, there were two press briefings at launch complex 39, followed by a tour of pad A. The afternoon mission briefing took place at the news center itself. John W. King, chief of the Public Information Branch, provided countdown commentary, starting five hours before liftoff. Loudspeakers carried this commentary to the press site at LC-39, the VIP site on the opposite side of the vehicle assembly building, the visitors information center, the KSC news center, all cafeterias throughout KSC, and the main buildings in the industrial area. The Manned Space Center in Houston took over the commentary after liftoff.

The Cape Royal auditorium was available to contractors for presentations at times not in conflict with NASA requirements. Contractors' representatives could schedule such events in advance with the approval of the KSC news center manager. The contractors also had space for displays and a liaison desk for their public relations representatives.

At least equally important, but more complicated than preparations for representatives of the media and the contractors, was the task of caring for the dignitaries who would descend on the area as long as viewing an Apollo launch would be a socially and politically prestigious event. NASA Headquarters had its own list of invitees, as did the three centers (Kennedy, Marshall, and Houston). Naturally many names were duplicated on the lists. The centers settled the overlapping among themselves, and each center director invited his guests personally. The distinguished visitors viewed the launch from uncovered bleachers northwest of the assembly building, which could accommodate 1000 guests.

Protocol representatives from NASA Headquarters, KSC, Marshall, and the Air Force Eastern Test Range set up a joint protocol center at the Sheraton Cape Colony Inn in Cocoa Beach, five days before liftoff. With the usual foresight, KSC had a contingency plan that did not have to be used on Apollo 4. In case of postponement or delay of a launch, the guests automatically had a valid invitation for the rescheduled time. In the meantime, the Protocol Office would provide further tours of the Kennedy Space Center until launch. NASA and contractor employees at KSC could view the Apollo 4 mission from a convenient area near their place of duty. Their dependents watched from Avenue E in the industrial area, south of the Apollo training facility. The Security Office provided badges, car passes, and instructions five days before the launch. Some contractors and range organizations chartered buses to bring dependents to the viewing site. Throughout all viewing areas, KSC provided emergency first aid and ambulance service. Security handled parking of vehicles and controlled traffic with an ease that was to grow with each launch.[36]

KSC Learns about Government Accounting

While KSC was wrestling with the protracted checkout of Apollo 4, its top management had to divert considerable attention to a U.S. General Accounting Office (GAO) audit that had started two years before and now reached a climax. Back in the spring of 1965, while KSC was still in the process of readying launch complex 39, the GAO began an audit of the contract for a second crawler-transporter. It wanted written studies substantiating the need for two vehicles. Close to five months after beginning its audit, the GAO informed KSC Director Debus that the evidence had "thus far eluded" it.[37] Debus replied on 15 October 1965 that such documentation did not exist—a surprising fact in view of the detailed documentation of almost every activity at the Cape. Debus pointed out that a second crawler was critical to the unit's performance because of the possibility of a breakdown, and that NASA had informed Congress and the Bureau of the Budget of its plans to purchase two transporters.[38]

Six months later, on 14 April 1966, the GAO notified Debus of its plans to examine other duplicate facilities on LC-39. Stanley Dyal, the auditor-in-charge, met with KSC officials and toured LC-39. On 6 July the GAO asked Debus about the decision to construct mobile facilities for six launches a year, in view of NASA and contractor studies that showed the mobile concept to be more economical only above a rate of 12–18 launches a year. Since NASA Headquarters had made the decisions the GAO questioned, Debus referred the letter to Washington. On 16 August 1966, Associate Administrator Mueller replied to the GAO, pointing to the presidential declaration on space of 1961 and the desire for sufficient flexibility to meet future needs, and adding that complete documentation of all studies, analyses, and conferences was not available.[39]

After a visit to KSC in late August 1966, the Associate Director of the GAO, Clerio P. Pin, decided to send Dyal from the Atlanta office to review what studies were available. For two months, Dyal read the extensive documentation pertaining to alternate methods of developing launch complex 39. When he departed on 28 October, Dyal stated that the GAO office in Washington would handle further inquiries at NASA Headquarters.[40]

Upon completion of the audit in June 1967, the GAO sent a 39-page draft report of its findings to NASA Administrator Webb, requesting comments and inviting discussion. A week later, NASA Headquarters transmitted the report to KSC for comments. Even though preparations were under way for the first flight of the Saturn V and the first operational use of LC-39's mobile facilities, KSC undertook an intensive month-long analysis of the validity of the GAO statements. Background documents relating to early decisions were researched and cost figures were reviewed. Almost every office involved in any way with launch complex 39 participated in the analysis. The GAO report was based on records at NASA Headquarters and KSC, as well as numerous industry and in-house studies made between 1961 and 1967. The report also indicated a shift from the initially announced purview of the audit—the concern about duplicate facilities—to focus on launch complex 39 and an analysis of whether or not NASA had "adequately considered the relative operational and cost merits of both mobile and conventional launch facilities." It accused NASA of failing to keep Congress informed on important matters and of using obsolete data in supporting its budget proposals.[41]

In preparing a response to the GAO report, Debus wrote NASA Headquarters on 19 July 1967 that the GAO had worked from inaccurate assumptions and arrived at erroneous conclusions. In the report, computed savings were based on outdated analyses. Debus asserted that KSC had kept Congress informed on all major issues, and that members of Congress had concurred with KSC's rejection of conventional launch facilities in favor of the mobile concept. At a meeting between representatives of NASA and GAO on 8 September 1967, NASA agreed to hold its response in abeyance until the GAO was able to review new documents.[42]

Queries continued to come in, and during October threatened to disrupt some KSC operations. NASA Headquarters asked KSC to comment on a statement that would be submitted to GAO the following month. Admiral R. O. Middleton, KSC's Apollo Program Manager, said that some of the key personnel who were being called on to answer the GAO queries were working overtime in preparation for the Apollo 4 mission. He recommended that KSC make no response until two or three weeks after the launch. KSC acceded to the request, however, completing a reply one week before the Apollo 4 launch. Headquarters used some of the comments and added many of its own in a communication to GAO two months later. In a letter of 24 January 1968 accompanying NASA's response, Harold B. Finger, Associate Administrator for Organization and Management, pointed out that the Apollo 4 flight had tested Saturn V–Apollo launch complex 39 and demonstrated that the United States now had a capability to launch large rockets. The choice of the lunar orbital rendezvous, the stress on ground tests in lieu of flight tests, and the successful flights of the Saturns had greatly reduced the number of launches anticipated at the time launch complex 39 had been planned.[43]

On 27 March 1968, after almost three years of reviewing the mobile concept facilities at LC-39, the GAO informed NASA that it planned no further investigations. Although this meant that it would not submit the report to Congress, the GAO offered NASA some precautionary advice. The review of some of KSC's planning studies convinced the auditors that fixed facilities could have been constructed at a saving of $55 million for a launch rate of 12 or less per year. The GAO would continue to make "reviews of this nature" in view of NASA's large expenditures. It suggested that NASA document all major decisions in a manner that would show clearly the basis for the actions at the time of the decision, and that NASA make all related files available at the start of future reviews. Finally, the GAO expressed hope for better communications with NASA.[44]

Although the audit diverted much effort to actions that contributed little if anything to the accomplishment of the manned lunar landing, it was not unproductive. By re-emphasizing the need for thorough documentation to support management decisions, the audit increased awareness that, in spite of the pressure of KSC's mission, the center had to remain responsive to periodic audit by the "government's financial watchdog" and had to manage its records toward this end. NASA Headquarters, on its part, hoped that the GAO had become more aware that NASA based its decisions on significant technical factors, in addition to management and cost aspects. NASA also hoped that GAO's constructive suggestions might help prevent such time-consuming, expensive reviews in the future.[45]

The GAO had questioned NASA's judgment on administrative decisions such as the evaluation of costs, technical decisions dealing with operational effectiveness, and communications between NASA and Congress. The complaints arose chiefly from the GAO's reliance on early planning documents, while NASA had used an evolving series of planning studies that kept pace with new developments. In other words, the GAO saw the early planning studies as an end product, whereas NASA saw them as the first step in a process. Also GAO had the advantage of hindsight. The decisions concerning the mobile concept had been made in the light of contemporary knowledge, at a time when experts were calling for upwards of 50 launches a year.

NASA's reply clearly indicated that it viewed the manned lunar landing program as an embarkation point, not a terminus. It cited 1963 statements to Congress by James E. Webb wherein the Administrator asked for a position of preeminence in space, and the more explicit one by D. Brainerd Holmes that called for "landing on other astronomical bodies."[46] The National Aeronautics and Space Act of 1958 lent support to this interpretation. The GAO, on its part, drew a distinction between what the government had authorized and what NASA planned or anticipated. In the broad sense, the GAO had zeroed in on the time-honored practice of government organizations trying to expand beyond immediate authorizations. A rigid adherence to authorized programs without thinking of the future might well have placed NASA in a strait jacket. It would have forced NASA to revert to the abandoned practice of constructing a special facility for each type of launch vehicle, something that members of Congress had hoped to prevent. It would also have restricted the speculation and experimentation necessary for progress.

KSC came out unscathed, except as regards its documentation of management decisions. This was paradoxical in view of KSC's extraordinary devotion to technical documentation. One could argue, as Harold Finger did, that a new organization must give priority to accomplishing its mission and defer paper work to a later date. But this could hardly satisfy the GAO, which by one means or another habitually reminded government organizations that the appropriation of funds is a beginning and not an end, and that they must one day answer for the use of those funds.

20

Man on Apollo

Two More Trial Flights—Apollo 5 and 6

With the success of Apollo 4, NASA had recovered much of the ground lost the previous January. An ambitious schedule was set up for the new year. The last of the Surveyors, the unmanned spacecraft that were photographing the lunar surface and analyzing the lunar soil, would go up 7 January 1968 from Cape Kennedy. The same month the unmanned Apollo 5 would be launched from LC-37 on a Saturn IB to test the lunar module. Two months later NASA would launch the second unmanned Saturn V, Apollo 6, on what was intended to be its final qualification flight. If both missions proved successful Gen. Samuel C. Phillips, Apollo Program Director, planned to advance the manned Apollo 7, a Saturn IB mission, to July or August. Assuming Saturn V was man-rated by then, Apollo 8 in October would have astronauts on the giant rocket for the first time. Should either Apollo 5 or 6 fail to meet its objectives, alternate plans provided for an additional lunar module test on a Saturn IB or a third unmanned Saturn V mission.[1]

The lunar module stood center stage on the Apollo 5 mission. The flight would verify operation of the subsystems of the lunar module, conduct the first firings in space of the ascent and descent stages, and test the capability of the ascent stage to fire while still attached to the descent stage—a procedure that would eventually be used on the lunar surface. Test engineers would monitor the lunar module's performance for six hours in near-earth orbit.

General Phillips's office originally planned to launch the first lunar module aboard Apollo-Saturn 206 in April 1967. Anticipating six months of checkout on the lunar module, Debus had requested a delivery date of September 1966. Development took longer than expected, however, and delivery slipped from month to month. The lunar module's arrival was still uncertain in January 1967 when KSC erected AS-206 on pad 37. In March AS-206 was taken down and replaced with AS-204, the launch vehicle from the ill-fated Apollo 1 mission.[2]

Lunar module 1 finally arrived on 23 June 1967. In the meantime NASA and Grumman engineers had built a plywood mockup on LC-37 to be

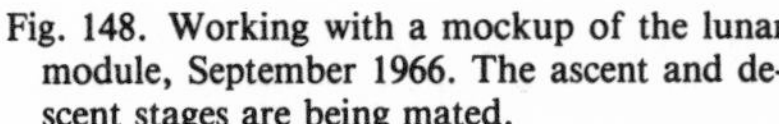

Fig. 148. Working with a mockup of the lunar module, September 1966. The ascent and descent stages are being mated.

used for facilities verification. For a simulation of the cable hookup, they bought hundreds of feet of garden hose at a hardware store and routed the garden hose "cables" down from the complex interfaces through the spacecraft lunar adapter. Since the first model did not carry all the extensive electrical systems of later lunar modules, checkout mainly concerned the propulsion system.[3]

The summer and fall were filled with problems. Both the lunar module and its ground support equipment required extensive modifications. The week of 13 August was typical: engineers replaced helium regulators and the water glycol accumulators on the ascent stage, corrected four deficiencies in the spacecraft acceptance checkout system, contended with leaks in the support equipment, and located the source of contamination in the gaseous nitrogen facility. On the 18th, the test office reported a "significant misalignment" at the juncture of the fuel inlet elbows and the spacecraft's propellant line. The elbows, built to the specifications of the original engine, did not fit the new engine, which had a slightly different configuration. Grumman fabricated new elbows and had them at KSC within three days.

The lunar module was mated to the launch vehicle on 19 November 1967. Without the command-service module, the vehicle stood 52 meters high. A protective covering that would detach in orbit shielded the 14 400-

kilogram lunar module and its adapter. The flight readiness test on 22 December came off satisfactorily. Preparations for the countdown demonstration test started on 15 January 1968. Following simulated liftoff on the 19th, the launch team began the actual countdown.[4]

Early in the month, Petrone had announced that Apollo 5 would be launched no earlier than 18 January. The indefinite date allowed for unforeseen problems with the lunar module, which lived up to expectations: problems in loading the hypergolic propellants delayed the terminal countdown until the 22d. At T – 2.5 hours, Test Supervisor Donald Phillips called a hold because of failures in the power supply and ground computer systems. These were corrected in time to launch the vehicle before dark. The S-IVB engine shut down ten minutes into flight and Apollo 5 went into orbit ten seconds later. The Saturn had performed entirely according to plan. The lunar module did likewise until a few seconds after the first ignition of the descent propulsion engine. The engine started as planned, but when the velocity did not build up at the predicted rate, the guidance system automatically shut down the engine. Experts analyzed the problem and recommended an alternate mission plan. The flight operations team carried this out successfully. As a result, Apollo 5 accomplished all its primary objectives.[5]

Apollo 6—A "Less Than Perfect" Mission

Besides its primary function as a flight-test vehicle, Apollo 6 (AS-502) served as a milestone in the site activation of LC-39. The Site Activation Board's second flow required that high bay 3, mobile launcher 2, and firing room 2 be in operation for the second Saturn V launch. Delays in the arrival of flight hardware and setbacks to the Apollo 4 schedule helped the board meet its schedule in time for Apollo 6. In April 1967, Boeing officials estimated that modifications on the swing arms, hold-down arms, and the tail service masts would require another 12 000 manhours. The mid-July date for the completion of this work was seven weeks behind schedule and threatened to delay a mid-August rollout. As events would eventually unfold, Apollo 6 did not reach pad A until February 1968, several months after the swing arm work was completed.[6]

The S-IC first stage arrived at KSC on 13 March 1967, and erection of the booster on mobile launcher 2 came four days later. Since the delivery of the S-II stage was another two months off, the Boeing crew substituted the S-II spacer again. The S-IVB stage and the instrument unit followed on the same day. The launch vehicle team quickly discovered that the high bay's environmental control system could not support the checkout. Portable high capacity air conditioners, used originally to protect Pegasus spacecraft on LC-37, were pressed into service. Even so, the humidity approached the maximum allowable for certain pieces of ground support equipment.[7]

During the month of April, a number of tests on Apollo 6 were postponed because of Apollo 4 support requirements, illuminating one of the limitations of the mobile concept in its early days. Although the facilities could physically accommodate two vehicles at the same time, their checkout could not proceed without the removal of men and equipment from one vehicle for temporary use on the other.

The S-II stage arrived on 24 May. It was mated with the interstage and moved to a low bay the next day. Further delays in the launch vehicle tests forced a postponement of several procedures including the launch vehicle overall test 1 (plugs in). Although propellant dispersion and power transfer tests were completed by the end of the month, the plugs-in test did not get under way until 13 June. The restacking of Apollo 4 in mid-June delayed the movement of the S-II to a horizontal position in the transfer aisle, and threatened the latter's erection date of 7 July. By the end of June, a new schedule for Apollo 6 was in hand, based upon the arrival of the command and service modules on 29 September.[8]

Apollo 6 operations in July and August continued to be marked by frequent delays. Several postponements were caused by hardware problems such as a request from Marshall that the launch team x-ray all liquid-hydrogen lines on the S-II stage. Vehicle tests were interrupted by the Apollo 4's plugs-in test on 1 August and again by ordnance installation on AS-501 during the week of the 14th. By September rescheduling had become a way of life for the checkout team.[9]

Another revised schedule in mid-September placed Apollo 6's countdown demonstration test in late January. Within a week the validation of swing arm 1 was four days behind schedule. Work on the service arms halted altogether on 26 September when most of the Apollo 6 crew was detailed to work on problems on mobile launcher 1. Support for Apollo 4 continued on an "as required" basis. Although the tests of the service arms for mobile launcher 2 fell three weeks behind schedule, this was not critical, because the delivery of the spacecraft was also postponed—this time by two months.[10]

With Apollo 4 launched and the spacecraft for Apollo 6 on hand, operations picked up. The swing arm tests were finally completed on 11 December, a day after the command and service modules joined the Saturn stack. During the remainder of the month, the launch team contended with a variety of problems: late flight control computers and flight program tapes, faulty memory in the RCA 110A interface unit, and glycol spilled on the outer surface of the spacecraft and S-IVB stage. The troubles of another plugs-in test on 21 December were typical: failure of a printed-circuit board in a digital events evaluator, a false fire alarm in the assembly building, failure of the emergency detection system test program, and a faulty battery that put an early end to the test.[11]

Problems with flight hardware continued to consume much time. During the plugs-out test on 28 December, the launch team had a premature cutoff of engine 5 in the S-II stage. An investigation indicated that the culprit was the engine control actuator. On 5 January 1968, North American began a three-day operation to replace the actuator. Just as this was being completed, a crack was discovered in the weld of a 2.5-centimeter LOX fill and drain purge line that paralleled a similar line inside the second stage. By the time the replacement line was cleaned and installed, the S-II crew had lost another three days. Unfortunately, the problems of the S-II stage on Apollo 6 were not limited to the checkout; they were precursors of malfunctions that would occur in flight.[12]

The space vehicle electrical mate, emergency detection system check, and overall test 1 were run during two days in mid-January, and the space vehicle swing arm overall test was completed on 29 January. As the launch crew reextended the swing arms after the test, the retract latch mechanism on arm 1 failed and the first stage took a blow. A gimbal joint in the support system was damaged, but the dent in the launch vehicle proved superficial.

Apollo 6 was transferred to the pad on 6 February. Under cloudy skies the crawler with its load paused briefly just outside the assembly building for the erection of the communications antenna and lightning rods on the mobile launcher. Winds and rain hit the area, and the crawler stopped when the storm disrupted communications with the launch control center. After two hours, with contact restored, the control center gave orders to proceed. The four double-track trucks moved ahead in the driving rain. A rainbow formed above the glistening height of Apollo 6 shortly before the crawler reached the foot of the pad. Two diesel engines began leveling the platform as the transporter negotiated the incline to the top of the pad. The sun had sunk behind a low bank of clouds, and the rocket inched up the pad in semidarkness. By the time the crawler reached the top of the pad shortly after 7:00 p.m., the clouds had scudded away, the winds had died down, and the stars glistened in a rain-washed sky. The mobile service structure could not be moved to the pad for two days because of high winds.[13]

The flight readiness test for Apollo 6 was completed early on the morning of 8 March. Three days later the flight readiness review was held at KSC. The meeting included representatives of all the major supporting elements for the mission. Apollo 6 was cleared for flight subject to the satisfactory completion of space vehicle testing and the closeout of action items identified by the review. The launch was set for 28 March. The next week the hypergolic loading team ran into some minor problems, and the stabilized platform in the instrument unit was replaced. The latter meant an extra 18 hours to reestablish the guidance system's integrity. The launch was changed twice again, first to 1 April, then 3 April.[14]

Preparations for the countdown demonstration test ended 23 March, and the precount began on schedule at 1:00 p.m. on the following day. The test was completed within a week. The launch countdown was picked up on 3 April at 1:00 a.m., the T – 8 hour mark. There were no unscheduled holds. At the mission director's informal review held at KSC on 3 April, Apollo 6 received a "go" for launch the next day. Launch day dawned warm and humid with scattered clouds. The prelaunch countdown and liftoff, in the words of Rocco Petrone, "followed the script"; but the script included one cliff-hanger, again in the S-II stage. During the countdown demonstration test, four propellant pump discharge temperatures had been a few degrees above redline values at the engine inlets. This threatened to convert the liquid hydrogen and oxygen into gases before reaching the injector. If this happened, Petrone told a prelaunch press conference, the pumps could malfunction and upset the ratio of fuel to oxidizer. After the test, steps had been taken to improve the insulation, and the LOX redline was raised two degrees to 98 kelvins (– 175 °C). Whether these changes would correct the condition would not be known until the countdown went into automatic sequence a little more than three minutes before liftoff. If the temperatures exceeded the new redline, the sequencer would be halted at T – 22 seconds. As it developed, the launch readings were within the new tolerances.[15]

Two Engines Out but Still Running

After liftoff Apollo 6 ran into a sea of troubles. In the closing seconds of the first stage burn, the vehicle went through 30 seconds of severe longitudinal oscillation—the pogo effect, it was called, because the space vehicle vibrated up and down like a child's pogo stick. As George Mueller later explained in a congressional hearing:

> Pogo arises fundamentally because you have thrust fluctuations in the engines. Those are normal characteristics of engines. All engines have what you might call noise in their output because the combustion is not quite uniform, so you have this fluctuation in thrust of the first stage as a normal characteristic of all engine burning.
>
> Now, in turn, the engine is fed through a pipe that takes the fuel out of the tanks and feeds it into the engine. That pipe's

> length is something like an organ pipe so it has a certain resonant frequency of its own and it really turns out that it will oscillate just like an organ pipe does.
>
> The structure of the vehicle is much like a tuning fork, so if you strike it right, it will oscillate up and down longitudinally. In a gross sense it is the interaction between the various frequencies that causes the vehicle to oscillate.[16]

The pogo effect had not been significant on Apollo 4. On Apollo 6 it started about 30 seconds after maximum dynamic pressure or "Max Q"—between 110 and 140 seconds after liftoff—and produced unacceptable g loads in the spacecraft.

Simultaneously, the spacecraft lunar module adapter was experiencing trouble. Made of bonded aluminum honeycomb, the adapter not only housed the lunar module but connected the command-service module to the Saturn launch vehicle. At T + 133 seconds, sizable pieces of the outer surface, more than 3 square meters, flaked off. Telemetry data and airborne cameras verified the damage. Nevertheless, the adapter performed its function without impairment of the overall mission.[17]

More was to come. Despite the pogo effect, the first stage completed its task and the S-II took over. At T + 206 and T + 319 seconds, the performance of engine 2 fluctuated. At T + 412 seconds, the engine shut down. Engine 3 cut off about two seconds later. The control system kept the vehicle stable for the remainder of the burn, 427 seconds or about 58 seconds longer than normal. This resulted in a deviation from the S-II flight pattern, and the third stage had to burn 29 seconds longer.[18]

In a postlaunch press statement, Phillips acknowledged, "there's no question that it's less than a perfect mission." However, he took comfort in a "major unplanned accomplishment"—the ability of the second stage to lose two engines and still consume its propellants through the remaining engines.[19] At the launch site Mueller described the mission as "a good job all around, an excellent launch, and, in balance, a successful mission . . . and we have learned a great deal . . . with the Apollo 6 mission."[20] The flight had tested altitude control, the navigation and guidance systems in conjunction with the service module engine, and the command module's heat shield. In spite of all difficulties, Apollo 6 had gone into orbit. Nonetheless, Mueller admitted later that Apollo 6 "will have to be defined as a failure."[21] The Apollo team set out to find what had gone wrong and why.

A week after the launch, Marshall issued an initial report. In relation to the malfunction of the J-2 engines, there was some speculation that the wires that carried cutoff commands to them had been interchanged. Although the basic source of the difficulties in the second stage had not yet been determined, this at least appeared to explain the premature cutoff. Later the trouble was identified as ruptures in small-diameter fuel lines that fed the engine igniters. The lines were redesigned to eliminate the flexible bellows section where the break occurred; the fix was then verified by tests at the Arnold Engineering Development Center.[22]

Coordinated plans for the resolution of the Apollo 6 anomalies, presented to the Apollo Program Director in a teleconference 2 May, included the fixes related to pogo. Prior to the launch of the first Saturn V, the longitudinal stability of the vehicle had been analyzed extensively. The results indicated that any pogo effect could be suppressed by detuning the natural frequencies of the propellant feed system and the vehicle structure. NASA had ruled that any modifications to existing hardware must be minimized. Now, from a screening process in which many solutions were considered, the corrective action emerged—it involved filling a series of cavities with helium gas. This required little change in hardware, but effectively changed the Saturn's resonant frequency. On 15 May a review of the oscillation problem determined that the fixes could be verified in an acceptance firing about mid-July. A final decision would be made at a planned August delta design certification review* for AS-503 (Apollo 8). All aspects of the problem were reviewed in June during a day-long teleconference among the Apollo Program Director and his staff, Marshall, Houston, KSC, and contractors. Tests and analyses had demonstrated that the modifications to 503 and subsequent vehicles had dampened the oscillations. The second of the major mechanical obstacles to man-rating had been successfully overcome.[23]

At the Manned Spacecraft Center, work on the spacecraft lunar module adapter's structural failure was concentrated in two areas: launch vehicle oscillation and spacecraft structures. No provision had been made to vent the honeycomb cells between the inner and outer surfaces of the adapter during launch. Pressures induced by aerodynamic heating of trapped air and free water in the cells could have ripped loose some of the adapter surface during the flight of the first stage. During the summer, North American engineers in Tulsa studied the effects of pressure on unbonded sections of the honeycomb panels. Dynamic tests at Houston verified a mathematical model of the spacecraft. At KSC the adapters for the Apollo 7 and 8 missions were inspected. Minor areas of unbonding were found and corrected. To equalize internal

---

*The delta design certification review was a programmatic review of all hardware changes in the Apollo-Saturn since the previous mission. With KSC engineers replacing many items of hardware on the space vehicle, these conferences served an important function. The name came from the widespread practice of using the Greek letter *delta* to stand for *difference*, hence by extension, *change*.

and external pressures during boost, holes were drilled through the adapter surface; and to reduce thermally induced stresses, a layer of thin cork was applied to all areas that had not been previously covered. The additional inspection at KSC and these two modifications were approved for subsequent missions, and as of late September no further changes were anticipated. It was generally agreed that the failure of the adapter had not been directly related to the pogo effect.[24]

NASA's efforts to resolve the Apollo 6 problems satisfied the Senate Committee on Aeronautical and Space Sciences, which in late April reported that NASA had analyzed the abnormalities of the flight, identified them with dispatch, and undertaken corrective action.

Apollo Astronauts at KSC

Before the end of February 1968, 18 Apollo astronauts had gone through exercises in the flight crew training building at Kennedy Space Center. This included both prime and backup crews for the first two manned Apollo missions. They used the mission simulators and the emergency egress trainer and were schooled on functional and operational aspects of the spacecraft.[25]

The saga of the astronaut as a superman had begun and ended with the first seven astronauts, not from their doing, but because the public demanded a space legend. With the Apollo program, it became clear that the astronauts were exceptional men, but human. Even though selection policies tended to produce a type, the crews included diverse personalities. Some were informal and convivial, some serious and tending toward the scientific in outlook, some difficult to deal with, others easy of access. Some astronauts were extremely courteous to the ground crew, totally cooperative; others were not. Some challenged the test teams to softball games or went fishing with them, while others remained aloof. But while the men on the pad knew this, the nation as a whole and the world at large saw a different picture—a group of all-Americans who, if not supermen, had "nary a failing among 'em."

In an article in the *Columbia Journalism Review* a few years later, Robert Sherrod attributed this stereotype to an unfortunate contract that *Life* magazine had made to tell portions of the astronauts' stories.[26] Sherrod told of a visit with a team of astronauts. He found them freely available. One cooked steaks for the *Life* crew. Another told of his Lincoln-like rise from obscurity. A third made flapjacks for his son's Cub Scout pack. "These three astronauts . . . went sailing together," Sherrod wrote, "though they didn't really like each other very much. . . . It took some time for the truth to sink in: these famous young men were doing handsprings for *Life* because they were being paid for it. . . . My story never came off, except as a picture story; the astronauts came out, as usual, deodorized, plasticized, and homogenized, without anybody quite intending it that way."[27]

In actuality they were distinct and interesting human beings, and, at times, major problems for the men who had to deal with them.[28] One of the heroic astronauts, for instance, was extremely rough in his language with the men on the ground—so much so that one of his most respected colleagues called a meeting of the ground crew to apologize for the man's conduct. One member of the launch team thought the tantrums deliberately contrived for two purposes: to get maximum efficiency out of the ground crew and to release personal tension. He said: "I would trust that astronaut to function perfectly in any tense situation. There is nothing I feel he couldn't do." The majority, however, agreed with their pad partner who remarked after listening to a recording of one outburst: "I hope they burn that tape."

The veteran astronauts were able to get one of their favorite pad men of Mercury and Gemini days, Gunter Wendt, transferred to Apollo. Gunter, a former Luftwaffe flight engineer, had emigrated to Missouri, where his father lived, after World War II. He had worked as a mechanic until he gained his citizenship papers and then joined McDonnell Aircraft Corporation. Sent to Florida, he had served on every spacecraft close-out crew from the launch of the monkey "Ham."[29] Wendt had a commanding way, a heavy accent, and a wiry frame—all of which brought him the nickname among the astronauts of "Der Fuehrer of the Pad."[30] The entire country was to hear his name in a few weeks. When Gunter looked in the window to make his final check of the Apollo 7 spacecraft, Wally Schirra quipped: "The next face you will see on your television screen is that of Gunter Wendt." Gunter retorted: "The next face you fellows better see is that of a frogman—or you're in trouble." Shortly after liftoff, Schirra asked Eisele what he saw out the window of the spacecraft. Eisele recalled the incident on the pad. As he looked out the window at endless space, he imitated Gunter's accent with words that went out to the television and radio audience: "I vunder vere Gunter vendt." This was to become the title of a chapter in a book of reminiscences by astronauts and their wives a few years later.[31]

Long before he "vundered vere Gunter vendt," Donn Eisele and his fellow crewmen of Apollo 7, Walter Schirra and Walter Cunningham, had gone through almost endless practice flights in the Apollo command module and lunar module simulators in the flight crew training building. Houston provided the management and operational personnel and KSC the facility support. After a series of lectures, the astronauts entered the simulators to practice all types of docking and rendezvous maneuvers, mission plans, malfunctions, and other situations that the pre-programmed computers threw at them. Gradually simulator work took precedence over briefings, and the astronauts concentrated on specific procedures for rendezvous and reentry.[32]

Each simulator consisted of an instructor's station, crew station, computer complex, and projectors to simulate stages of a flight. Engineers served as instructors, instruments keeping them informed at all times of what the pilot was doing. Through the windows, infinity optics equipment duplicated the scenery of space. The main components of a typical visual display for each window of the simulator included a 71-centimeter fiber-plastic celestial sphere embedded with 966 ball bearings of various sizes to represent the stars from the first through the fifth magnitudes, a mission-effects projector to provide earth and lunar scenes, and a rendezvous and docking projector which functioned as a realistic target during maneuvers.[33]

Two years later, when simulated moon landings had become commonplace for the astronauts and the simulator crews, they invited important guests to participate. Surprises were occasionally arranged for special guests. When French President Georges Pompidou moved the module toward the moon, he found the Eiffel Tower in the Sea of Tranquility. Another time, Chancellor Willy Brandt of the Federal Republic of West Germany landed the simulator on a Volkswagen symbol.[34]

Fig. 149. Spacecraft simulator in the flight-crew training building.

While the astronauts continued their repetitious exercises in the simulators, crews prepared two altitude chambers in the manned spacecraft operations building, adjacent to the flight crew training building, to test the Apollo spacecraft before its first manned flight. One chamber would serve the command and service modules, the other the lunar module. The program called for manned sea-level tests of the command-service module with astronauts on board, an unmanned altitude test, and two manned altitude tests, one with Schirra's prime crew and one with the backup crew of Thomas Stafford, John Young, and Eugene Cernan. These tests were principally designed to prove the machines at very low pressures. Mercury and Gemini flights had already demonstrated man's capabilities.

During the final 90 days prior to their flight, the astronauts lived on a relatively permanent basis in crew headquarters on the fourth floor of the manned spacecraft operations building. From here, they could "big brother" their flight hardware as each system went through its tests. The quarters

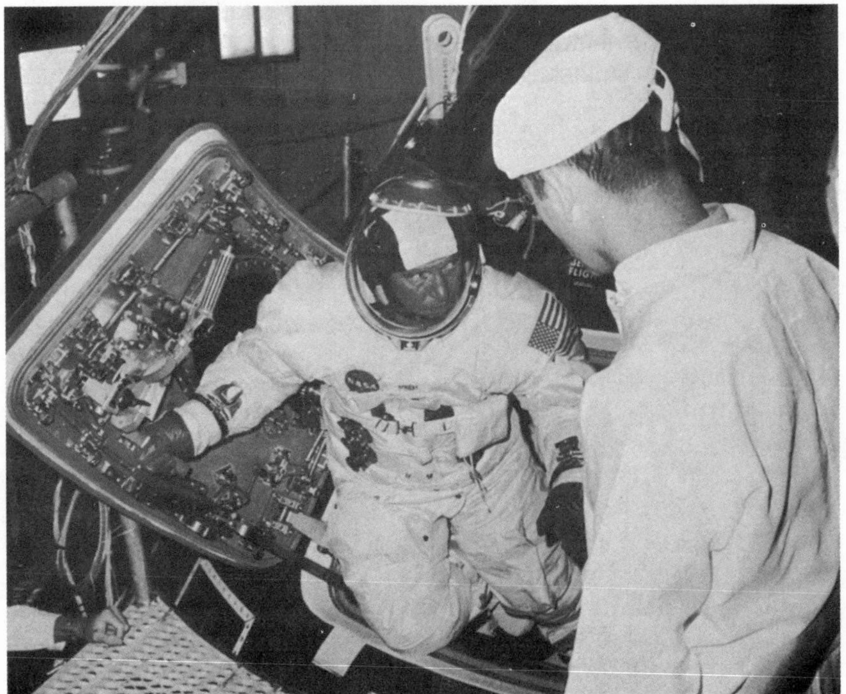

Fig. 150. Walter M. Schirra emerging from the spacecraft in an altitude chamber of the operations and checkout building, July 1968. Escape training was in progress.

consisted of three 3-man apartments, a small gymnasium, a lounge, and a kitchen, as well as a small but fully equipped medical clinic.

Apollo 7 Operations

Apollo 7, the first manned mission, was also the last Saturn IB flight in the Apollo program. Originally scheduled for late 1966, the launch had been delayed about 20 months by the fire and its repercussions. In mid-1967 while NASA was scrambling to recover from the disaster, the mission was tentatively set for March 1968. On the eve of AS-501, the Apollo Program Office scheduled the mission for October 1968. If the lunar module test on Apollo 5 went well, Phillips planned to proceed to the first manned flight in July 1968. Apollo 5 had accomplished its objectives, but because of extensive modifications, the command and service modules for Apollo 7 arrived at KSC more than two months late—on 30 May, three weeks after the launch vehicle had been erected on LC-34. In his operations schedule of 3 June, Petrone planned to stack the spacecraft on 19 July and launch in mid-September.[35]

Despite the best intentions, North American could not meet Petrone's schedule. The new block II command module was substantially different from the earlier model; there had been nearly 1800 changes to systems and procedures since the fire. The unmanned altitude run, scheduled for 1 July, was not completed until the 23d. The following week the astronauts made the manned altitude runs. The prime crew of Schirra, Eisele, and Cunningham spent more than nine hours in the spacecraft on 26 July, most of the time at a simulated altitude of 68 900 meters. They performed many assigned tasks to test their ability to work in their pressurized spacesuits. Technicians first purged the cabin, using a mixture of 65% oxygen and 35% nitrogen. Then the test team "dumped" the cabin's atmosphere, the astronauts relying on their spacesuits as the pressure dropped to nearly zero. After about an hour's work in near-vacuum conditions, the cabin was repressurized to 0.4 kilograms/square centimeter (5 psi) of pure oxygen—the normal atmosphere used in orbit. Three days later, the backup crew of Stafford, Young, and Cernan spent eight hours in the spacecraft at a simulated 61 000 meters altitude.[36]

While launch team and astronauts tested the command-service module, other KSC engineers tried out a slidewire that would serve as an alternate route of escape from the 65-meter level of LC-34's service structure. The 360-meter wire, designed by Chrysler, increased the options open to the astronauts and launch crew. If the hazard were a fire at the base of the service structure or any immediate threat, the slidewire offered a better means of

Fig. 151. The Apollo 7 flight crew (Schirra, left; Donn F. Eisele, entering the spacecraft in the background; and Walter Cunningham) during a test in September 1968. They are in the white room atop launch complex 34.

escape than the high-speed elevators. Inside the spacecraft, of course, the astronauts could employ the launch escape system. On 16 August after a successful dummy run, the engineer in charge strapped his harness to the slide mechanism and rode safely to the ground. The next test, a mass exit of dummies, revealed some problems. With a strong wind behind them, the 89-kilogram dummies sailed down the wire faster than expected; two overshot the embankment. The mass exit was tried again two weeks later, using a different brake setting on the slide mechanism. Five dummies and then five men rode the slidewire safely to the ground. The system was ready for Apollo 7.[37]

Petrone revised the Apollo 7 schedule on 1 August, laying out the remaining milestones at an Apollo Launch Operations meeting:

• Space vehicle erection	10 August
• Space vehicle electrical mate	28 August
• Plugs-in test	30 August
• Countdown demonstration test	11 September
• Flight readiness test	24 September
• Final countdown	7 October

There were no serious delays during the last ten weeks of the operations. The flight crew's presence gave the mission extra meaning for many members of the launch vehicle team who had not launched an astronaut since the Mercury-Redstone days. The countdown began at 2:34 p.m. on 10 October 1968 with the launch scheduled for 11:00 the following morning. After a smooth countdown, with only one brief unscheduled hold, the Saturn IB lifted off.[38]

Apollo 7 went into a circular orbit about 242 kilometers in altitude. The spacecraft, consisting of command and service modules, but no lunar module, separated from the Saturn's second stage nearly three hours after liftoff. The crew practiced docking maneuvers by bringing their spacecraft to within a few feet of a target circle painted on the S-IVB stage. In 11 days the crew demonstrated that three men could live and operate in the Apollo spacecraft for the period of time needed to get to the moon and back. The astronauts appeared to millions around the world via seven live television transmissions from "The Lovely Apollo Room High Atop Everything."

Splashdown was close to home. At 7:11 a.m. on 22 October, less than 30 seconds off the scheduled time, the astronauts hit the squally Atlantic south of Bermuda. The command module tipped over after the splashdown, but inflation devices soon righted it. Helicopters from the prime recovery carrier *Essex* brought the bearded trio onboard for medical assessment. They returned to Kennedy Space Center for further debriefing.[39]

"The Apollo 7," von Braun stated flatly, "performed . . . as nearly perfect as one can rightfully expect a development flight to be."[40] The Director of NASA's Apollo Program Office, General Phillips, agreed. "Apollo 7 goes in my book as a perfect mission," he stated. "Our official count is that we have accomplished 101 per cent of our intended objectives."[41]

Apollo 7 evoked more lines from budding poets than most previous launches from the Cape, as well as three memorable letters from youngsters. One small boy volunteered to "ride on a space ship to Mars," and listed three outstanding qualifications he had: he weighed only 27 kilograms, he was very observant, and he would not marry any of the women up there because he was "not fond of girls of any kind or shape." Another asked if he could train for interplanetary space travel, stating: "I have a very high eye cue and am smart." A 14-year old commented, "I would like to congratulate you on your progress. As I see it, you have only two problems remaining to conquer space—how to get there and how to get back."[42] No one at KSC disagreed!

Apollo 8 Launch Operations—Early Uncertainties

When AS-503—the third Saturn V—was erected on 20 December 1967, it had been scheduled for the unmanned launch of a boilerplate Apollo in May 1968. By late January the launch team had stacked the remaining stages on mobile launcher 1. Despite the success of Apollo 4, the flight hardware still carried considerable research instrumentation. As the Apollo 6 mission neared, KSC hastened to complete the integrated testing of AS-503 in the assembly building. Admiral B. O. Middleton, KSC's Apollo Program Manager, had informed Phillips that, if Apollo 6 failed and another unmanned

Saturn V were needed, AS-503 could roll out to the pad within ten days. Final preparations for the move were held pending analysis of the Apollo 6 flight test data and the decision whether AS-503 would be manned or not. KSC's chance to demonstrate the relative speed and economy of the mobile concept disappeared in the ripples created by pogo.[43]

Despite the disappointment of the Apollo 6 flight, NASA was reasonably confident in its analysis of the Saturn V problems. On 23 April, Mueller recommended a revised Apollo schedule to Administrator Webb, including provisions to man Apollo 8. The next day in a press briefing at NASA Headquarters, Phillips stated that, in spite of the problems, Apollo 6 had been a safe mission. He supported Mueller's recommendation by advocating that NASA prepare for a manned flight late in 1968 on the third Saturn V with the option to revert to an unmanned mission if corrections did not meet the requirements felt necessary to ensure crew safety.[44]

The revised schedule was approved by the Administrator on 26 April in a note endorsing the planning, design, fabrication, development, and proof-testing necessary for a manned AS-503. The Administrator did not, however, authorize such a mission at that time. The decision would come later and would be subject to several restrictions. Specifically, manning the mission was contingent upon the resolution of the Apollo 6 problems and the results of the Apollo 7 (AS-205) flight.[45]

KSC work schedules reflected the ambivalence of the Apollo 8 mission. If the vehicle was to have the unmanned boilerplate aboard with a lunar module test vehicle, the launch date would be 10 July. Allowances for a slippage to 15 October were built in for testing the fixes. If 503 was to be manned, it would fly CSM-103 and LM-3 no earlier than 20 November. As the manned alternative took precedence, KSC moved quickly to meet its demands. One requirement was an additional cryogenic proof pressure test for the S-II stage at the Mississippi Test Facility. By 30 April the launch team had taken the Saturn V apart and put the S-II aboard a barge. At Mississippi Test Facility the second stage, in addition to cryogenic testing, underwent modifications to the spark igniters. The J-2 engine of the third stage received the same modifications at KSC. Phillips hoped to increase the chances of meeting a manned launch in November by spreading out the necessary modifications among the various centers.[46]

In early May a problem in the first stage added to KSC's hardware difficulties. On 7 May, during a leak check on the turbopump of an engine on the first stage of AS-503, about 0.6 liter leaked from the main fuel seal in a period of 10 minutes. After evaluation, a decision to change the engine was made. The new engine was shipped on 20 May and arrived at KSC the following day. It took the remainder of the month to install and check out the replacement.[47]

By the middle of June, all approved modifications to the stages and the ground support equipment at KSC were in work. Several expected modifications, however, had not yet been approved or received. Consequently, KSC officials had some doubts that the planning schedule could be maintained. One anticipated change was the modification to suppress pogo in the first stage. Although KSC had not received approval for the modification, the work had to be done and it would probably delay the internal power tests on the stage. On 13 June the RCA computers in firing room 1 and in mobile launcher 1 malfunctioned. They, too, were undergoing modifications. The troubles were isolated to two printed-circuit cards and an open circuit in the mobile launcher's computer.[48]

After two Saturn V missions, operations at LC-39 were still not what might have been hoped. As one participant later observed about the period after Apollo 6: "Few working here on a daily basis really thought we were going to be able to make it by 1969. Everything took too long." This observation was directed largely at the Apollo spacecraft.[49] The same mood was evident at a closed meeting held at Grumman Aircraft and Engineering Corporation in July 1968. At that time, Phillips noted that "the lunar landing next year is within our grasp, but we don't have a hold of it because of the [contractors' disregard of planned delivery dates]." Mueller noted that "the rate of changes in the [lunar module] was three times that of the Apollo command module, whose rate of changes, in turn, was four times that of the Saturn V rocket. . . . The changes placed added burden on [KSC] technicians who should be concentrating on launching operations, not on vehicle modifications."[50]

By the summer of 1968, problems at Apollo factories were stretching KSC's workload beyond its capabilities. Furthermore, the preparation of lunar module 3 for the Apollo 8 mission was only the second mission for the Grumman team, and its inexperience showed. Charles Mathews, former Gemini program director, expressed concern about launch operations after a two-day visit to KSC. "The amount of rework [on LM-3] necessary at KSC was more than should be required in Florida." While acknowledging the overload, Mathews criticized Grumman engineers for reacting too slowly. They in turn complained about a lack of support from Bethpage. Mathews believed that neither North American nor Grumman had sufficient knowledge of manufacturing requirements. He recommended that both contractors appoint spacecraft managers to direct operations from factory to launch—"someone with as much authority within the Cape organization as he has at the factory."[51]

In mid-July Debus addressed the problem of KSC's handling three Apollo–Saturn V missions concurrently. A letter to Mueller noted an apparent misunderstanding between headquarters staff members and KSC.

Debus pointed out that, prior to the issuance of Apollo Program Directive 4H in November 1967, no schedule had shown more than two Saturn V vehicles at KSC simultaneously. Since then, he continued, discussions with Phillips had indicated that KSC should be able to process three vehicles concurrently. Funding constraints, however, had hampered efforts to enlarge the stage contractors' operations team.[52]

In a reply to Debus the following month, Phillips stated that the schedule was not, in fact, being met by KSC. To carry out the flights that were programed for the next year, KSC had to be able to process three vehicles concurrently. Phillips emphasized the efficient use of available resources and authorized KSC to provide crews for some phases of work on three vehicles simultaneously.[53]

Lunar Module Problems and Another Change of Mission

The uncertainty about the Apollo 8 mission, temporarily relieved by the progress on Apollo 6's deficiencies, reappeared in June when KSC began testing lunar module 3. Although it was to have arrived in flight-ready condition, KSC soon found out otherwise. The ascent and descent stages were delivered separately during the early part of June. Several leaks appeared during early tests of the ascent stage; one of them required a redesign and valve change. Early in July, a damaged flight connector in the rendezvous radar of the spacecraft caused a delay in its final installation. A week after this, there was a meeting at KSC of Houston, Grumman, and KSC officials to resolve the modification requirements. KSC estimated that it would take four days to complete the approved modifications prior to altitude chamber operations. An additional three to four days might be required if other pending modifications were approved. While work proceeded around the clock, engineers began a combined systems test for the spacecraft on 17 July. Problems with the radar, guidance, and communications systems delayed completion of the test for three days.[54]

During July, KSC was also investigating an electromagnetic interference problem in which the rendezvous radar locked onto the telemetry signal. Filters sent from the Grumman plant did not correct the problem. Attempts to tune the coaxial connection between the radar dish and the electronics package lessened the interference with the telemetry system, but resulted in a new interference with the abort guidance system. On 2 August when the spacecraft internal systems were activated, electromagnetic interference increased and further investigation began. As George M. Low later recalled, it was about this time that a circumlunar mission without a lunar module first

Fig. 152. A lunar module arriving at KSC aboard the Super Guppy, June 1967.

appeared as a real possibility. Difficulties encountered at KSC were having their impact on decision-making at headquarters.[55]

The S-II second stage had gone immediately to the low-bay transfer aisle after its return on 27 June. Between 1 and 11 July, the augmented spark igniters in the five engines were changed. When the second stage was erected on 24 July, the third stage was still undergoing modification. Forecasts that the instrument unit's flight control computer would not arrive on time threatened the schedule. Between delays in the delivery of launch vehicle hardware and difficulties with the lunar module rendezvous radar, the period of late July and early August was critical. Without a firm decision from headquarters, KSC could not move effectively, and difficulties at KSC tended to preclude firm decisions.[56]

At a Management Council Review in Houston, 6–7 August, Low presented the details of the lunar module problems and asked the Houston mission director, Christopher C. Kraft, to look into the feasibility of a lunar orbit mission without a lunar module. Low noted that the KSC work schedule was currently headed for a January 1969 launch and that insistence upon the use of lunar module 3 could result in a delay of up to two months. At a second meeting on 9 August, Kraft reported that the lunar orbit mission was feasible. Debus indicated that KSC could support such a launch as early as 1 December. Only two items remained open: the location of a suitable substitute for the lunar module and the approval of the Administrator, who was overseas at the time. Within three days after the meeting, the command and service modules for Apollo 8 had arrived at KSC.

At a meeting in Washington on 14 August, NASA substituted a test article for the lunar module. Since the circumlunar mission depended on KSC's ability to support a 6 December launch, Debus was asked to assess the launch team's chances. The KSC director replied that he had no technical reservations. Although Mueller expressed a reluctance to decide before Apollo 7 results were evaluated, he conceded the necessity of doing so. The overall review of the circumlunar mission plan resulted in an informal "go." KSC's response was immediate and positive: the following day, the spacecraft facility verification vehicle was erected on the instrument unit.

Administrator Webb agreed on 17 August to man Apollo 8 for an earth-orbital mission, but postponed the decision on a circumlunar mission until after the Apollo 7 flight. The launch of Apollo 8 was set for 6 December. On 19 August, General Phillips announced the earth-orbit mission to the press in Washington. He ascribed the change to the problems with the lunar module, then six weeks behind schedule.[57] To expedite prelaunch operations for Apollo 8, Phillips relieved KSC of much of the burden for hardware modification. The appropriate development centers were given the responsibility with the understanding that only changes necessary for crew safety would be accomplished.[58]

In mid-September KSC completed the first ten parts of the launch vehicle malfunction test satisfactorily; part 11 was scrubbed because of a failure in the RCA 110A computer. A modification of the computer in the launch control center delayed the plugs-out test until 18 September. At this point the spacecraft was approximately 5 days behind the 10 September schedule.[59]

NASA conducted a delta design certification review on 19 September by means of a teleconference. Since Boeing had not yet completed the testing and analytical work associated with pogo, Phillips asked MSFC to recommend a date in November for the final review of the Saturn V. Two days after the spacecraft was added to the launch vehicle stack, Apollo 8 rolled out to the pad on 9 October.[60] During the remainder of the month, the launch team conducted a series of space vehicle tests. The flight crew participated in several, such as verifying the performance of the command, control, video, and optical systems in support of the abort advisory system. They were also active in emergency egress training. Unlike earlier programs in America's manned space effort, the crew did not spend a great amount of time with the actual flight vehicle.[61]

The Apollo 7 mission ended with splashdown on 22 October. Six days later, NASA outlined the steps that would lead to a final decision on the next manned Apollo during the week of 11 November. Dr. Thomas O. Paine, acting Administrator, said: "The final decision on whether to send Apollo 8 around the moon will be made after a thorough assessment of the total risks involved and the total gains to be realized in this next step toward a manned lunar landing. We will fly the most advanced mission for which we are fully prepared that does not unduly risk the safety of the crew."[62] On 12 November NASA made its decision public—Apollo 8 would fly a lunar-orbital mission beginning 21 December.[63]

Launch Countdown for Men on Saturn V

Problems with the Sanders display unit (see page 338) in the firing room forced a postponement of the flight readiness test on 15 November. The second attempt on the 19th proved successful. The presence of a crew led to some alterations in the launch procedures. The commander could call a "hold" if he felt it necessary, or he could initiate an inflight abort. Weather restrictions for the launch were supplemented to meet the danger of impacting on land after a pad abort. The presence of a thunderstorm cell within 20 miles of the pad could force crew egress, and under no circumstances could a launch take place during or through a thunderstorm. These contingencies were the province of the flight director (who took control of the flight once the vehicle had cleared the tower of the mobile launcher), the launch operations manager, and the test supervisor.[64]

The countdown demonstration test for AS-503 began early on 5 December. The spacecraft slipped approximately 14 hours behind schedule because of problems in the astronaut communications and cryogenic systems. On 8 December the wet test progressed to T−9 hours when a problem in a data transmission system caused several hours delay. Later in the day an error in the memory of a digital events evaluator and a malfunction in a helium regulator terminated operations. The launch team resumed the following morning after the problems were resolved. A defective heat exchanger in the third stage's ground support equipment halted operations at T−2.5 hours. Once again the test conductor recycled the test clock to begin at T−9 hours the morning of the 10th. After completing the test by mid-afternoon, the launch team concluded the demonstration test with a dry run the following day. Problems with the astronaut communication system and ground support equipment were grim reminders of the 204 disaster.[65]

The launch countdown for Apollo 8 began at 7:00 p.m. on 15 December and headed for a launch on the 21st. The following day, a three-hour physical examination found the crew in good health. Both the men and the machine appeared ready.[66]

Apollo 8—A Christmas Gift

Activation of the fuel cells and the loading of cryogenics heralded the final count on the night of 19–20 December. The added tension of a manned launch began to show. Debus expressed the general mood on the afternoon of the 19th: "To go to the moon is symbolic of man's leaving earth, the opening of a vast new frontier. If we hadn't gained confidence in what we're doing, it would be an unendurable stress."[67] According to Paul C. Donnelly of the Test Operations Office, the astronauts "do not make it more difficult. They make it easier, because people respond better; everyone does a little better than he did when they were unmanned."[68] When night came, huge searchlights made Apollo 8 visible for miles. Poised on its pad, ready for man's first trip to the moon, it was a Christmas scene of rare beauty. Before dawn of the 21st, the sightseers already clogged the roads. The air was chilly, the dark sky filled with stars. Buses brought newsmen through the gates, and helicopters carried VIPs above the traffic. The distant Atlantic was the pale blue of predawn. With the morning light, Apollo 8 held everyone's gaze. People stopped their nervous prelaunch chatter, and stood in front of their cars. Radios announced "T – 30 minutes and counting." Astronauts Frank Borman, James Lovell, and William Anders had long before taken their cramped, temporarily supine positions. On ignition, a jet of steam shot from the pad below the Saturn. The crowd gasped. Then great flames spurted. Clouds of smoke billowed up on either side of the giant, completely hiding its base. From the midst of this fiery mass, Apollo 8 rose, slowly at first, as if unsure it could really lift free.

Suddenly the noise rolled across the three intervening miles, and vibrations struck the VIP and press bleachers. Flocks of ducks, herons, and small birds rose frantically from the marshes and filled the sky; and then came the most memorable noise of all, a triumphant cheer. A cloud blurred the view. Something fell out of the cloud, cartwheeling toward the blue ocean—the first stage had cut off. The giant second stage reappeared above the cloud, a bright star, diminishing second by second, until it faded from sight. People again turned their attention to their radios, listening attentively until the news came that Apollo 8 was in earth orbit.

Apollo 8 will be remembered for its demonstration of a great advance in space technology, for the incredible perfection that men and machines achieved throughout the mission, and for its television exploits. By television, people saw the earth from a distance of 313 800 kilometers. They saw the moon's surface from a distance of 96.5 kilometers and watched the earth rise over the lunar horizon. The astronauts described the dark Sea of Tranquility—an area designated as a landing site for a later Apollo mission. The television cameras measured the long shadows of the sunrise on the moon.

Then followed on Christmas Eve one of mankind's most memorable moments. "In the beginning God created the Heaven and the Earth." The voice was that of Anders, the words were from Genesis. "And the Earth was without form and void and darkness was upon the face of the deep. And the spirit of God moved upon the face of the waters and God said, 'Let there be light,' and God saw the light and that it was good, and God divided the light from the darkness."

Lovell continued, "And God called the light day, and the darkness he called night. And the evening and the morning were the first day. And God said, 'Let there be a firmament in the midst of the waters. And let it divide the waters from the waters.' And God made firmament, and divided the waters which were above the firmament. And it was so. And God called the firmament Heaven. And evening and morning were the second day."

Borman read on, "And God said, 'Let the waters under the Heavens be gathered together in one place. And the dry land appear.' And it was so. And God called the dry land Earth. And the gathering together of the waters he called seas. And God saw that it was good." Borman paused, and spoke more personally, "and from the crew of Apollo 8, we close with good night, good luck, a Merry Christmas and God bless all of you—all of you on the good Earth."[69]

On Christmas Day in the morning, Borman reignited the Apollo engine to break free of lunar gravity. Mission Control Center soon announced that Apollo 8 was on course, on time, at the correct speed. It landed in the Pacific shortly before midnight on 27 December.

The men of Apollo 8 had many firsts to their credit: they were the first to navigate the space between earth and moon, the first to experience the gravity of a body other than earth, the first to show live television transmission of the full earth disk, the first to exceed speeds of 38 625 kilometers per hour, the first to view the moon close-up with the naked eye, and they had set a distance-from-earth record for manned flight of approximately 359 000 kilometers. Somewhere in the list, but with a high priority at KSC and throughout the world of NASA and its contractors, they were the first men to ride the Saturn V.[70]

During the outward flight and to a lesser extent on the return, Borman suffered some form of sickness that appeared to be related to sleeping pills and led to a feeling of nausea. As there had been a slight epidemic of influenza at Kennedy Space Center, there was some concern that the astronauts might be suffering from this illness, as had the Apollo 7 crew members. Fortunately this proved false, and the crew completed the mission in good physical shape.

The official objectives of the mission went beyond flying ten orbits around the moon and included navigating the command-service module, communicating with earth, making corrections in mid-course, determining food needs, and controlling temperatures inside the spacecraft. In addition, engineers had tested in detail the systems and procedures directly related to lunar landings and other operations in the vicinity of the moon.

NASA planned this mission, like all others, on a step-by-step "commit point" basis. This allowed Mission Control to decide whether to continue the mission, return the craft to earth, or change to an alternative mission before each major maneuver, depending upon the condition of the Apollo and its crew. Thus, Control could have returned the spacecraft to earth direct by way of an elongated, elliptical path in space, in effect still an earth orbit, instead of entering lunar orbit. The flight was a faultless demonstration of the command and service modules, particularly the restart capability of the main service module engine on which the return journey depended. Besides television, the crew carried cameras loaded with color film. These yielded dramatic pictures of the earth viewed from the vicinity of the moon and color photographs of the moon's surface.

The Apollo 8 mission also highlighted KSC's tremendous achievement in the managerial task of assembling the equipment, controlling the ground support facilities, and achieving a liftoff within 1/6 of a second of the time scheduled months before. In the technical debriefing of the Apollo 8 astronauts at the Manned Spacecraft Center in Houston on 2 January 1969, Borman, Lovell, and Anders had little to suggest for improvement of preflight procedures at KSC. Their only recommendation was for a later date for the emergency egress test at the pad.[71] Other recommendations dealt with in-flight procedures of immediate concern to the experts at the Manned Spacecraft Center, not KSC. NASA granted awards to 12 key spaceport officials for their contribution to the Apollo 8 launch. Among those honored was George F. Page, Chief of the Spacecraft Operations Division. The recognition pleased the spacecraft team, several of whom spoke about Page as "one of the unsung heros, an outstanding intermediate management man. He applies the pressure that makes others perform."[72]

In briefing the Subcommittee on Manned Space Flight on 28 February, Debus said of the flight:

> The impact of Apollo 8, in my opinion, is something that defies quantitative measurement. Following the launch of Apollo 8, the Kennedy Space Center received over 5000 telegrams, phone calls, and letters from all over the world. This was by far the greatest volume of messages . . . following a launch. A similar theme ran through all of these communications. For Americans, it was one of intense pride in their country and its achievements. From friends all over the world—and letters came from 28 countries—it was one of pride in the human race and a feeling of gratitude to America. These letters came from men, women, the elderly, the young, the black, the yellow, the Christian, the Jew, the Moslem, the Heads of State, the laborer, the engineer and the underprivileged. Apollo 8 had some specific significance for everyone. This reaction certainly must be evaluated in terms of world prestige, technological accomplishments, and power.[73]

21

Success

The Launch Complex Becomes "Operational"

The achievements of Apollo 8 obscured some of the limitations of that flight. Most important from KSC's point of view, Apollo 8 was not a complete moon-landing vehicle. A test article had done duty for the real lunar module. In the launch vehicle, the S-II stage had carried extra insulation, and research and development instrumentation had been flown on all stages. Final confirmation of the LC-39 launch procedure would have to wait on a fully operational Apollo-Saturn. Apollo 9 (AS-504) would bring the space vehicle much closer to operational status. It would be the first test of the mated command-service and lunar modules. The 10-day mission in earth orbit would check out combined spacecraft operations and run the lunar module through a series of solo flights.[1] Some viewed the mission as a relatively mundane exercise in earth orbit except for the checkout of the lunar module's docking capabilities; but in General Phillips's words, Apollo 9 was "certainly one of the most vital missions that we've had in our mission sequence [and the risks] a little greater than the risks which we knowingly accepted in committing the Apollo 8 mission."[2] Moreover, Apollo 9 was to become the standard for processing subsequent Apollos through KSC.

Early schedules had listed Apollo 9 as the first manned Saturn V mission after three unmanned development flights. In the letter of 19 August 1968, which removed the lunar module from the Apollo 8 configuration, the Apollo 9 mission was redefined as a test of the lunar module in earth orbit. The crew slated for a later flight—James McDivitt, David Scott, and Russell Schweickart—was moved up to Apollo 9, and launch date was set for late February 1969.[3]

Launch operations began in May 1968 with the arrival of the S-II stage—first on hand this time after holding up three previous Saturn V missions. In August the North American team began modifying the S-II stage, not without complaint that Huntsville and the home office were not providing adequate direction. This dereliction, the daily status report for 28 August warned, might once again delay the high-bay testing of the S-II.

X-ray reports in mid-September gave the forward skirt splices a clean bill. At the same time the team made extensive changes in the propellant utilization and instrumentation systems to accommodate the S-II's new engines, which had been uprated to nearly one million newtons (230 000 pounds of thrust). Thanks to its early arrival and the team effort, the S-II stayed close to schedule. The third stage S-IVB arrived 12 September, followed in late September by the instrument unit, flight control computer, and S-IC first stage with its pogo modification. After inspection in the transfer aisle, the first stage was erected on 1 October; stacking of the entire vehicle was completed on 7 October. Erecting launch vehicles was becoming routine. Testing of the Saturn systems progressed according to plan during October, and faulty accumulators on two swing arms were replaced without delaying the schedule.[4]

Early in November a problem developed that involved both the vehicle and the ground support equipment. During the S-IC fuel pre-pressurization leak and functional test, a significant amount of RP-1 was spilled in the mobile launcher. Pressure in the Saturn fuel tank had forced fluid from the engine supply and return lines into a hydraulic pumping unit reservoir. The back pressure caused an overflow. An additional failure of a check valve on the gaseous nitrogen purge line allowed RP-1 fuel to back up into the electrical system of the hydraulic pumping unit. Accumulators from launcher 3 were borrowed for use on launcher 2. This type of problem illustrates the close interrelation of the rocket and ground support equipment. In effect, they formed a single unit, and malfunctions in one frequently caused damage to the other.[5]

The boilerplate spacecraft was removed from the stack on 2 December and the flight spacecraft replaced it the following day. At this point, the countdown demonstration test and launch countdown for Apollo 8 halted the testing of Apollo 9. The preliminary flight program tapes for the launch vehicle arrived at KSC on 20 December and the electrical mate of the space vehicle was finished six days later. After a plugs-in test in the assembly building on the 27th, ordnance installation was completed on New Year's Eve. The processing of Apollo 9 was on the schedule set in September and the space vehicle was ready for the trip to the pad. Despite problems, both vehicle and launch complex schedules had been maintained in a way hitherto unknown for Saturn V. Experience was beginning to show results.[6]

The Slowest Part of the Trip

Apollo 9, like every Apollo–Saturn V, started its epochal journey with the trip from the assembly building to launch pad 39. Eventually astronauts would travel at speeds in excess of 40 000 kilometers per hour, but 1.1 was about as fast as the crawler crew dared move the transporter with the Apollo-Saturn on its mobile launcher—an unwieldy 5715 metric tons rising 137.5 meters above the ground. "You can't imagine the difference between .7 and .9 miles per hour with this weight," one of the hydraulic engineers said. "At .7 the ride is very smooth, at .8 the vibrations may be noticeable but tolerable, and at .9 it might be difficult."[7]

Fred Renaud, a crewman on the crawler, had called it a "Texas tractor" in conversation with Representative Robert Price of Texas.[8] But a local newspaper was to refer to it as "one of the strongest, slowest, biggest, strangest, and noisiest land vehicles ever devised by man." With pardonable exaggeration, the newspaper spoke of the 5.6-kilometer trip as "nearly as important as the 500 000 miles [870 000 kilometers] to and from the moon."[9]

Each transporter had two cabs containing the usual controls found in an automobile: an accelerator, foot and parking brakes, speedometer, air conditioner, adjustable seat, and windshield wiper, plus radio for two-way communications. While the accelerator on the family car controls a single engine rated at around 250 horsepower, the crawler's accelerator controlled 16 motors with a capacity of more than 6000 horsepower. But starting a car, even on a winter morning, was easy compared to getting the crawler-transporter ready to move. It took an hour and a half for the crew of 14 to warm up the six diesel engines, energize several dozen electrical circuits, start up three hydraulic systems, one pneumatic system, a fuel system, and two lubricating systems, and make a series of checks called for by the 39-page "Start-Up Procedure Manual."

Handling such a monster required a cool head, extreme patience, and much teamwork, especially while loading and unloading at either end of the trip. Inside the assembly building, the crew had to steer the transporter with the aid of gauges, guidelines, and the judgement of technicians stationed at strategic points with walkie-talkie radios; and to bring it to within 5 centimeters of a set of pedestals ranging across the 45.7-meter width of the mobile launcher, so that the load could be firmly bolted down.

"When a man stands next to the crawler, the crawler looks big," Bruce Dunmeyer, supervisor of the transporter team, said, "but when you see the crawler under the mobile launcher, the crawler looks incapable of lifting such a big load." Spectators, and sometimes the crewmen themselves, were to feel that at any moment spacecraft and launcher could tip over and crash to the ground.

Renaud described a typical run down the level part of the crawlerway:

> This part of the move is not particularly hard . . . the main concern is just staying on the road, and if you have to stop quickly,

don't lean on the brake. The small jolts and jerks down here are sledge hammers at the top. One of the hazards is you tend to over-control the machine because it takes things so long to happen. You come up to a curve, put in a steering signal, and about 25 minutes later you come out of the curve. The tendency is to put all the steering on at once."[10]

The transporter had a crew of as many as 30, most of them with walkie-talkie radios, to monitor the last stage of the trip, the 365-meter incline with a grade of about 5%. The control room engineers and the head engineer supervised the critical task of keeping the Apollo-Saturn on an even keel while ascending the grade. This meant an endless chain of orders to systems of the transporter, including the cab engineers. William Clemens, one of the control engineers, felt that negotiating the grade was easier on the way up because there was so much excess power. But coming down, the driver could not allow the crawler to move too fast. "She wants to free wheel and coast," he stated, "and if you overspeed too far the diesel engines will shut off—which spells trouble! You must keep the speed under control."[11]

In Supervisor Bruce Dunmeyer's view, connecting the mobile service structure to the Saturn V at the pad was the trickiest and most delicate maneuver of all. The service structure towered 122.5 meters above the ground and provided access platforms for final checking of the Apollo spacecraft and the booster stages. "You have only a few inches of clearance when you are mating the structure to the pad," said one of the hydraulic engineer chiefs. "There are clamshelled doors that hinge and close around the bird, and if you run into it, there will be no shot. It is as simple as that."[12] Just before launching, the crawler-transporter would take the mobile service structure back to its parking area. The crawler crew's work represented hours of extreme tension between days of routine. In spite of this, the original crew was to see little turnover, with only two men leaving over the years.

The Launch of Apollo 9

Apollo 9 took seven hours to travel to pad A on 3 January 1969. The next three days were devoted to moving the mobile service structure to the pad. This included mating the mobile launcher to the pad, hookup and checkout of the data link and RCA 110A computer, final validation of swing arm 9, an integration test for the environmental control system, and moving the mobile service structure. On 6 January vehicle power was applied, two days later the Q-ball* was installed.[13]

The Manned Space Flight Management Council, which consisted of the major figures in the NASA manned spaceflight program from all the centers, met at KSC early in February. The meeting was followed by the flight readiness review for Apollo 9. At the time, the space vehicle was going through hypergolic loading, RP-1 loading (for the S-IC stage), and the main fuel valve leak test. During the electromechanical test of the service arms, oxidizer fumes were detected externally at the S-IVB aft interstage area. Examination revealed a vapor leak in the LOX system. The problem was solved by a decision to plug the leak detection port and to launch in that configuration.[14]

The countdown demonstration test began early on the morning of 12 February at T – 130 hours. As a practical matter, this test was the start of the countdown for the lunar module. System and subsystem checks as well as full servicing and close-out of much of that spacecraft left little to be done beyond loading the crew equipment. Crew participation during the "dry" demonstration test required only activation of systems needed to support the spacecraft-crew interface. Swing arm 9 retracted to its park position at the proper time, but instead of remaining retracted, the arm moved back to the command module. Activation of the fuel cells was simulated, since they were not required for crew support. The test was completed successfully on the morning of 19 February. Tests of the RF telemetry systems of the space vehicle and the return of the mobile service structure to the pad marked the beginning of the precount preparations for the launch itself.[15]

The countdown for Apollo 9 began the following week, aiming toward a launch on 28 February. While matters went smoothly for the launch team, the flight crew developed colds. The day before launch, at T – 16 hours, NASA officials postponed the mission until 3 March. KSC recycled the countdown to T – 45 hours so that the spacecraft team could replace the supercritical helium in the lunar module. The liquid oxygen and liquid hydrogen tanks required only a topping-off. For the launch vehicle team the delay meant charging a new set of flight batteries to install on 1 March. The principal effect of the flight crew's "malfunction" was to give the KSC team its first lengthy respite in a Saturn V countdown. The machine had proven more reliable than the men.[16]

---

*A 16-kilogram, cone-shaped instrument 36 centimeters high, the Q-ball was located above the Launch Escape System on top of the Saturn rocket. Unequal pressures on the four holes of the Q-ball indicated a change in trajectory.

At 11:00 a.m. on 3 March 1969, Apollo 9 lifted off on its flight into earth orbit. With an almost flawless performance, the Saturn V emerged as a proven piece of space hardware. Launch damage to the ground support equipment was slight compared to prior launches. During the countdown there had been no significant failures or anomalies in the ground system. As the first Apollo–Saturn V space vehicle in full lunar mission configuration, Apollo 9 demonstrated not only its own capabilities, but those of the ground facilities as well. The first comprehensive test of the vehicle, the complex, and the philosophy was a very satisfying success.[17]

The Contractors Receive Their Due

During a visit by the Teague Subcommittee on 28 February 1969, the congressmen inquired into KSC relations with the contractors who were now playing so large a role in launch operations. With its lunar module atop Apollo 9, Grumman Aircraft and Engineering Corporation had joined the list of major contributors at KSC. The Boeing Company was supervising ground support equipment for all stages of the Saturn launch vehicle, a task involving design and logistics engineering for 17 launch support systems. In addition, Boeing had the S-IC stage and technical integration and evaluation of the total program. North American, besides the tricky job of developing the S-II, was building rocket engines for all three stages at its Rocketdyne Division, as well as constructing the spacecraft.* McDonnell-Douglas was building the third stage at its Huntington Beach plant. IBM, with over 900 employees at KSC, had a dual role in launch operations: the installation and flight readiness checkout of the IBM-built Saturn instrument unit and maintenance of the Saturn ground computer complex, the hub of the semiautomated system that had been designed and built by RCA. This was not the only instance of a computer company operating products built by a rival firm. General Electric was sharing in the operation of Honeywell-produced computers.

Congressman Teague and his committee members directed most of their questions to basic principles of the KSC-contractor relationship. Deputy Director of Management Albert Siepert answered for KSC. The congressmen wanted to know why KSC had not developed an in-house work force instead of contracting the work out. The answer was that since Apollo employment had rapidly risen to 300000, then dropped back to 153000 in

*North American merged with Rockwell Standard Corporation on 22 September 1967 to become North American Rockwell, later renamed Rockwell International.

18 months, any attempt to handle such a short-term buildup with government employees would have disrupted the civil service. Committee members also asked why contractors were using KSC space and facilities instead of maintaining their own. Siepert explained this avoided duplication. KSC could wrap up the services needed by all the major contractors and handle them in a single service contract. Examples (not cited by Siepert): the Wackenhut Detective Agency's security contract and LTV Aerospace Corporation's contracts for audio-visuals, graphics, library, data management, and publications.

Siepert made the contractor representatives happy with his forthright answer to one question: Were contractors better informed than KSC personnel? He said that a contractor, who had designed and manufactured a piece of equipment, was the best authority on how it should perform. This warmed the hearts of some contractor employees who felt their NASA counterparts had been wanting in such appreciation.[18]

Even KSC's paperwork was testifying to the steady elimination of rough spots, technical and organizational. The first three Saturn V vehicles had been accompanied by a veritable mountain of printed documents that dealt with almost every conceivable topic and contingency within the purview of KSC. As each mission unfolded, numerous revisions in these materials took place before the launch itself: the final document not infrequently varied considerably from its predecessors on the same mission and topic. A change in this pattern began with the Apollo 9 mission. From September 1968 onward, the documents relating to any given mission (or to Apollo missions in general) became increasingly uniform. In this respect, the paper system was an outward manifestation of the increasingly "operational" character of both the vehicle and the facilities. The mobile concept had demonstrated its feasibility with the first Apollo–Saturn V mission, and the results showed up in the paperwork after AS-504.[19]

The test and checkout requirements document provides a good index of the operational complexities involved in the launch operations. Issued for each mission, this manual delineated the path of the vehicle through KSC facilities. As the program developed, the test and checkout requirements were modified. Examples of such changes were the elimination of the plugs-out overall tests and the rescheduling of the flight readiness test to precede the countdown demonstration.[20]

By the time of Apollo 9 an increase in the number of automated programs devoted to test and checkout was also apparent. The first two Saturn V missions had used 21 Atoll programs (pp. 355–56). A sharp increase had occurred on the AS-503 checkout, and this was countered by a drop in the use of programs written in the more difficult machine language. With Apollo

9, the total number of automated programs increased to 78, of which 36 were written in Atoll. Although this growth did not end with Apollo 9, it was clear that Atoll had proved its utility as a checkout tool. It made possible the launching of progressively more complicated missions from LC-39.[21]

Changes in the Telemetry System

The success of the Saturn V flights depended in large part on the performance of the telemetry system. A characteristic of all spacecraft programs, telemetry transmitted prelaunch and flight performance data from the vehicle to ground stations. From the start of the planning for Apollo, NASA realized that many more varied and sophisticated demands would likely be placed on the system. The development of launch operations at KSC was, in part, conditioned by those demands (pp. 356–58).

For prelaunch operations, the ability of the launch vehicle to check itself out was limited by requirements for ground support. A digital computer in the Saturn V instrument unit was primarily intended for guidance and navigation. It had triple redundancy throughout, except in memory and power sources, and its self-check capability was limited primarily to flight. Support for prelaunch operations came from the digital data acquisition system located in each stage. Tailored to the specific needs of its stage, this system transmitted data either through the data link or by means of pulse-code-modulated radio transmissions. The radio link could be used either on the ground or in flight.[22]

The Saturn V launch vehicle had 22 telemetry links carrying more than 3500 instrumentation measurements during flight. In prelaunch checkout each link and instrumentation channel was tested to assure operation within specified tolerances. Since the vehicle instrumentation system was used to acquire data during tests on other vehicle systems (such as pneumatics and control), frequent prelaunch checks of the instrumentation were required.[23]

The S-IC contained six very-high-frequency links, including the single-sideband-frequency-modulated telemetry system, one of the Saturn V components that had evolved throughout the launch vehicle development program. Because data during staging might be concealed or lost due to the effects of the engine exhaust, a tape recorder was included in the stage to collect that information, which was subsequently recovered by playback over radio. Range-rate data for the tracking of the vehicle was provided by the offset doppler transponder in the stage. Two other telemetry links used ultra-high-frequency receivers for range safety purposes. If the safety officer on the Cape issued a destruct command to these receivers, they would trigger the explosive network.[24]

The second stage had systems similar to those on the S-IC, but had one less single-sideband link. The S-IVB (third stage) carried no tracking transponders; otherwise, its telemetry equipment was identical to that of the first stage. The instrument unit carried an offset doppler, an Azusa (or Mistram) transponder, two C-band beacons, and a command and communication system. It had no range safety receivers.[25]

Long before the first Saturn V flew, the configuration of the vehicle allowed the use of either the Mistram or Azusa tracking systems, but not both at once. To reduce the complexity of the system, Phillips in 1965 directed that Azusa be used on future Saturn flights. Real-time support at the Cape would be required at least through the AS-503 mission. Experience in tracking early Saturn vehicles indicated a need for only one beacon, and some viewed even that as possibly unnecessary. It was later confirmed that the Saturn V was large enough to reflect enough radar energy to be visible on ground indicators to the limits of safety responsibility. Though a beacon might not be required for tracking purposes, range safety personnel considered it desirable.[26]

To receive telemetry from its vehicles, NASA maintained three ground networks. One of these, the Manned Space Flight Network, was under the operational control of Goddard Space Flight Center, Greenbelt, Maryland, during Apollo missions. In order to operate effectively for the lunar landing program, the system had to be able to control the spacecraft (both the command and lunar modules) at lunar distances. While the equipment had been adequate for earth-orbit missions, the greater distances, as well as the complexity of Apollo, led to the introduction of the unified S-band system.[27]

The term *S-band* derived from the period of the Second World War when letters were used to designate bands of frequencies. The band selected for Apollo lay between 1550 and 5200 megahertz. For use with its unmanned space probes, the Jet Propulsion Laboratory (JPL) had developed equipment that operated on these frequencies. A useful feature of the JPL equipment was the combination of several radio functions into a single transmission from only one transmitter to a given receiver. For Apollo, these functions included tracking and ranging; command, voice and television communications; and measurement telemetry. The versatility of the system was inherent in its structure.[28]

For the lunar mission the unified S-band offered the twin advantages of simplicity and versatility. The line-of-sight signal lost little of its strength when it passed through the atmosphere, and transceiver and power supply equipment could be relatively small. In providing direct communications between the spacecraft and ground stations, the unified S-band worked equally well in near-earth operations or circling the moon.

Apollo's tracking system required close, continuous communication among the major centers and the Manned Space Flight Network. This was accomplished by means of digital data, teletype, and voice links which were the responsibility of the NASA communications system centered at Goddard. A combination of land lines, undersea cables, high frequency radio, and satellites linked more than 100 locations throughout the world. For Apollo, the system had to be augmented. Major switching centers ensured maximum sharing of circuits, while giving Houston priority for real-time data during Apollo missions.[29]

During Apollo operations, the three manned spaceflight centers were connected outside the Goddard system by two links—the launch information exchange facility and the Apollo launch data system. Operated by Marshall during launch operations, the former was primarily an information transfer link between Huntsville and KSC with connections to Houston. It carried real-time telemetered data, closed-circuit television, facsimile, classified typewriter, voice, and countdown information. The Apollo launch data system was the primary information link from KSC to Houston. It had four independent subsystems that handled telemetry, television, countdown and status data, and launch trajectory data during prelaunch and launch operations. By using the Apollo launch data system, personnel in Houston could conduct closed-loop tests of the spacecraft while it was at KSC. During powered flight, the system transmitted trajectory data from the impact predictor for the information of the flight director at Houston.[30]

The Apollo program significantly increased the tracking and data acquisition requirements for KSC and the Air Force Eastern Test Range. To ensure uniformity, the Office of Tracking and Data Acquisition, NASA Headquarters, was designated in August 1964 the "single point of contact" with the Department of Defense for such coordination. Although heavily involved in the development of the unified S-band system for Apollo operations, the Jet Propulsion Laboratory and Goddard were directed to support the planning and operations.[31] The agreement that resulted between NASA and the Defense Department emphasized colocation of KSC and Air Force Range facilities whenever possible "to achieve a maximum of mutual assistance, to avoid unwarranted duplication, and to realize economies where practical and consistent with mission requirements . . . ."[32] To support Apollo, range facilities needed considerable modernization. During 1965 about 85% of the existing Air Force tracking equipment was modified. Over three years, the cost exceeded $50 000 000, including the updating of telemetry stations downrange as well as at the Cape.[33]

The entire Apollo tracking and data acquisition network, including ships, planes, and unified S-band ground stations, was integrated with the Manned Space Flight Network between November 1966 and June 1968. The AS-202 mission in August 1966 provided the first test under actual operating conditions. By the launch of Apollo 9 the new system was operational at stations in Texas, Mexico, Ascension Island, the Canary Islands, Bermuda, Spain, Hawaii, Australia, Wales, and California.[34]

There was no major change in tracking and data acquisition comparable to the introduction of the mobile concept. The primary alteration in tracking was the increasing sophistication of the hardware.[35] From early Saturn I missions through Apollo 9, development of hardware had tended to proceed steadily, dependent largely upon launch vehicle requirements. At the same time, less and less direct control over telemetry was allowed to KSC. In this respect, the attempt of NASA to spread the R&D among several centers had led to an unexpected constraint upon launch operations at LC-39. In the end, the Saturn V was measured and tracked by a telemetry system largely outside the control of KSC.

At Long Last

The liftoff of Apollo 10 on 18 May 1969 would mark KSC's fourth manned Apollo launch in the short space of seven months. The mobile concept was proving its efficiency. Before Apollo 8 moved out of the vehicle assembly building on 9 October 1968, the crews had already stacked AS-504 for Apollo 9. Before Apollo 9 was subsequently moved out, crews had stacked AS-505 for the Apollo 10 flight. Apollo 10 rolled out on 11 March to pad B, the more distant of the two pads on LC-39. This would prove the only use of pad B for an Apollo mission and the only use of firing room 3.

On the flight of Apollo 9, McDivitt, Scott, and Schweickart had checked out the lunar module in flight and docking maneuvers with the command-service module. On Apollo 10, the crew of Thomas Stafford, John Young, and Eugene Cernan took the spacecraft to the vicinity of the moon where the lunar module closed to within 16 kilometers of the surface before redocking with the orbiting command module. At long last, KSC was set for the Apollo 11 mission that would put men—Neil Armstrong and Edwin Aldrin, Jr.—on the moon while Michael Collins waited for them in the command module.

Apollo 11 stages had been arriving at KSC since the beginning of the year, the second stage undergoing rigorous inspection on account of a stormy barge voyage from California via the Panama Canal. During March

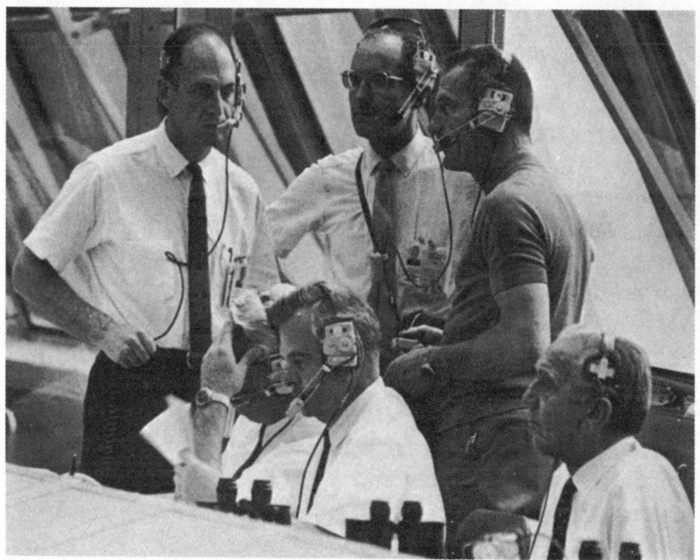

Fig. 153. Apollo officials in the launch control center during the countdown for Apollo 10. From left, standing: George M. Low, Apollo Spacecraft Program Manager; Lt. Gen. Sam Phillips, Apollo Program Director; and Donald Slayton, Flight Crew Operations Director, MSC. Seated, John Williams, Spacecraft Operations Director; Walter Kapryan, Launch Operations Deputy Director; and Kurt Debus, KSC Director.

the prime and backup crews participated in the spacecraft tests, with mid-April bringing the docking tests in the altitude chambers. During a checkout of the lunar module descent stage, technicians discovered faulty actuators in the machinery that would push out the legs of the lunar module for the moon landing. The repair area was inaccessible to men of average build, and Grumman scoured its rosters for two qualified technicians who were "very slim." The two men—William Dispenette and Charles Tanner—squirmed into the narrow space and replaced the actuators.[36]

The lunar module on Apollo 11 differed in several respects from that on Apollo 10. A very-high-frequency antenna would facilitate communications with the astronauts during their extravehicular activity on the moon's surface. The lunar module would also have a lighter-weight ascent engine, increased thermal protection on the landing gear, and a packet of scientific experiments. The only change in the command-service module was the removal of a blanket of insulation from the forward hatch. On the launch vehicle, the first stage was stripped of its research and development instrumentation. Insulation was improved on the second stage, and slight

Testing the Apollo 11 spacecraft in the operations and checkout building

Figure 154

Fig. 154. The command and service modules in an altitude chamber. Segments of a workstand, used to work near the top of the spacecraft, have been lifted and pulled back against the circular walls of the chamber. Fig. 155. Testing the landing gear of the lunar module. Fig. 156. Mating the command and service modules with the spacecraft-lunar module adapter, April 1969.

Figure 155

Figure 156

changes were made in the connections between the third stage and the instrument unit.[37]

The crawler-transporter picked up the 5443-metric-ton assembly and started for pad A at 12:30 p.m. on 20 May while Apollo 10 was still on its way to the moon. The countdown demonstration test got underway 27 June with vehicle and spacecraft fueled, powered up, and counted down for simulated launch on 2 July. On the following day, with the fuel tanks drained, Armstrong, Collins, and Aldrin participated in a dry test.[38]

Meanwhile, KSC was preparing for the hundreds of thousands of people who wanted to see the men off to the moon. Special guests, members of the press, and dependents of Apollo team members would number close to 20000. Some 700000 people were expected to watch the liftoff, possibly the largest crowd to witness a single event in the history of the world. The anticipated traffic jam prompted KSC to arrange for helicopters to fly in key personnel, should they be otherwise unable to reach their work. Guests included Vice President and Mrs. Spiro Agnew, former President and Mrs. Lyndon Johnson, Army Chief of Staff General William Westmoreland, four cabinet members, 33 senators, 200 congressmen, 14 governors, and 56 ambassadors. Close to 3500 accredited members of the news media were occupying the press site. Over two-thirds were American; 55 other countries, including three Iron Curtain nations, sent representatives, with Japan's 118 leading the way. All western European countries except Portugal were represented, and all western hemisphere nations except Paraguay.[39]

Brilliant lights illuminated the launch area and Apollo 11 during the night of 15 July. The crawler-transporter carried the mobile service structure to its parking area a mile away. In the early hours of 16 July, the tanks of the second and third stages were filled with liquid hydrogen. More than 450 people occupied the 14 rows of display and control consoles in firing room 1. Sixty-eight NASA and contractor supervisors occupied four rows; seated at the top, nearest the sloping windows that looked out toward the launch pads, were the KSC chiefs, the Saturn V program manager for Marshall, and the Apollo program manager for the Manned Spacecraft Center. One hundred and forty Boeing engineers occupied consoles linked to the Saturn IC stage and mechanical ground support equipment. North American Rockwell had 60 engineers at consoles connected with the S-II stage, while 45 McDonnell-Douglas engineers monitored the S-IVB stage. Ninety IBM engineers manned three rows of consoles hooked up to the instrument unit, IBM stabilization and guidance systems, and flight control. About 8 kilometers to the south two automatic checkout stations in the operations and checkout building monitored the spacecraft.[40]

The fueling of the launch vehicle was completed more than three hours before liftoff. Then the closeout crew of six men under the direction of Gunter Wendt and Spacecraft Test Conductor Clarence Chauvin returned to the pad. They opened the hatch and made final cabin preparations. The backup command pilot, Fred Haise, Jr., entered the spacecraft at 3 hours and 10 minutes before liftoff. With the assistance of Haise and a suit technician, Neil Armstrong entered Apollo at 6:54 a.m. Michael Collins joined him five minutes later in the right couch, and Edwin Aldrin climbed into the center seat. The closeout crew shut the side hatch, pressurized the cabin to check for leaks, and purged it. At two hours before liftoff Houston participated in a final checkout of the spacecraft systems. At one hour before liftoff, the closeout crew left the pad. Almost a kilometer to the west, protected by a sand bunker, 14 rescue personnel stood watch. Equipped with armored personnel carriers and wearing flame protective gear, they could move to the pad quickly if the astronauts needed help.[41]

To make the occasion more memorable, the day was ushered in by a beautiful dawn. A few fleecy clouds scarcely cut the warm sun. The slight wind cheered the assemblage. As the moments ticked off, loud speakers reported that everything was moving according to schedule.

The countdown became automatic at 3 minutes, 20 seconds, when the sequencer took over. Ignition commenced at 8.9 seconds with a wisp of white

Fig. 157. The departure of Apollo 11.

smoke indicating that the first engine would soon come to life. All five engines built up full thrust with an awesome roar. For a moment Apollo 11 seemed to stand still; then at 9:32 a.m. on 16 July 1969, the moon rocket rose slowly and majestically. A voice broke the tension: "The vehicle has cleared the pad." Apollo 11 had gone beyond KSC's control and the men in firing room 1 turned for a moment from their consoles to view the rocket rising over the Atlantic.

Many people moved away from the viewing sites as soon as the vehicle disappeared from view. Others stood silently, or chatted quietly, or sat on the grass if they were not among the privileged visitors in the stands. Exhaustion held some—others simply did not want to fight the traffic. A cameraman asked how the launch looked. He had not seen it, because he had been busy photographing the reactions of the VIPs.

"Eagle Has Landed"

Cleared to proceed to the moon, the astronauts fired the S-IVB engine again, increasing their velocity to 38 400 kilometers per hour. On 20 July, Sunday in the United States, Armstrong and Aldrin occupied and powered up the lunar module, Eagle, and deployed its landing legs. The two craft separated at 1:46 p.m. (KSC time). Collins fired the command module rockets to move about three kilometers away. Flying feet first, face down, Armstrong and Aldrin fired Eagle's descent engine at 3:08 p.m. Forty minutes later, as the command module emerged from behind the moon, Collins reported: "Everything is going just swimmingly." The two astronauts guided the Eagle into elliptical orbit. Armstrong throttled the engine at 4:05 p.m. to slow its descent.

As the moonscape came into clearer view, Armstrong saw they were approaching a crater almost as large as a football field. He took over manual control and steered toward a less formidable site. At Mission Control physicians noted his heart beat had increased from a normal 77 to 156. While Armstrong manipulated the control, Aldrin called out altitude readings: "750 feet, coming down at 23° . . . 700 feet, 21 down . . . 400 feet, down at nine. . . . Got the shadow out there . . . 75 feet, things looking good . . . lights on . . . picking up some dust . . . 30 feet, 2½ down . . . faint shadow . . . four forward . . . drifting to the right a little . . . contact light . . . O.K. Engine stop." As the probes beneath three of Eagle's four footpads touched the surface, a light flashed on the instrument panel. The world heard Armstrong's quiet message: "Houston. Tranquility Base here. Eagle has landed."[42]

Later the crew explained that at some distance from the surface, fine dust had blown up around the spacecraft and obscured their vision. They felt no sensation at the moment of landing, and set to work telling people on earth what they could see from Eagle's windows. At 6 p.m. Armstrong recommended that the walk on the moon should begin about 9 p.m., earlier than originally planned. Later than he proposed, but still five hours ahead of schedule, Armstrong opened the hatch and squeezed through it at 10:39 p.m. He wore 38 kilograms of equipment on his back, containing the portable life support and communications systems. On the moon, the weight amounted to only 6.3 kilograms. Wriggling through the hatch, Armstrong cautiously proceeded down the nine-step ladder. He paused at the second step to pull a ring to deploy a television camera, mounted to follow his movements as he climbed down. At 10:56 p.m. he planted his left foot on the moon. Then the words that were to take their place among the great phrases of history: "That's one small step for a man, one giant leap for mankind."[43]

At 1:54 p.m. 21 July, after 22 hours on the lunar surface, Aldrin fired the ascent stage engine. It functioned perfectly. They docked with the command module at 5:35 p.m. Collins touched off the main engine at 12:55 a.m. 22 July, while on the back side of the moon, and the astronauts headed for home. Because of stormy seas, they adjusted their course to a new landing area 434 kilometers from the original site. They splashed down in the Pacific at 12:50 p.m. 24 July. President Nixon greeted them on the aircraft carrier *Hornet*.[44]

The Apollo program had achieved its objective five months and ten days before the end of the decade.

One of the most perceptive writers of our time, Anne Morrow Lindbergh, probed the deeper meanings of these amazing engineering accomplishments. In *Earthshine*, she spoke of the "new sense of awe and mystery in the face of the vast marvels of the solar system," and the feeling of modesty before the laws of the universe that counterbalanced man's pride in his tremendous achievements. Many had remarked that mankind would never again look on the moon in the same way. She thought it more significant that people would never again look at earth in the same way. We would have a new sense of its richness and beauty. She concluded: "Man had to free himself from earth to perceive both its diminutive place in the solar system and its inestimable value as a life-fostering planet."[45]

22

A Slower Pace: Apollo 12–14

With the arrival of Apollo 12's spacecraft in late March 1969, four months before the first moon landing, Kennedy Space Center again had three Apollos in the operational flow. On 30 April Grumman mated the lunar module ascent and descent stages, while North American readied the command-service module for a cabin leak test in the altitude chamber. Launch vehicle stacking awaited the arrival of the S-IC first stage, which arrived from Michoud on 3 May and was placed on mobile launcher 2 four days later. Operations on Apollo 12 halted for two days while ordnance was installed in Apollo 11. The remaining Saturn stages were erected on the 22d. At the operations and checkout building, a number of hardware problems delayed operations by a week. Although the Launch Operations office postponed until 30 June the transfer of the spacecraft to the vehicle assembly building, the launch team continued under a tight schedule. If Apollo 11 failed, KSC faced a possible September launch for Apollo 12.[1]

Testing went well during the next six weeks and the space vehicle stack was complete by 1 July. The successful lunar landing later that month relaxed the pace. After the splashdown on the 24th, General Phillips announced a 14 November launch of Apollo 12 to the moon's Ocean of Storms. Among the mission goals, NASA hoped to improve its landing techniques and secure low-orbit photographs of sites for further exploration. During extra-vehicular periods, the astronauts would gather lunar samples and deploy the first Apollo lunar surface experiments package.[2]

With the accomplishment of Apollo's primary objective, a number of key program officials decided to move elsewhere. In August 1969 Phillips left his position as Apollo Program Director to command the USAF Space and Missile Systems Organization. Rocco Petrone moved up from KSC to fill the vacancy in the Office of Manned Space Flight. The following month Rear Admiral Roderick O. Middleton vacated the Apollo Program Manager's Office at KSC to assume command of Cruiser-Destroyer Flotilla 12. In December Dr. George Mueller resigned as NASA's Associate Administrator for Manned Space Flight and was replaced by Dale Myers, an executive from North American Rockwell. That same month Albert Siepert announced his retirement from KSC.[3]

Petrone's departure from KSC brought Walter J. Kapryan to the post of Director of Launch Operations. A native of Flint, Michigan, and a graduate of Wayne State University, Kapryan had served as a B-29 flight engineer in World War II. In 1947 he had entered the field of hydrodynamic research at Langley Research Center in Virginia. When NASA absorbed Langley, Kapryan became a member of the Space Task Group. He came to the Cape in 1960 as a project engineer for Mercury-Redstone, worked for a time in Houston, and then headed Houston's Gemini Program Office in Florida. Kapryan came to Apollo in late 1966, first as Assistant Apollo Spacecraft Program Manager and then as Petrone's deputy. Although less imposing physically and less assertive in manner than Petrone, Kapryan enjoyed wide respect within Apollo program ranks. Middleton's successor, Edward R. Mathews, was a veteran of the Missile Firing Laboratory. As Chief of the Saturn IB Systems Office, Mathews had played an important role in LC-34 and 37 modifications. He had served as Deputy Apollo Program Manager since September 1967. Siepert's duties as Deputy Director for Center Management were taken on by Miles Ross when the latter became Deputy Center Director in June 1970. Fortunately there was little turnover during the remainder of the program.[4]

Apollo 12 rolled out in the early daylight of 8 September. The prime crew for the mission—Commander Charles Conrad, Command Module Pilot Richard Gordon, and Lunar Module Pilot Alan Bean—joined hundreds of other spectators. During September and October, the checkout proceeded in routine fashion. In the local jargon, it was a nominal operation. The countdown demonstration test ended on 29 October without incident, although rain and high winds stormed through the complex at simulated liftoff.[5]

The launch team started precount procedures one week before launch day. The 70 hours of activities moved along smoothly until Wednesday, 12 November. That morning technicians began filling the service module's liquid hydrogen tanks, which fed gaseous hydrogen to the spacecraft's fuel cells. There the hydrogen mixed with oxygen to provide electrical power and drinking water. Within minutes the North American test team knew it had a problem: one tank was not chilling down. When the team stopped the hydrogen flow, the fuel level dropped off rapidly. A crew member, looking through a panel window, detected frost on the tank. It was found that a leak in the outer shell had destroyed the vacuum insulation. Less than 40 hours remained on the countdown clock, and the problem was a new one for the North American crew at Merritt Island. After consulting with the Manned Spacecraft Center, John Williams, Spacecraft Operation Director, decided to replace the faulty tank with its corresponding unit from Apollo 13. Judging from experience at Downey, there was ample time for the operation. If the replacement could not be accomplished within the remaining hold time, KSC would delay the launch one month. The exchange involved removing an access panel and the cryogenic service lines leading through the panel, disconnecting a series of cryogenic feed lines and electrical connections between the tank and the hydrogen subsystem shelf, exchanging tanks, and refastening all the parts. The North American crew worked deliberately since spacecraft power was on, but still managed to complete the work within 24 hours. Meanwhile Launch Operations rescheduled the spacecraft cryogenic loading for Thursday morning at T-17 hours.[6]

The launch team had planned one other major change in the countdown—the installation of the fuel capsule that would provide power for the package of experiments to be left on the lunar surface. The experimental instruments received power from a small (45 centimeters high, 40 centimeters in diameter) atomic generator. The 3.8 kilograms of plutonium 238 that fueled the generator rode to the moon inside a graphite cask. Understandably, the plutonium was one of the last items placed aboard.[7]

Lightning Strikes

Scattered rain showers, forerunners of a cold front, marked the approach of Apollo 12's launch day. A broad band of clouds and precipitation, punctuated by numerous thunderstorms, moved into central Florida on Thursday afternoon. By nightfall, the thunderstorms ended and the rain slackened. The next morning, radar displays of precipitation echoes placed the cold front about 80 miles north of the Cape. Despite the weather, large crowds were on hand to watch the liftoff. President and Mrs. Nixon headed the list of 3000 guests, marking the first and only appearance of a Chief Executive at an Apollo launch. Other names on the VIP list included Vice President Agnew, Henry Kissinger, Roy Disney, Jr. (of Walt Disney Productions), Arnold Palmer, and James Stewart.[8]

As the space vehicle underwent final preparations, the approaching cold front pushed large banks of clouds toward the Cape. Cold rain drenched the spectators. Up in the command module, Yankee Clipper, Commander Conrad noticed water leaking between the boost protective cover and the spacecraft. He later recalled:

> I could see water on my two windows—window 1 and 2. We experienced varying amounts passing across these windows, dependent on how heavily it was raining. These [rain and wind] were the only things noted up to liftoff.[9]

With a half hour to go, Merritt Island was experiencing peak winds of 14 knots, light rain showers, broken clouds at 240 meters, and overcast skies at 3000 meters. But the ceiling exceeded the minimum requirement of 150 meters, and the ground winds were within limits. The Apollo design permitted launch during rain. The possibility of lightning concerned Launch Operations Director Kapryan, however, and he considered a hold. As he explained at the postlaunch briefing:

> We were within our minimums. . . . The only consideration as far as launching under what apparently are adverse conditions—they are really twofold. Number 1, we would not launch into a thundercloud; number 2, we would not launch when we had lightning in the system. There was some concern. We had very unpredictable weather predictions. The weather was deteriorating. . . .[10]

A weather report from the Eastern Test Range helped Kapryan make up his mind. An Air Force plane reported only mild turbulence and no indication of lightning within 32 kilometers of LC-39. Air Force 1, bringing the President to the launch, experienced no turbulence while flying through the front. Astronauts Slayton and Stafford told Kapryan the weather was satisfactory. The launch operations director also had to weigh a "now or scrub" situation: the liquid oxygen replenish pump had failed at T-1 hour and 22 minutes, and everything depended on a backup pump. With the launch rules and available evidence giving him an affirmative, Kapryan opted for an 11:22 a.m. launch.*[11]

Apollo 12 lifted off on schedule. Thirty-six seconds later, as the space vehicle reached 2000 meters, spectators observed two parallel streaks of lightning flash toward the launch pad. The Yankee Clipper experienced a power failure. As Conrad later recalled:

> I was aware of a white light. I knew that we were in the clouds; and although I was watching the gauges I was aware of a white light. The next thing I noted was that I heard the master alarm ringing in my ears and I glanced over to the caution and warning panel and it was a sight to behold.[12]

The spacecraft sustained a second lightning discharge 16 seconds later at an altitude of 4400 meters. Conrad reported to Mission Control: "We just lost the [stabilizing] platform, gang: I don't know what happened here; we had

---

*After the launch some newspapers suggested that President Nixon's presence influenced Kapryan's decision. The launch director denied it.

everything in the world drop out."[13] Fortunately, the spacecraft automatically switched to a backup power source, and the astronauts soon restored primary power.

That Apollo 12 had been hit by lightning was a matter of dispute for some time. At the postlaunch briefing, one hour after liftoff, reporters asked Stafford, Apollo 10 commander, and Kapryan about reports of lightning. Stafford dismissed the reports as only speculation. Kapryan said, "I think we're pretty certain that it was not lightning. If the vehicle had been struck by lightning the damage would have been quite severe rather than a momentary dropout." When reporters pressed the matter, Stafford and Kapryan responded that NASA had quite a few people watching after liftoff and no one reported a sighting. Subsequently, the lightning reports from numerous viewers were substantiated by space vehicle data and KSC cameras.[14]

President Nixon chose not to mention the incident in his postlaunch remarks at the launch control center. He commented on the "great experience and awe" of an Apollo launch. He repeated the remarks made to him by astronauts "that those on the ground, the engineers, and the technicians, and the scientists, and all of those who work in the program, that they are really the heart of this great, successful experience for the American people and for all the people of the world."[15] Nixon promised to keep the United States first in space.

After the unnerving lightning incident, the mission moved smoothly. Apollo 12 went into earth orbit 11 minutes and 43 seconds after liftoff. By 2:15 p.m., it had accelerated to 38 000 kilometers an hour and was headed for the moon. There was a significant change in the trajectory. Three earlier Apollos flew a course that permitted looping the moon and returning to earth if the spacecraft failed to attain lunar orbit. Apollo 12, by a midcourse maneuver, entered a trajectory that did not allow free return. This was necessary to reach the desired landing site.

On 19 November 1969, Conrad and Bean landed in the Ocean of Storms, within 180 meters of the unmanned Surveyor 3 that had been there for two years. The two astronauts spent 7 hours and 45 minutes on the lunar surface, setting up scientific instruments, collecting pieces from the Surveyor, gathering materials, and photographing the landing craft, the Surveyor, and other objects of interest. They lifted off on the morning of 20 November and splashed down in the South Pacific on 24 November.[16]

With plans afoot for a world tour, the crew first returned to KSC on 17 December for a reunion with the launch team. Debus led them into the transfer aisle of the vehicle assembly building as a Navy band played "Anchors Aweigh" and 8000 members of the government-industry team applauded. He complimented the crew on leaving as commanders and returning as U.S. Navy captains.

"The crew didn't consider the flight over until we got back here," Conrad said. "We forgive the weather man for his job, but had we to do it again, I'd launch exactly under the same conditions." Gordon pointed out that

> the real guts of these flights, after their formative, opening stages, are really put together here. The hardware is brought here, it's mated here in the VAB, and a great amount of testing is done. But more importantly, the crew is here most of the months before launch. And this is really the way it ought to be. This is really our home.[17]

The astronauts received enlarged color photographs of the Apollo 12 liftoff, plus a stone from the crawlerway over which their vehicle began its journey. Then they walked through cheering crowds along the transfer aisle, exchanging handshakes and signing autographs. They lunched with the KSC Management Council and contractor managers where they regaled the party with some lighthearted comments about their achievement. The astronauts were presented with such trinkets as whiskbrooms to remove lunar dust, tiny parasols to ward off the intense sunlight on the moon, and joke books to while away the time on lunar journeys. It was a happy family reunion.[18]

Whys and Wherefores of Lightning

The strike on Apollo 12 led to another study of lightning protection, this one focusing on the atmospheric conditions that might threaten a launch. At a meeting of the American Geophysical Union in December 1969, experts discussed the incident and offered NASA some observations. The scientists generally agreed that Apollo 12 triggered the lightning discharges. There were no other signs of lightning or thunder for six hours before and six hours after the launch. However, readings on electrical field meters in the Cape area indicated disturbed weather conditions. Apparently Apollo 12 had entered an electrical cloud and distorted the field sufficiently for breakdown to occur. The 110-meter space vehicle and its 500-meter ionized exhaust plume then formed an excellent conductor. The space vehicle had probably triggered a lightning stroke from an electrified cloud incapable of producing lightning on its own. Although the launch vehicle's design incorporated safeguards against electrical discharges, lightning could damage components in the spacecraft such as solid-state electronic devices. The Apollo 12 experience prompted NASA officials to reexamine the space vehicle and the weather criteria for a launch.[19]

The lightning investigation team opposed any modifications to the spacecraft. They recommended, instead, further launch restrictions to reduce the possibility of touching off another lightning strike. The new "severe weather restrictions" appeared in the launch rules for Apollo 13. The space vehicle would not be launched if the nominal flight path would carry the vehicle within 8 kilometers of a thunderstorm, through cold-front or squall-line clouds, or through cumulus clouds with tops at 3050 meters or higher.[20] The additional weather limitations would have a moderate effect on winter and spring launchings; in those seasons, high winds would more often cause delays. On a February afternoon, the probability of delay would increase from 10% to 18% with the new restrictions. In the summer, the probability of a scrub would jump from 3% to 18%. Despite the new rules, the odds for acceptable weather were still better than nine out of ten for most three-hour launch windows.[21]

Apollo 13 Launch Operations

The launch vehicle and spacecraft for the Apollo 13 mission arrived at KSC in June 1969. Following the Apollo 11 success, NASA set a March 1970 launch date for Apollo 13. More planning time was added in January, moving the launch to 11 April. Prelaunch operations went smoothly through the fall and winter months. The work in high bay 1 marked its last use for Apollo; subsequently the area would be used for Skylab operations. The Bendix crawler team transferred the space vehicle to pad A on 15 December. The flight readiness test, scheduled before the January program change, was run on 29 January as a "confidence test" and rerun on 26 February. After four days of hypergolic load tests in mid-March, the launch team began the countdown demonstration test on the 18th.[22]

A strange accident punctuated the last day of the test. Early on 25 March, Graydon Corn's propellants crew started the chill-down of the LOX pumping system. The operation required a 760-liter-per-minute flow to the replenishing pumps (which could handle five times that rate) and a lesser amount through a bleed line that had been added to the LOX system after the 500-F spill in August 1966 (pp. 343–44). During the 40 minutes of precooling, the launch team emptied 39000 liters of LOX into a drainage ditch outside the perimeter fence. Normally ocean breezes dissipated the oxygen fog. On the morning of the 25th, however, there was no wind and a pronounced temperature inversion. A dense fog built up in the drainage ditch; at a culvert where the road to the slide wire bunker crossed the ditch, the invisible oxygen overflowed onto the bank. At 6:00 a.m. the closeout crew and safety personnel left the LOX storage area. First-stage loading could begin after a three-minute chill-down of the 38 000-liter-per-minute main pumps. A security team completed its job of clearing the pad area and proceeded in three cars to the perimeter gate southwest of the LOX sphere. The driver of the first car, Patrolman Nolan Watson, drove through the gate and parked. As he walked back to Earl Paige's car, an order over the radio directed the team to clear the slide wire bunker area. Paige turned his ignition on and heard a loud pop. Soon flames sprang up from beneath the hood. Watson ran back to his car, only to find it also on fire. About the same time, the third car burst into flames. The three guards quickly ran for cover. A fire and rescue crew arrived in five minutes but took no action until the oxygen cloud dissipated. It was nearly 7:00 a.m. before the fire was under control, leaving three burnt hulks and a shaken crew.

Debus called for an immediate investigation. The preliminary report, rendered a week later, blamed the accident on the enriched oxygen atmosphere. Spontaneous ignition resulting from the engine heat, combustibles (oil and grease on the engine covers and gas around the carburetors) and the oxygen vapor cloud caused two of the fires, the third apparently starting when the driver turned the ignition switch. The report criticized the practice of dumping large quantities of cryogenics and termed the resulting vapor a hazard. Recommendations included immediate studies of the drainage system leading from the LOX storage area and its dump reservoir, of entry and exit routes at pad 39 A, and of KSC's safety training course. The major change brought about by the accident was to extend the LOX drainage pipes beyond the perimeter ditch to a marshy area farther from the pad.[23]

Another anomaly during the demonstration test appeared insignificant at the time; in fact, it was the beginning of what was to prove Apollo's most nerve-wracking hours. On 24 March the North American launch crew finished loading the cryogenics into the service module. Tank testing had gone smoothly and nothing about the loading operation presaged troubles ahead. The first sign came when the launch crew partially emptied the two liquid oxygen tanks. While the first tank performed normally, emptying half of its contents, the second tank released only 8% of its LOX. The crew prepared an interim discrepancy report and postponed further action until the end of the demonstration test.

The spacecraft team resumed detanking operations on the 27th, after discussing the matter with Houston, Downey, and Beech Aircraft Corporation. The problem centered on a possible leak between the fill line and the quantity probe because of a loose fit in the sleeves and tube. A second failure of the detanking procedure strengthened this view. After additional attempts at higher pressures proved unsuccessful, the KSC team decided to "boil off" the remaining LOX. The tank heaters, energized by 65 volts of direct current, were turned on; 90 minutes later the tank fans were also activated. The solution proved to be a slow one. After 6 hours the quantity of LOX in the tank still stood at 35%. The team continued to run the heaters and began pressurizing the tank for a few minutes and then venting the fill line. After two more hours of alternately heating and venting, the tank emptied.

Apollo officials faced a difficult decision. Replacement of the oxygen shelf in the service module would take two days and posed the possibility of damaging other equipment. If the problem were a loose fill tube, the shortcoming would not threaten the mission. The LOX tank would still supply the fuel cells properly and any electrical short at the capacitance gauge would be insignificant. After further discussions with Washington, Houston, and Downey, KSC undertook a partial fill on the 30th. Both tanks reached the 20% level without any trouble, but emptying the second tank again required heating and pressure cycling. Apollo technical and management personnel weighed the possible hazards of flying with a loose fill tube against the problems of shelf replacement. The decision was to keep the defective tank.[24]

A second cryogenic tank problem received more publicity in the closing days of the prelaunch operations. Liquid helium from a tank in the lunar module was used to assure a steady flow of propellants to the descent engine. The tank's design allowed for a slow increase of pressure as the helium warmed, but during the countdown demonstration, pressure in the tank began rising too fast. If a faulty vacuum allowed heat to build up too rapidly, the increased pressure would blow the tank's burst disk and prevent a lunar landing. Over the first weekend in April, newspapers reported the helium tank as a serious problem. A test conducted on Monday the 6th, however, indicated that the "heat rate loss was well within parameters and acceptable for launch."[25]

Apollo operations continued to attract famous people from around the world. In early March, French President and Mme. Pompidou spent a day at the center. The following week, 50 members of the U.S. Congress and the Canadian Parliament got a close view of Apollo 13; 60 German and Japanese astronomers visited Merritt Island on 9 March, after viewing a solar eclipse in north Florida. Later that month the British astronomer, Sir Bernard Lovell, and his wife were guests. The VIP list for the 11 April launch included Willy Brandt, Chancellor of West Germany, Vice President Agnew, and Secretary of State William Rogers.[26]

A Case of Measles

Three notables in residence—the crew of commander James Lovell, command module pilot Thomas K. Mattingly, and lunar module pilot Fred Haise—kept busy in the simulators and altitude chambers. While Lovell and Haise trained for two moon walks, Mattingly studied his photographic assignments which included the moon, sun, and other astronomical subjects. Training went smoothly, the hectic pace of previous launches seemingly a thing of the past. The situation changed dramatically, however, when NASA's Medical Director, Dr. Charles A. Berry, reported on 6 April that the prime crew had been exposed to measles. Backup lunar module pilot Charles Duke had a case of german measles (rubella) and Jeffrey Lovell, the commander's son, was down with the red measles (rubeola). Although the three astronauts were in good physical condition, blood samples were taken to determine their immunity. Initial tests showed satisfactory antibody levels in all three astronauts, but a recheck cast doubts on Mattingly's condition. Further tests indicated that Mattingly had no immunity and would likely experience the illness about the middle of the lunar mission.

At a press briefing on 8 April, Dr. Berry indicated that he would recommend against Mattingly's flying. NASA's preventive medicine program was questioned and Berry acknowledged the need to re-examine the subject. Previously crews had been restricted to essential contact during the last 21 days of prelaunch operations. This still included many people—training personnel, workers at the crew quarters, even younger members of the immediate families. The astronauts' schedule kept them in KSC's crew quarters much of the last three weeks, but risks were inevitable. Berry noted that some loopholes in the isolation program were necessary; others might be eliminated. He mentioned the likelihood of more antibody testing and immunization, even for such unlikely adult diseases as measles.[27]

Mattingly's health posed a difficult decision for NASA. Duke's illness ruled out the substitution of the alternate crew. Delaying the launch a month would lessen confidence in the space vehicle and add $800 000 in costs. Another alternative was to replace Mattingly with his backup, John Swigert. The longer time between missions had permitted extensive simulator training with the backup crew. Although a late substitution for the other two crew members was out of the question, the command module pilot was more on his own. A last-minute switch might work. Thursday morning, 9 April, a new crew of Lovell, Haise, and Swigert entered the flight crew simulators.[28]

The Flight Crew Operations Branch concentrated on situations that required rapid teamwork. First they tested the crew's ability to handle various abort situations. Then the crew practiced the mission's critical maneuvers: the translunar injection, transposition and docking, lunar orbital insertion, descent orbit insertion, rendezvous and docking, and transearth injection. Mechanical failures were cranked into each of the maneuvers, forcing Swigert to make corrections. One situation required a decision and response within two seconds.[29] The major concern was communications between crew members. As Riley McCafferty, branch chief, put it:

> From the standpoint of putting these three guys together and these three guys accepting each other and these three guys establishing confidence in each other, that wasn't our concern. Our concern was, did we have the proper communication, so when Jack Swigert said, "that's good," did they really know what "good" meant to Swigert versus what "good" meant to Mattingly?[30]

The Flight Crew Operations Branch had striven for compatibility in training the prime and backup crews. With Apollo 13 came the first fruits of their labor. By Friday afternoon Deke Slayton and Riley McCafferty were convinced that Swigert could work with his new crew mates.

More important, Lovell was satisfied with the new arrangement. Paine, after discussing the matter with Lovell, Slayton, and other Apollo officials, gave the mission a go-ahead at KSC's prelaunch press conference Friday afternoon. Paine told reporters there was never any question about Swigert's ability as a command module pilot: "Jack literally wrote the book on the malfunctions and how to overcome them." NASA's concern had been whether the astronauts could work together effectively, and the 12 hours of intensive tests had removed all doubts. Slayton praised the crew training group: "They got the equipment on the line for the last 36 hours in A number 1 shape, came up with a beautiful plan, and we in fact did it. I guess we were all surprised also that the crew did integrate as well as they did." A reporter asked whether the change had caused extra crew fatigue. Slayton noted that the tests had not exceeded the normal work schedule. If the crew had not been ready by Friday noon, NASA was prepared to postpone the launch.[31]

A Fragile Lifeboat

The Apollo 13 countdown proceeded without a major incident, and liftoff came at 2:13 p.m. on 11 April. When the S-II stage's center engine shut down 132 seconds early, an extra 34-second burn from each of the four outboard engines made up most of the difference. An additional nine-second burn of the S-IVB stage brought the vehicle to within 0.4 meters/second of the planned velocity and left sufficient fuel to boost the space vehicle out of the earth's gravitational field.[32] Aside from the S-II problem, the first two days of the mission went according to plan. The crew started the third day in space by inspecting the lunar module. Lovell and Haise read a supercritical helium pressure well under the danger line. Fifty-five hours into the mission the crew began a television transmission from the command module, Odyssey. Fred Haise demonstrated movement through the tunnel into the lunar module, Aquarius, and remarked: "There's a little bit of an orientation change that, even though I'd been through it once, in the water tank, is still pretty unusual. I find myself now standing with my head on the floor, when I get down into the LM." For the next half hour the crew described their temporary quarters in a space version of "Person to Person." The television interview ended on a light note as Lovell showed off a floating tape recorder. Musical selections included "Aquarius" from "Hair" and the theme from "2001, A Space Odyssey."[33]

The good cheer came to a sudden end a few minutes later when the warning system indicated low pressure in hydrogen tank 1. Mission control asked the crew to turn on the cryogenic fans and heaters. Ninety seconds after the fans started up, Mission Control lost all telemetry for two seconds. The crew heard a loud "bang" and observed a low voltage condition on d.c. main bus B.* Swigert reported, "Okay, Houston. Hey, we've got a problem here."[34] The full extent of the problem, however, was not immediately apparent. Voltage on main bus B recovered momentarily. The quantity gauge for oxygen tank 2 fluctuated and then returned to an off-scale high reading. Repeated firings of the attitude control thrusters on the service module added to the confusion. According to a later NASA report the thrusters were probably firing to overcome the effects of venting oxygen and a blown panel. Within minutes, the electrical output from fuel cells 1 and 3 dropped to zero. At first the mission controllers focused on the electrical systems, postulating a possible disconnect between the fuel cells and their respective buses. Upon realizing that the fuel cells were not working, mission control directed an emergency powerdown of the command module. With indications of a pressure loss in oxygen tank 1, Houston directed a switch in electrical power to obtain a reading from the number 2 tank's instrumentation. The tank was empty. The reading substantiated a crew report that the spacecraft was venting something into space. As the pressure in oxygen tank 1 continued to drop, Lovell's crew abandoned the mission and sought refuge in the lunar module.

---

*An electrical bus is a conductor that serves as a common connection for several circuits.

A subsequent investigation pieced together the probable sequence of events. Apparently the start of the fans in oxygen tank 2 caused an electrical short circuit. Damaged Teflon insulation around the fan motor wires caught fire. Although the Teflon burned slowly, increasing heat and pressure soon ruptured the tank. The escaping oxygen either ignited with combustibles in the oxygen shelf compartment or blew an access panel off by itself. The panel struck the spacecraft high-gain antenna, disrupting telemetry signals momentarily. The pressure in oxygen tank 1 began dropping immediately after the telemetry loss. Apparently the same force that blew off the panel also damaged tank 1. The sudden and possibly violent failure of tank 2 may have broken a line to tank 1 or caused a valve to leak.[35]

The plight of the astronauts reawakened world interest in the Apollo program. Television carried the drama to millions. Foreign countries offered their services for a recovery outside the intended Pacific splashdown area. At the manned spaceflight centers, concern was matched by a determination to return the astronauts safely. Two major activities dominated the remainder of the mission: planning and conducting the propulsion maneuvers with the lunar module so as to bring the spacecraft back to earth, and managing the vital resources—oxygen, water, electricity, and the canisters of lithium hydroxide used to remove carbon dioxide from the cabin atmosphere. Open communications lines between KSC and mission control at Houston carried advice and test requirements. The two centers simulated the various maneuvers and conservation measures before directions were given to the flight crew. A team under Charles Mars, lunar module project engineer, devised a means of recharging the command-service module's re-entry batteries from the lunar module's electrical system. Another KSC recommendation turned off the radar heaters to save electricity. North American and Grumman engineers at KSC helped devise ways to transfer water from the portable life support systems into the lunar module's water coolant system. One of the biggest problems was the removal of carbon dioxide from the crowded Aquarius. KSC engineers, again duplicating activities at Houston, rigged a system that carried the CO_2-rich air from the lunar module, through a hose, into the command-service module's lithium hydroxide canisters. Over in KSC's flight crew training building, the Houston team simulated in advance the various situations to be encountered by the astronauts.[36]

Apollo 13 looped around the moon on 14 April 1970. While the lunar module barely provided room to turn around, the crew preferred its narrow confines to the chilly 11°C of the powerless command module. Respect for Aquarius increased as its systems continued to function well past their two-day mission expectancy.[37] Splashdown came in the South Pacific on 17 April.

While the dramatic rescue earned plaudits for the entire Apollo team, the mission had failed. Paine took steps to determine the cause of the accident as the astronauts were returning to earth; on the 17th he announced the appointment of an Apollo 13 review board under the leadership of Langley Research Center's Director, Edgar M. Cortright. The board conducted an intensive investigation during the next six weeks and the positive reception of its report contrasted sharply with the earlier Apollo fire investigation. The board concluded "that the accident was not the result of a chance malfunction in a statistical sense, but rather from an unusual combination of mistakes, coupled with a somewhat deficient and unforgiving design."[38]

Oxygen tank 2—the one that first ruptured—had undergone acceptance tests at the Beech Aircraft Corporation factory in 1967. The tank was installed in SM-106 (Apollo 10) and later removed for modifications. During the operation, the oxygen shelf was jarred and fell some 5 centimeters. North American officials analyzed the incident and concluded that there was no damage. The review board found the likelihood of tank damage from the incident "rather low," but listed the accident as a possible cause of the loose-fitting fill tube. The oxygen shelf was retested after the modifications, but no cryogenics were used. As the components worked satisfactorily, the shelf was installed in SM-109 (Apollo 13) on 22 November 1968.

Unfortunately, when the tank arrived at KSC in June 1969, it had an even more serious shortcoming. The two protective thermostatic switches on its heater were built to 1962 specifications for 28-volt d.c. power. In 1965 North American had issued a revised specification—the heaters would operate on 65 volts for tank pressurization. Beech did not change the thermostatic switches, and both North American and NASA documentation reviews overlooked the error. Subsequent qualification and acceptance tests did not require complete switch cycling, and so they too failed to reveal the incompatibility. During tank pressurization the 28-volt switches could accommodate the 65 volts from KSC's ground support equipment because the thermostats remained cool and closed. However, if the tank temperature rose considerably, as it did for the first time during KSC's special detanking, the 28-volt thermostatic switches would fail. When the switches started to open at their upper limit of 300 kelvins (27°C) on 27 March, the current in the ground equipment welded them permanently closed. The review board estimated that, after the switches failed, temperatures in the tank reached 811 kelvins (538°C) in spots during the eight hours of detanking. The intense heat would have severely damaged the Teflon insulation on the fan motor wires.

As the board indicated, the special detanking on 27 March 1970 did not violate KSC procedures. However, the launch team could have detected

The Apollo 13 accident

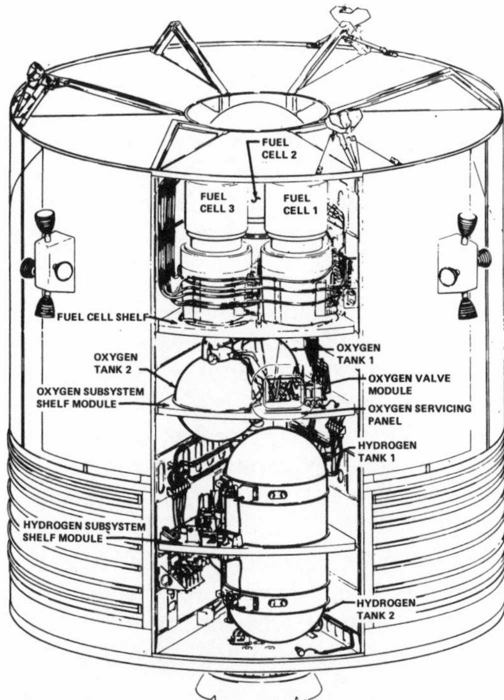

Figure 158

Fig. 158. Cutaway of the fuel cells and cryogenic tanks in bay 4 of the service module, Apollo 13. Fig. 159. The hydrogen tank shelf, with an oxygen sphere, upper left. Fig. 160. Schematic of the oxygen tank.

Figure 159

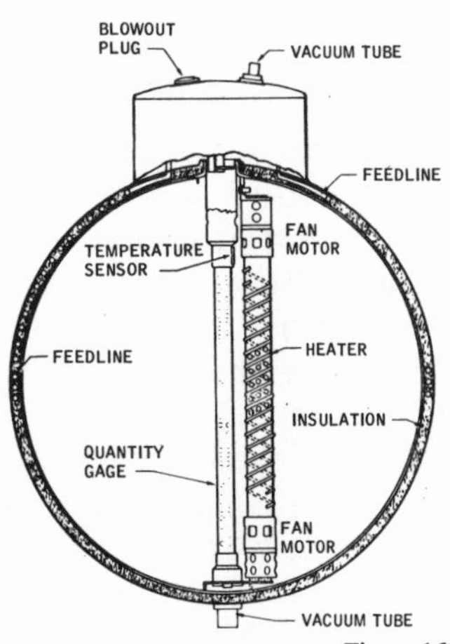

Figure 160

the failure of the thermostatic switches to open by observing heater current readings on the control panel. The tank temperatures indicated that the heaters had reached their temperature limit and a switch opening should follow. There was also an apparent communications gap while the oxygen tank problem was under debate. Attention focused on the loose fill tube, and many individuals at Houston, North American, and Beech were unaware of the extended heater operations. Those aware of the special detanking procedure failed to consider the damage that might result from the excessive heating and did not alert Apollo management to the possible consequences.[39]

The board summed up the lesson of Apollo 13 in the preface to its report:

> The total Apollo system of ground complexes, launch vehicle, and spacecraft constitutes the most ambitious and demanding engineering development ever undertaken by man. For these missions to succeed, both men and equipment must perform to near perfection. That this system has already resulted in two successful lunar surface explorations is a tribute to those men and women who conceived, designed, built, and flew it.
>
> Perfection is not only difficult to achieve, but difficult to maintain. The imperfection in Apollo 13 constituted a near disaster, averted only by outstanding performance on the part of the crew and the ground control team which supported them.[40]

Apollo 14 Launch Operations

The Apollo 14 launch, originally scheduled for February 1970, was postponed to July, then to October, then to December, and finally, after the Apollo 13 review board, was set for 13 January 1971. As Stuart A. Roosa remarked to a press conference, his "was the only crew that's been six months from launch four times."[41] The delays allowed the launch team to check and recheck the Apollo hardware, and ensured ample training time for all concerned. Roosa, for example, logged more than 1000 hours in the command module simulator.

The Apollo 14 command-service module had arrived at KSC on 19 November 1969 and moved into the altitude chamber the following week. The ascent stage of the lunar module was flown down from Bethpage, Long Island, on the 21st; the descent stage followed three days later. When tests in early December revealed a faulty oxidizer flow control valve on the descent stage, a new engine was substituted before Christmas. Operations continued during the holiday season; on the 29th and 30th, North American technicians replaced a defective hydrogen tank in the service module. The first major exercise of the new year involved a successful command-service–lunar module docking test on 9 January 1970. Two days later the S-IC stage sailed into its slip. The Boeing team erected the booster the following day on a mobile launcher. A week later the S-II and S-IVB stages arrived and were placed in low bay stalls. At the operations and checkout building, the descent stage entered the altitude chamber on 16 January. The Grumman team moved the ascent stage to the chamber on the 19th and began a three-day mating operation.

There were no significant problems for the next ten weeks until an accident in mid-April caused about a month's delay. North American's work schedule for the 15th included installation of a new inertial measurement unit with an improved gyroscope. A technician accidentally punched a hole in the unit and a pint of water glycol spilled out over the command module's lower equipment bay. Spacecraft officials initially estimated the cleanup and modification would take nine days. As the North American crew removed wires and equipment, the damage proved more extensive. However, there was no need to rush; the Apollo 13 accident had, by this time, delayed the launch of 14 indefinitely. When the cleanup was completed, a special altitude chamber run was conducted on 15 May to dry out the command module.[42]

The instrument unit arrived at KSC on 6 May; the following month North American engineers finished modifying the S-II stage's center engine. During the Apollo 13 flight the pogo effect had reappeared, this time on the second stage. Severe oscillations had forced an early shutdown (two minutes ahead of schedule) of the inboard engine. Although the outboard engines had burned longer and compensated for the loss, NASA officials did not want any more pogo. Marshall and North American engineers devised three changes to the second stage. They installed a helium gas accumulator in the LOX line of the center engine. This reservoir served to dampen fluid pressure oscillations, keeping them out of phase with the vibrations of the thrust structure and engines. North American added a cutoff device to shut down the center engine in case the accumulator failed to control the oscillations. Finally, simplified propellant valves were installed on all five J-2 engines. The valves controlled the propellant mixture to the engines, providing a rich mixture for high thrust during the early portion of the burn and a leaner mixture later.[43]

In the operations and checkout building, the primary and alternate crews conducted altitude chamber runs in the lunar module and simulated the command-service module altitude run. Grumman engineers traced noise problems in the VHF communications system to the VHF transceiver and the signal processor assembly. After replacing the defective parts, the lunar module team scheduled another run for 10 July.

The date for the altitude test slipped several times. Excessive leakage in a propellant quantity gauge caused the first delay. Tests at Houston and Bethpage resolved the problem by late July, and the rerun was set for 13 August. A conflict with the flight crew training schedule led to a second postponement, this time until the 18th. On 11 August, Houston asked KSC to check the ascent stage's ball valves, and on the 17th the Office of Manned Space Flight ordered ball valve leak checks on both stages. The test involved the removal of the lunar module from the altitude chamber to the high bay work stands and separation of the two stages. The spherical valves, located above the ascent and descent engines, controlled the flow of hypergolic fuel and oxidizer into the thrust chamber. The launch team used gaseous helium to test the valve seals. The leak checks and other propulsion system tests were completed by the end of August. On 2 September the lunar module returned to the altitude chamber, where all systems were reverified with an altitude run on 18 September.[44]

A stretch-out in the Apollo 14 launch schedule had prompted the decision to revalidate the ball valves. In early July, NASA Headquarters released a new flight schedule moving the launch date from 3 December 1970 to 31 January 1971. The postponement was caused by the Apollo 13 review board's recommended modifications for the command-service module. The board added a third cryogenic oxygen tank (placed in a previously empty bay of the service module), an auxiliary battery as a backup to the fuel cell, and an emergency supply of drinking water. The modifications recommended for the oxygen tanks—for example, replacing the Teflon insulation on internal wires with stainless steel conduits—required more time.

Following manned altitude runs in early September, North American removed the command-service module from the chamber on the 17th for cryogenic modifications (to comply with the Apollo 13 review board recommendations). The lunar module remained in the chamber until 13 October, when Grumman began installing the landing gear at the high bay stand. In late October, the launch vehicle team detected a condition that might inhibit S-II separation from the booster; paint on a mating flange had bonded it to the second stage. After the upper stages were removed and the area cleaned, the Saturn was restacked on 2 November. The spacecraft was added on the 4th and rollout followed five days later. Milestones passed in routine fashion the last two months:

- 7 December Launch readiness review
- 14 December Space vehicle overall test 1 completed

- 17 December Flight readiness review
- 19 December Flight readiness test completed*
- 19 January Countdown demonstration test completed
- 25 January Launch countdown begun.[45]

One major change in operations during the last month concerned the flight crew's health program. In December strict rules were instituted to preclude a recurrence of Apollo 13's measles. At KSC more than one hundred individuals were designated "primary contacts," i.e., people who had direct contact with the astronauts during the performance of essential duties. The primary contacts underwent immunization against nine diseases and were required to report any illnesses in their families. No one else was permitted in the astronauts' presence. Beginning at T-21 days, the Medical Surveillance Manager maintained a 24-hour command post near the astronauts' quarters on the third floor of the operations and checkout building. The astronauts were restricted during the last three weeks to their quarters, the flight crew training building, the flight line, and pad A's white room. Within these areas, bells and horns warned secondary contacts of approaching astronauts. Certain facilities were also modified. Air filtration units, installed in the air conditioning systems of the operations and checkout building and the flight crew training building, screened out 97% of airborne bacteria. Airtight doors and positive air pressure in the two buildings provided additional protection. Inside the training building an airtight glass partition protected the crew from secondary contacts. Communication with the crew was by intercom. Despite the elaborate precautions, the cooperation of the KSC work force was essential to the success of the program. Posters, adorned with comic strip characters, reminded workers to report any sign of illness. During the first week, five primary contacts reported in sick—all with a respiratory illness. But there was no recurrence of the problems that had beset Apollo 9 and 13.[46]

The Apollo 14 mission attracted widespread interest, in part because of its predecessor's near disaster, but also because its popular commander, Alan Shepard, was making a comeback after ten years. Following his 15-minute Mercury flight in May 1961, Shepard had been grounded for a minor ear disorder. He had continued in the program, serving for a while as chief of the Astronaut Office at Houston. Flights had passed him by, however, until surgery corrected his ear problem in 1969. Navy Captain Shepard's presence on the team bothered some junior members of the Flight

---

*During the flight readiness test, NASA officials received an anonymous telephone call threatening to blow up the launch center. KSC increased its security measures for the countdown and launch. *Washington Star*, 20 Dec. 1970.

The difficulties of practicing for work on the moon

Figure 161

Figure 162

Figure 163

Fig. 161. Training aids were built much lighter than the actual equipment, to approximate the effects of lunar gravity; but those effects could never be duplicated exactly. Fig. 162. As the scientific tasks became more elaborate on the later missions, so also did training for making observations on the moon. Fig. 163. Even the simplest tasks were difficult to perform.

Crew Operations Branch. They feared that Shepard would "pull rank" and prove uncooperative. But as Riley McCafferty recalled:

> Shepard eased himself in and, over a period of about four weeks, he had a relationship with the young engineer on the floor that was good. They had a lot of confidence in each other and they talked back and forth; and the instructor, the young engineer, felt like he could tell Al Shepard, "you fouled up, buddy."[47]

There were no mishaps during the last week of Apollo 14 operations. The countdown, begun on 25 January, included 102 hours of scheduled tasks and five holds totaling 48 hours. The amount of intended hold time, representing rest periods and contingency planning for unforeseen problems, had changed little in four years. The holds proved largely unnecessary for Apollo 14. On launch day, 31 January, overcast skies gave Walter Kapryan some anxious moments. Light rain was falling on the large Sunday afternoon crowd when Kapryan halted the count at 3:15 p.m., only eight minutes from launch. Within 40 minutes, the cloud peaks had moved from the flight path. Launch officials changed the flight azimuth of the space vehicle from 72° to 75.6° and sent Apollo 14 on its way.[48]

Pruning the Apollo Program

While 1969-71 were the harvest years—four missions that put men on the moon, and the safe return of Apollo 13 after its breakdown in space—they were not so kind to Kennedy Space Center and the men who worked there. Congress cut the NASA budget, NASA cancelled Apollo missions, KSC and its contractors laid off thousands of employees—not in one fell swoop but in a succession of smaller blows. Space enthusiasts had hoped to go on to a manned landing on Mars in the mid-1980s; it was not to be. American public opinion was shifting its priorities to other matters: civil disorders, Vietnam, decaying cities, campus unrest, and inflation. And Apollo was a victim of its own success. For laymen, one moon landing after another was a little boring. Noting the public's limited interest in Apollo 12, the *New York Times* concluded that a collective sense of anticlimax was "perhaps predictable considering the intense national emotion spent on the first moon landing four months ago."[49]

Probably the biggest reason for Apollo's decline was the detente in American-Soviet relations. In 1961, amid cold war animosities, the United States was trailing the Soviet Union in the world's most widely publicized form of competition, manned spaceflight. Eight years later, the United States had clearly demonstrated its superiority. Despite the Russian invasion of Czechoslovakia, relations between the two nations had improved. Americans seemed less eager to spend "whatever it took" to surpass the Russians in space. Agreement on a U.S.–U.S.S.R. rendezvous mission (the Apollo-Soyuz flight of 1975), signed before the end of the Apollo program, clearly indicated a new policy of cooperation in space.

NASA budgets marked the contour intervals of Apollo's descent. Appropriations had exceeded $5 billion in the mid-1960s; in fiscal years 1969 and 1970 they fell below $4 billion. Apollo research and development funding declined from $2.9 billion in FY 1967 to $2 billion in FY 1969. Initially, NASA's follow-on programs to Apollo—Skylab, an earth orbital laboratory; Voyager, an unmanned Mars mission; and Nerva, a nuclear rocket engine—bore the brunt of the cutbacks. Funding for space programs to follow Apollo appeared in the Johnson administration's 1968 budget. Congress sharply reduced Nerva and Apollo Applications (Skylab) appropriations, cutting the latter from $454.7 million to $253 million. Voyager was eliminated entirely, while Apollo funds fell by less than 2%. For FY 1969 the Johnson administration budgeted $439.6 million for Apollo Applications, $38 million for 1971 and 1973 unmanned missions to Mars, and $41 million for Nerva. Again all three programs were cut sharply: Skylab eventually received $150 million that year. Apollo received all but $14 million of its $2.039 billion request. After the first lunar landing, however, Apollo lost its immunity to cutbacks, and further tight budgets brought reductions there as well.[50]

The Apollo flight schedule that was published on the eve of the first lunar landing called for nine additional flights before June 1971—a launch every 11 weeks. Apollo 12–15 would develop man's capability to work in the lunar environment; 16–20 would extend the astronauts' stay time on the moon to three days and increase their range of exploration. A primary purpose of the latter missions was to study the technological requirements for a potential lunar base.[51]

American lunar scientists opposed the rapid pace of the launches. They wanted 6–12 months between flights to study moon samples and plan future experiments. Dr. Lee A. DuBridge, Presidential Science Advisor, expressed the scientists' viewpoint in congressional testimony on the FY 1970 NASA budget: "Nothing can do more harm to support for the space program than to have a series of missions for which there are no clear objectives—such as a series of manned revisits to the moon without providing the capability to perform new scientific experiments and to exploit interesting new lunar features."[52] Three weeks after the first lunar landing, John Noble Wilford, space correspondent for the *New York Times*, publicized the dispute over Apollo's future. The scientific community, according to Wilford, sought a larger role in mission planning and more scientist-astronauts, as well as more time between missions.[53]

The July 1969 schedule had included an alternate plan that extended the nine remaining launches by 18 months and provided a launch interval of 4–5 months. Following the success of Apollo 11, NASA officials approved the compromise schedule. In defending the choice, George Mueller acknowledged the scientific arguments but cited other major factors. Among these, Mueller included "operational considerations in keeping a steady workload through the Cape" thereby "minimizing the cost."[54]

While NASA debated the pace of the remaining Apollo missions, a Space Task Group examined the future of America's space program. What lay beyond Apollo was the subject of their September 1969 report, "America's Next Decades in Space." The report's sponsors, a panel including Vice President Spiro T. Agnew and NASA Administrator Thomas O. Paine, recommended a balanced manned and unmanned space capability. The group listed three possible NASA programs leading to a manned landing on Mars before the end of the century. The most ambitious plan called for a lunar orbiting station by 1978, a lunar surface base and a 50-man, earth-orbiting station in 1980, and the first Mars mission in 1983. The cost of all this would reach an annual $8 billion by 1976. The least ambitious plan postponed the lunar base and earth-orbiting station by three years and left open the date for the initial Mars expedition. The funding estimates for this second plan ran slightly more than $4 billion a year during the 1970s. Apollo missions would lay the groundwork for the lunar surface base. The report generated little support, and NASA's budget slipped to $3.3 billion the following year.[55]

The decline in Apollo funding was even more severe; a reduction of nearly 50% dropped the program's budget below the $1 billion mark for the first time in eight years. While much of the decline represented an expected slowdown in costs, the shortage of funds forced drastic program changes. Edward Mathews, KSC's Apollo Program Manager, notified Debus in March 1970 that FY 1971 funding constraints had eliminated the Apollo 20 mission. There would be an average interval of six months between launches, with Apollo 18–19 put off until 1974 after a year of Skylab missions. Further budget cuts in September included a $50 million reduction for Apollo. NASA officials reluctantly cancelled missions 18 and 19. The flight of Apollo 17 in late 1972 would bring the program to a close.[56]

The Impact of the Apollo Slowdown on KSC

NASA budgets translated into people at KSC. Center employment peaked at 26000 during Apollo 7 operations in 1968, the same year that KSC's budget reached a high of $490 million. America's space program provided over 40% of Brevard County's employment. By the following spring, KSC faced sharp reductions in both money and manpower. Debus let community leaders know what was coming at a 30 April 1969 briefing. A revised FY 1970 budget, prompted by the Nixon administration's concern over inflation, lowered KSC's appropriation from $455 million to $410 million. The entire reduction came out of the $345 million earmarked for Apollo and reflected the intent to slow the program from five to three launches per year. In terms of manpower, the lower budget would reduce KSC's work force from 23500 to 18500 by 30 June 1970. A five-day work week would replace the six- and seven-day weeks that had been typical. Instead of three-shift operations, KSC would employ two, with only enough people on the second to continue necessary tests. Debus took an optimistic view of the cutbacks. The 20% reduction in force affected both stage and support contractors and could probably be met in large part through attrition. Contractor turnover rates at KSC in this period varied from an average of 14% annually for stage contractors to as high as 25% for some support contractors. He thought that "others will see the first lunar landing as a logical milestone in their career plans and move into other programs elsewhere." It would be "a difficult but orderly retrenchment."[57]

The reduction took a greater toll than Debus had predicted. By mid-1970 KSC's work force had fallen to 16235. The numbers engaged in Apollo launch operations showed an even steeper decline, 50% from the 17000 high of 1968. KSC civil service employment dropped less sharply in FY 1970, from 2920 to 2880. One reason NASA had contracted a large amount of Apollo work had been to avoid an excess of civil service personnel at the end of the program. Subsequently civil service enrollment at KSC was forced down to 2425 by the end of the program. Newspapers captioned the plight of Brevard County: "Cocoa Beach Boom Reaches Perigee"; "Most of Brevard in Gloomy Mood"; "Depressed Brevard Banks on Space Shuttle." Reporters described long lines at the employment office and a buyer's market of empty homes and stores. The articles were exaggerated; unemployment never exceeded 6.5%. Realtors and the Chamber of Commerce launched an aggressive campaign in metropolitan newspapers, describing Brevard homes as the best buys in Florida. Within two years an influx of retirees brought stability to the housing market. In similar fashion small businesses were encouraged to locate in the Cape area. Many members of the

Apollo team had found jobs in other parts of the country as stage and support contractors made a strong effort to relocate their personnel.[58]

Although KSC retrenched in orderly fashion, the atmosphere at the center showed a marked change. The pace slowed considerably as the time between launches stretched to eight months. Morale was jeopardized by the space program's uncertain future. As Alan Shepard, Apollo 14 commander, put it: "We kind of feel like the Wright brothers would have felt if they had been told there's not enough money for a second plane because there's no need for airplanes."[59] During Apollo's last three years, the launch team's esprit was of concern to center and contractor officials alike. The presence of the astronauts remained a positive factor. Launch Director Walter J. Kapryan made them as visible as possible, encouraging their visits with workers at the assembly building and on the pad. Efforts were made to keep everyone busy. That morale never became a significant problem is a tribute to effective civil servant and contractor leadership and to the personal pride of the launch team members.[60]

23

Extended Lunar Exploration: Apollo 15–17

A Change of Course for Apollo

Scientific investigations highlighted the last three Apollo missions. The cold war competition that had put men on the moon was fading. Congress and the American public now wanted tangible benefits from space expenditures. NASA adjusted its manned programs to the new climate. Skylab, the post-Apollo manned program, would focus on practical applications, most of them earth-oriented. Apollo, reduced by funding cuts to three more missions, would emphasize lunar exploration. Missions 15–17 did not disappoint American scientists; indeed those missions proved a fitting climax to one of the nation's great achievements.

NASA's plans for the concluding Apollo missions were announced on 2 September 1970. Modifications to the spacecraft and astronaut support systems would double the time the astronauts could stay on the moon. The weight devoted to lunar surface experiments would also double. A lunar rover vehicle—an appropriate gift to the moon from an America-on-wheels—would more than double the distance the astronauts could travel on the surface. Other new equipment included a lunar communications relay unit, which enabled the crew to maintain contact with earth while exploring beyond the lunar module's horizon. Transmissions from the portable relay station to Houston included voice, TV, and telemetry. Although the suitcase-sized device was normally mounted on the front of the rover, it could be detached and carried by an astronaut—a feature that ensured the crewmen a means of communication if they had to walk back to their spacecraft. The rover also mounted television cameras that were operated by remote control from Houston.[1]

The command-service module's lunar orbiting experiments, while less dramatic, were a vital part of the last missions. The scientific instrument module (SIM) in Apollo 15's service module included three spectrometers—gamma-ray, x-ray fluorescence, and alpha-particle—to measure the composition and distribution of the lunar surface. A mass spectrometer would measure the composition and distribution of the lunar atmosphere. A

Fig. 164. Discussing the lunar surface ultraviolet camera during an experiments review in the operations and checkout building, November 1971. From left, lunar module pilot Charles Duke; Rocco Petrone, Apollo Program Director; George Carruthers, Naval Research Laboratory, the principal investigator for the camera; and Apollo 16 commander John Young.

subsatellite, ejected from the SIM bay into lunar orbit, would beam earthward information about solar winds, lunar gravity, and the earth's magnetosphere and its interaction with the moon. Other equipment in the SIM, two cameras and a laser altimeter, would map about 8% of the lunar surface, in all some three million square kilometers.[2]

The extended missions on the moon required major modifications to the lunar module. Supplies of water, oxygen for the portable life support system, and electrical power were increased. Grumman enlarged the capacity of the propellant tanks by 7% and redesigned the descent stage to make room for the lunar rover. Altogether, the lunar module modifications and the SIM additions added about 2270 kilograms to the Apollo 15 spacecraft, bringing its total weight to over 48 metric tons.[3] This put a burden on Saturn engineers. Marshall and its contractors met the payload increase through minor hardware changes in the S-IC stage and by revising the Saturn V's flight plan. The hardware modifications reduced the number of retrorocket motors, rebored the orifices on the F-1 engines, and set the burning time for the outboard engines nearer LOX depletion. Better use of the Saturn's thrust was achieved by launching the AS-510 rocket in a more southerly direction (changing the launch azimuth limits from 72–96° to 80–100°) and by using an earth parking-orbit of 166 rather than 185 kilometers. Apollo 15 also stood to gain some advantage from the July launch date, when temperature and wind effects would be favorable.[4]

Interfaces with the First SIM

Apollo 15 launch operations got off to a slow start, impeded by spacecraft modifications. Checkout of the lunar module began in mid-June 1970, about the time the Apollo 13 review board announced its findings. Service module modifications, recommended by the board, delayed the launch date by five months. The September decision to enlarge the final missions brought further hardware changes. Spacecraft operations resumed in November with the arrival of the modified stages of the lunar module. Initial testing concentrated on the propulsion systems. Early in the new year Grumman engineers added three equipment pallets to the descent stage and brackets for the lunar rover. The new command-service module arrived in mid-January and went almost immediately into the altitude chambers.[5]

January also brought the first instruments for the scientific instrument module. By that time the Experiments Section had been at work on the SIM for more than a year. Preparations for the lunar orbiting experiments included the construction of a laboratory in the operations and checkout building and development of ground support equipment. When testing began, the 7-man Experiments Section supervised 25 engineers representing 8 contractors. An occasional visit from an experiment's scientist-author further complicated the three-shift operations. The contractor representatives proved invaluable from a logistical standpoint, securing minor design changes and spare parts. They did not always, however, seem to appreciate the need to meet a launch date.

From the very beginning the test engineers faced a familiar problem—hardware designed for use under conditions of zero gravity could not stand up to the rigors of earth gravity. The 7.5-meter extendable booms, which would deploy the mass and gamma-ray spectrometers, were built by North American for zero gravity. They could not support the spectrometers on

Merritt Island, Earth. North American designed a long rail to help carry the load for test purposes. The operation was generally unsatisfactory, however, since it introduced problems that would not occur in zero g.

The SIM work crew joined North American's spacecraft operation in late February and placed the SIM, with its eight experiments, inside the service module. Interface problems between the scientific instruments and the service module appeared almost immediately. The alpha spectrometer's data stream failed to synchronize with the spacecraft data-relay system. The Experiments Section had more trouble with the gamma-ray and mass spectrometer booms. When the engineers extended the boom, they received no indication that signals were being received. Investigation indicated that diodes in the boom circuitry were blocking the signal. North American subsequently modified the spectrometer booms.[6]

Test procedures caused nearly as much trouble as the hardware:

> The SIM bay complicated the checkout flow in every major procedure we ran. In some cases the vendors got the scientific instruments to us late. In other cases they would want to conduct a last-minute check at a very inconvenient time. Every time we powered up the ship for a major test somebody would come down with a special requirement for their instrument.[7]

The initial requirements for the calibration of the gamma-ray spectrometer called for halting all motor vehicles within 16 kilometers. NASA and the contractor negotiated the matter for several weeks, agreeing finally to a late night test with a traffic ban in nearby parking lots and roads. Following the weekend calibration exercise, the Experiments Section tested all SIM systems on 15 March and returned them to the factory for a month's rework by the responsible contractors.

The instruments arrived back at KSC in mid-April. While there were some minor problems, e.g., the mapping camera would not turn off, the test team closed out the SIM bay temporarily in late April for the move to the pad. When the subsatellite arrived a month later, the Experiments Section installed its batteries, checked out the transmitter, and tested the interface with the mechanism that would eject the subsatellite into a lunar orbit. Technicians entered the space vehicle stack on 9 June and added the subsatellite to the SIM bay.[8]

The Moon Gets an Automobile

For the public, the big feature of the Apollo 15 mission was its little lunar rover. Americans immersed in an automobile age contemplated with no small joy the beginnings of a stop-and-go traffic jam on the moon. And the rover was worthy of its homeland; it boasted bucket seats and power steering. The 207-kilogram vehicle would run for 65 kilometers on its two 36-volt batteries. As a safety precaution, NASA restricted travel to a 9.5-kilometer radius from the lunar module, the limit of the astronauts' ability to walk home. The rover's payload allowed about 363 kilograms for the two astronauts and their portable life support systems, 54 kilograms for scientific and photographic equipment, the same for communications equipment, and 27 kilograms to bring home lunar samples. All weights, of course, would be reduced by five-sixths when the little car operated in lunar gravity. To meet space limitations inside the lunar module, the rover folded into a wedge-shaped package less than half its operating size.[9]

The Boeing Company and its prime subcontractor, the Delco Electronics Subdivision of General Motors, designed and built the first lunar rover in 18 months—one of the major rush jobs of the Apollo program. While the forced schedule contributed to the $12.9 million cost, the high price was principally a result of the rover's unique engineering requirements. The harshness of the lunar environment—its extremes in temperature, lack of atmosphere, one-sixth gravity, and rough yet silt-soft surface—posed design problems in vehicle propulsion, stability, control, and wheel-soil interaction. Special wheels made of woven spring steel wire with titanium chevrons for traction were developed to meet the launch weight restrictions and still provide the support and mobility required on the moon. Each wheel had its own electric drive motor. The vehicle had independent steering motors for front and rear wheels so the driver could use front, rear, or both.

The two crewmen sat side by side on the vehicle. Control was provided by a T-bar "joy-stick" mounted on a console in the center. The joystick provided acceleration, brake, and steering control through complex electrical circuitry. It could be operated by either crewman. The console also provided electrical system control, monitor, and alarm capabilities.

Since a magnetic compass could not be used to indicate direction on the moon and because of other problems—such as having to go around craters—a special navigation system was built around a directional gyroscope, odometers, and a small computer. The system used the distance and direction traveled to determine range and bearing from the lunar module. With this information the astronauts could easily determine the shortest course back at any time. The navigational system also provided data for the location and placement of scientific equipment on the lunar surface.[10]

The launch team started preparing for the rover in late 1970 when the requirements document arrived from Marshall Space Flight Center. Arthur Scholz, Boeing's rover project manager at KSC, drew up the test and checkout plan describing the sequence of operations. The first events on the flow chart involved reception and inspection, activation, and calibration of the rover's ground support equipment. Meanwhile, Boeing engineers began preparing test procedures for the rover. They relied first on preliminary design data from the Seattle plant and then on the formal requirements document from Marshall. In January 1971 R. Dale Carothers, KSC's manager for rover operations, and a group of Boeing and government engineers journeyed to Seattle where they took part in the last two months of factory tests.

The action switched to KSC in mid-March when the rover arrived at the Cape's skid strip* aboard a C-130 Hercules aircraft. The first rover spent two days in the operations and checkout building undergoing inspections, first in its folded and then in its unfolded condition. During the next three days technicians installed simulators for the two 36-volt batteries and checked out the vehicle's power. The second week was taken up with electrical systems tests including front and rear wheel steering, the four drive motors, and the alarm system. During the tests the rover had to rest on a pedestal while the wheels turned in mid-air. The pedestal also supported the chassis when an engineer or astronaut entered the rover. The vehicle could support its own weight on earth, but no more. On one or two occasions, with the rover mounted on the pedestal, the test team witnessed a strange sight—the front wheels moving forward and the rear wheels in reverse. Boeing engineers said the drive motors were out of synchronization and that the phenomenon could not occur on the moon, where the wheels would be touching the lunar surface.

On 26 March the prime and backup crews went through the Crew Fit and Function Test, known in KSC parlance as CF^2. The test marked the astronauts' first opportunity at KSC to work with the rover. There were several operations: removing the rover's communication, television, photographic, and data-gathering equipment from the pallets in the spacecraft, placing the equipment in its proper place aboard the rover, and selecting items from the rover for further operations. The task was made more difficult by bulky gloves, the only part of their life support system the astronauts wore for the test. The exercise revealed a number of small problems such as recalcitrant strap fasteners and poorly fitting safety belts. As the rover's stowage date was only a month away, Scholz and Carothers sought immediate modifications. The paperwork took more time than the physical changes. Coordinating design modification with contractors and other NASA centers was always a slow process. On this occasion a money dispute threatened further delays. Marshall did not want to authorize additional

*The Cape's hard-surface 45 × 3050-meter runway earned its name in the 1950s when Air Force launch teams retrieved winged Snark missiles by landing them on skids.

funds to accomplish the changes. Houston wanted the modifications but did not want to finance the work. In the end, the astronauts' wishes prevailed; program managers from Marshall, Houston, KSC, and Boeing approved the proposed modifications and the work was under way in two weeks. The changes did not affect the stowage schedule.

The third week of rover testing began with a navigational systems check. The rover was mounted on the work stand, the wheels started turning in mid-air, and an engineer moved the steering handle. The test team observed the computer's performance as it assimilated driving data from the odometers and gyroscope. The following day the launch team tested the rover's mechanical brakes. Wheel and fender replacements and the closing out of discrepancy records took the remainder of the week. During Easter week technicians completed most of the modifications. A silicone-oil leak from the shock absorbers caused several days' concern before the test team declared the shock absorbers "acceptable for flight." On 16 April Boeing undertook one of the more difficult tasks—loading the rover aboard the lunar module. Technicians successfully deployed the rover the following day, using a landing platform to reduce the distance it fell, so that the impact was equivalent to what would be experienced under lunar gravity.

A second CF^2 test inaugurated the last week of operations. The exercise provided a check on the various modifications that had been made since the first test. The rover group joined Grumman engineers the next two days for the electromagnetic compatibility test. As its name implied, this test was to detect interference, primarily with the lunar communications systems. With radios, computers, radars, even the rover's wheels operating, no problems developed in the lunar communications relay unit. The launch team then moved on to the climax—simulated mission runs with the two astronaut crews.

The simulated missions gave Test Conductor Herman Widick some uneasy moments. Whereas lunar module tests usually attracted little attention, the novelty of the rover drew a large crowd of Apollo officials. The simulation involved a number of organizations: a Hamilton Standard representative for the portable life support system, NASA spacesuit technicians, Grumman engineers for the lunar module and rover storage, RCA and Goddard Space Flight Center communications experts, and Houston observers. While Widick had worked with most of these men, the Boeing engineers were new. Matters were complicated by two communications systems. The test conductor talked with the crew by radio through the portable life support system; communications with the rest of the test team were over the operational intercommunication system. The astronauts, their vision limited by the spacesuits, unwittingly interrupted Widick on several occasions.

The lunar rover

Figure 165

Figure 166

Fig. 165. Deployment of the rover watched by Apollo 16 astronauts Young (center) and Duke (right). Fig. 166. Apollo 15 astronauts training with the rover. David R. Scott (left) prepares to deploy the vehicle's antenna, while James Erwin considers a pile of equipment from a mockup of the lunar module.

Fig. 167. Apollo 17 astronauts Harrison H. Schmitt, geologist-pilot (left), and commander Eugene A. Cernan. The rover's antenna is fully deployed. Fig. 168. Rep. Olin Teague (D., Texas), chairman of the House Committee on Science and Astronautics Subcommittee on Manned Space Flight, and Mrs. Teague seated in the training vehicle. Duke (left) and Young answer questions. James C. Fletcher, NASA Administrator, is in the background, left of the antenna mast.

Figure 167

Figure 168

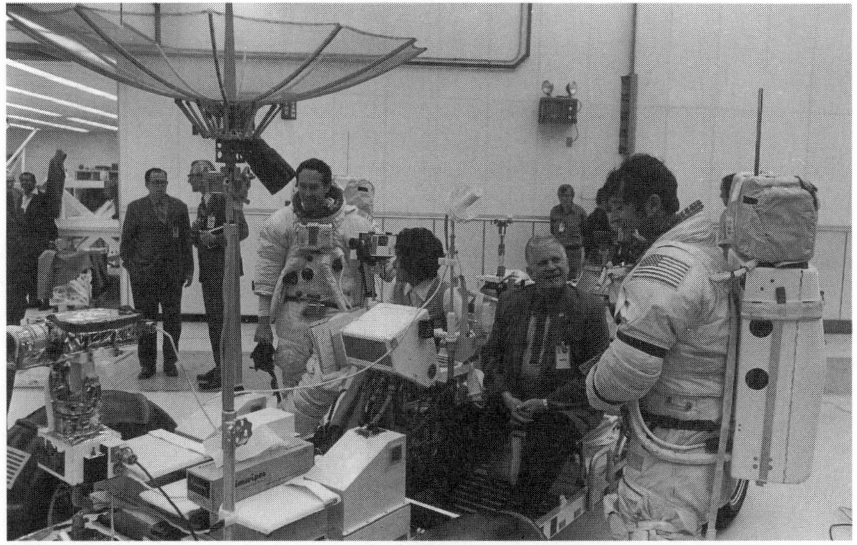

Spacecraft engineer Ernest Reyes had tried for several days to give the rover a sportier look, but Widick rejected every suggestion. As the test was about to begin, Commander David Scott pulled a fox tail from his spacesuit and attached it to the rover's low gain antenna.

The setting for the simulated mission resembled a science-fiction film of the 1940s. Sunlight gleamed off the lunar module's aluminum foil covering. Antennas stretched along the wall of the operations and checkout building's high bay. In the center of the scene the rover, fully deployed, rested on its pedestal. The astronauts, dressed in the lunar surface version of their Apollo spacesuits, completed the picture. Since the test employed the communications equipment within the portable life support system, the 38-kilogram unit was strapped to each astronaut's back. A reasonable load on the moon, it was too heavy to carry on earth, so a dolly with an overhead cantilevered arm supported the equipment. A technician guided each dolly as it moved behind the astronaut.

The mission simulation began with a communications check. While the astronauts stood in front of the rover, a test engineer switched on the drive motors. The wheels were noisy but produced no electrical interference. Technicians then removed the life support system from each astronaut's back and placed the packs on the rover seats. The crew moved to the back of the vehicle where they engaged the equipment pallet pin and disengaged the rear steering pin. The latter operation, a difficult maneuver, was only for emergencies. If the rear steering mechanism failed on the moon, the astronauts could lock the rear wheels in place and steer with only the front wheels. After these tests, the astronauts seated themselves in the rover. A wedge was placed between the life support system and the seat to give the astronaut the feel of lunar gravity with the pack. Wearing pressurized suits, the astronauts had considerable trouble with the seat belts. Finally ready, the astronauts began their lunar ride—simulating specific distances and directions until they returned to the lunar module. The astronauts also checked the TV signal that would return to earth over the relay unit.

After successful mission simulations with both crews, Boeing technicians folded the rover and reinstalled it. Two days of rover deployment followed. The exercise included possible malfunctions and appropriate responses. (On subsequent missions the launch team rigged various deployment malfunctions in the rover trainer and lunar module simulator.) On 25 April the launch team placed the rover inside the lunar module. The spacecraft joined the Saturn stack two weeks later.[11]

The rover was stowed away but not forgotten. On 5 May the training vehicle was demonstrated for the press. Scott and Irwin answered newsmen's questions and then drove the one-gravity rover trainer to the lunar simulation area. A few fortunate reporters tried their hand at the T-bar handle controls. The reporters, with instructors at their side, drove the rover through the astronauts' crater-pocked sandpile. Their enthusiastic response carried over to the next day's newspapers.[12]

Lunar Module Problems and More Lightning

The SIM and rover were the newest, but not necessarily the biggest, problems on Apollo 15; February and March 1971 were difficult months for the lunar module team. A combined systems test ran for ten days in February as test engineers attempted to resolve a series of discrepancies. Problem areas included the rendezvous radar, a frequent source of trouble in the past. The NASA–Grumman team discovered a malfunction in the radar's range-finding circuitry and called Bethpage for a replacement on the 24th. It arrived on the 26th and went immediately to the boresight range. Weekend tests disclosed another deficiency: the radar would not slue from its normal straight-ahead position to directly overhead, which would be needed for tracking the command-service module from the lunar surface. The fault was apparently in the mode position switch. On 1 March KSC asked Bethpage to send another radar. It arrived the following day and passed inspection at the boresight range; the self-tests (internal checks of electrical circuits), range determinations, and angle readout checks were all satisfactory. The angle readouts told what direction the radar was pointing and therefore the azimuth and elevation of the target. After installation, tests of the third radar looked all right except for some ambiguous range sensings. Then a technician reported an unusual grinding sound in the gyroscope assembly. The noise increased; a bearing had gone bad. With its fourth radar installed on 13 March, the launch team had a satisfactory rendezvous system.[13]

Herman Widick's test team uncovered more problems when the lunar module began altitude chamber runs in late March. During an unmanned run on the 26th, engineers noted unsatisfactory conditions in the communications system and the environmental control system. The radio problem involved extraneous noise from the transceiver. With no crew on board, the radios operated in a relay mode, i.e., signals went to the lunar module on the VHF uplink and came back immediately over the S-Band downlink. Harold Cockran's engineering team traced the problem to an improper setting on the VHF receiver squelch circuit.[14]

The possible malfunction in the environmental control system concerned a relief valve on the suit circuit assembly. Two demand regulators controlled the oxygen pressure to the assembly.* The relief valves protected the suit circuit from overpressurization. On the test of the 26th, the regulator's maximum pressure and the relief valve's minimum line were closer than the prescribed tolerance. It was realized, however, that the regulator pressure would drop somewhat during the manned runs. On the 29th, as the backup crew prepared for an altitude run, a technician inadvertently applied too much pressure to the commander's oxygen umbilical, damaging the hose. The rescheduled test failed when both demand regulators continued to pressurize the suit circuit after reaching the acceptable limit. Technicians removed the regulators the following day. Finally, on 6–7 April, the test team managed successful altitude runs with the prime and backup crews.[15]

The problems prompted James Irwin, Apollo 15's lunar module pilot, to speak out publicly. At a press conference in Houston, Irwin singled out the difficulties with the lunar module's environmental control system, the landing radar, and the rendezvous radar. Irwin attributed some of the problems to the extended shelf life of the Apollo equipment. Due to the stretch-out of Apollo flights, equipment was remaining in storage longer than manufacturers had expected. Irwin also noted that a lot of trained people were leaving the Cape and said, "I think maybe morale is slipping perhaps." Apollo 15's other two astronauts, David Scott and Alfred Worden, disagreed. Worden remarked: "I think the people there are more fired up about 15 than they have been before."[16] Concern about the lunar module lessened after the successful altitude runs.

Lightning strikes were the most significant events after the Saturn V was moved to the pad in early May. During the flight readiness test on 14–15 June, lightning struck the mobile service structure and mobile launcher. Although there was no apparent harm to the space vehicle, some ground support equipment was damaged. Schedules were revised to permit retesting of all spacecraft systems. On 25 June, the day following the flight readiness review, lightning struck again with the same results. Damaged electrical components were replaced and the spacecraft systems checked once more. Pad A experienced a third strike on the evening of 2 July during hypergolic loading. While there was no apparent damage, some tests were repeated during the countdown demonstration test.

Minor problems during the countdown demonstration test, 7–14 July, were corrected before the start of the countdown on 20 July. When more

---

*The suit circuit assembly included fans, a heat exchanger (cooling water), and lithium hydroxide to remove CO_2 from the air. The suit circuit provided an environmental control system for the cabin and life support for the astronauts' spacesuits. When the mission called for cabin depressurization, e.g., prior to an extravehicular activity, the astronauts hooked up to the suit circuit assembly. The portable life support system provided the same support on the lunar surface.

lightning struck LC-39A during countdown week, Kapryan delayed moving the service structure from the pad until the evening of the 25th.[17] Apollo 15 lifted off the next morning at 9:34 a.m. Commander Scott radioed back from space: "As we watch the S-IVB drift away here, how about passing along to Jim Harrington [Apollo 15 space vehicle test conductor] at the Cape congratulations from the crew to the launch team for a superior job."[18]

Apollo 16 Operations

While astronauts Scott and Irwin motored around Hadley Rille, KSC officials turned their attention to the Apollo 16 mission scheduled for March 1972. In early August, North American mated the command and service modules. Three weeks later Grumman joined the two LM stages for their altitude tests. September saw the start of lunar rover checkout and the erection of the S-IC stage. In October the launch vehicle team stacked the Saturn stages. Meanwhile the astronauts went through the crew compartment fit and functional tests and the altitude chamber runs. The spacecraft modules moved out of the chambers in November and landing gear was installed on the lunar module. In December the spacecraft team mated the Apollo spacecraft to the lunar adapter and moved the combination to the assembly building. Twelve days before Christmas Apollo 16 rolled out to the pad.[19]

The launch team had made relatively few changes to the Apollo 16 spacecraft during the first five months of launch operations. Malfunctions on Apollo 15 prompted two command module changes: replacing panel switches for the spacecraft propulsion system and replacing the main parachutes. One of the three main parachutes had failed to open for the splashdown of Apollo 15, and NASA officials suspected hydrogen embrittlement in the connector links of the suspension lines. After replacing the suspect parts with steel alloy links, North American shipped a new set of parachutes to KSC in mid-November. That same week the launch team replaced the water glycol accumulators in two fuel cells of the service module. When the fuel cells converted oxygen and hydrogen to electricity and water, considerable heat was produced. As it transferred this heat to a series of radiators, the glycol expanded and the excess liquid accumulated in reservoirs. The accumulators had been damaged in September when technicians overpressurized the glycol system during a vacuum-purging test.[20]

One of the few problem areas in the Saturn operations involved the engine actuators on the S-IC stage. These hydraulic actuators, 1.5 meters in length, swivelled the four outboard F-1 engines to change pitch, yaw, and roll. Actuator tests included the calibration of a recorder in the launch control center. As the actuators swivelled the F-1 engines, a potentiometer sent a voltage to the recorder indicating the direction and amount of movement. During November tests, excessive noise in one actuator interfered with the signal to the control center; the actuator was replaced on the 25th. The following week Boeing engineers inspected the S-IC LOX and RP-1 tanks for stress corrosion but found no problem.[21]

Early in the new year a spacesuit alteration and two spacecraft problems delayed the Apollo 16 launch to 16 April. Grumman engineers had increased the capacity of the lunar module batteries and wanted more time to gather test data. At Downey technicians discovered that an explosive device used to separate the command-service and lunar modules would malfunction under certain conditions; modification required additional time. The delay proved a godsend for KSC in late January when a fuel tank in the command module's reaction control system ruptured.[22]

The hypergolic propellants of the reaction control system, which controlled the attitude of the command module during reentry, were forced from their tanks by high pressure helium gas. Within each fuel tank, the fuel was inside a teflon bladder. As gas entered the tank, outside the bladder, rising pressure squeezed the bladder and forced the hazardous fuel from its tank. The flow of helium was tested during the integrated systems test. The primary and secondary regulators were checked to guarantee that an accurate flow was maintained, that the regulator shut off properly, and that after shutoff the pressure did not creep up, which would indicate internal leakage.

Problems with ground support equipment had put the launch team about two shifts behind schedule on 25 January when technicians completed the fuel-tank relief-valve checks and moved to the regulator tests. For these tests, the bladders were filled with helium gas instead of the hazardous monomethyl hydrazine. Human error brought the team grief: a technician failed to fully engage a quick-disconnect valve that controlled the flow of helium to a pressure regulator. Pressure inside a fuel tank, but outside the filled bladder, dropped quickly, and the bladder ruptured.[23]

The seriousness of the problem stemmed from its location. Replacement of the fuel tank involved removing the command module's aft heatshield, an operation that had to be conducted in the operations and checkout building. KSC faced a roll-back of the space vehicle from the pad to the assembly building—the first time this had happened since a hurricane threatened 500-F in June 1966. At first glance the accident seemed to preclude the April launch, and NASA officials announced a possible second month's delay; but after reviewing the work needed to replace the damaged fuel tank, Kapryan and Petrone concluded that the launch team could recover in time for the 16 April launch. The space vehicle was returned to the assembly building on 27 January. The following day the launch team transferred the spacecraft to the operations and checkout building where both fuel tanks were replaced, along with the descent propulsion system regulators. By working overtime and weekends,* KSC had Apollo 16 back on the pad in less than two weeks.[24]

While operations resumed their smooth course for most of the KSC team, the propellants section experienced more headaches. The spacecraft was undergoing another integrated systems test on 17 February when a leak developed in a quick-disconnect test point. A North American engineer closed off test points improperly and excessive pressure ruptured discs on both oxidizer tanks. While the launch team waited for replacements to arrive, the program office rescheduled the remaining propulsion tests. New burst discs were emplaced and x-rayed on the 22d, and propulsion tests resumed the following day.[25]

Spacecraft Stowage

Stowing equipment on the Apollo spacecraft grew more complicated with the lunar exploration missions. The SIM bay and the rover have been described. The modularized equipment storage assembly occupied another quadrant of the descent stage. These cargo pallets provided room for tools, the lunar communications relay unit, various cameras including the color television equipment, and other items to be mounted aboard the rover. Inside the command and lunar modules the astronauts required more of nearly all supplies: food, clothing, film, and life support items. During the latter missions the Manned Spacecraft Center placed a number of experiments aboard the command module, e.g., Apollo 16 carried 60 million microbial passengers in a small rectangular container, a light flash detector, a biostack, and a Skylab food package.†[26]

The launch team stowed the spacecraft cabins on three separate occasions during the Apollo 16 operations: first, in the chambers prior to the astronauts' altitude runs, a second time for the crew compartment fit and function test; and finally the day before launch. KSC had dropped the practice of stowing the cabin for the countdown demonstration test; instead

---

*On 16 April 1972 the *Washington Post* noted that the damaged fuel tank had added an extra $200 000 to the cost of Apollo 16. Most of the money had gone for overtime pay at KSC.

†NASA measured the effects of reduced oxygen, zero gravity, and solar ultraviolet irradiation on the microbes representing five strains of bacteria, fungi, and viruses. On one flight experiment a crewman donned an emulsion plate device or deflector while his mates wore eye shields. The purpose was to correlate light flashes, seen on each mission since Apollo 11, to cosmic rays. The biostack, a cylindrical aluminum container 10 cm. high, contained live biological material that was exposed to high-energy heavy ions in cosmic radiation. The Skylab food package included some experimental snap-top cans with dried peaches, puddings, peanuts, and other items.

technicians placed empty lockers inside the command module to give the astronauts the appearance of a flight-ready cabin. A team of nine normally stowed the command module. Inside the cabin two technicians secured each item in its proper place. A KSC quality control representative observed their work. Outside, two technicians unpacked the flight articles. A North American quality representative and engineers from Houston, KSC, and North American completed the team. While the six "outside members" of the team found the white room of the mobile service structure confining, they preferred it to the occasional use of the ninth swing arm from the umbilical tower, which had to be used in changing flight articles when Swigert replaced Mattingly on Apollo 13. Carrying equipment across a catwalk a hundred meters above the ground unnerved some members of the group. During the countdown, stowage of the command module began about 24 hours before launch and ran for seven hours. If no problems arose, the team could finish with several hours to spare.

The stowage exercise culminated two weeks of intensive preparations for KSC's Anne Montgomery. Her group checked many of the flight articles such as cameras, communications equipment, and the lithium hydroxide canisters. The items were tested individually and then in conjunction with other flight articles and command module systems. Some items required special packaging; all were weighed and recorded by serial number. Every flight article received a detailed quality inspection and each mission disclosed a number of discrepancies.

James McKnight directed a similar activity for the lunar module, the final stowage of which began just before the start of the formal countdown. At T - 55 hours Grumman technicians placed most of the articles aboard and checked out the lunar equipment conveyor. The astronauts relied on this moving clothesline to carry heavy items such as rocks inside the lunar module. The group completed stowage at T - 30 hours. After placing a portable life support system on the cabin wall and another on the floor, the technicians took pictures of their work and then sealed the hatch.

Houston prepared the stowage plans for each mission; these took into consideration when and where the astronauts would use a particular flight article. Emergency items received first consideration. The Manned Spacecraft Center was also responsible for the contents of the crew preference kits, the bags in which astronauts carried their personal mementos. Following the incident with unauthorized postal covers on Apollo 15, NASA tightened its restrictions on what the astronauts could take to the moon.[27]

After a successful flight readiness test on 1 March, officials met for the launch readiness review. The session covered all major aspects of Apollo

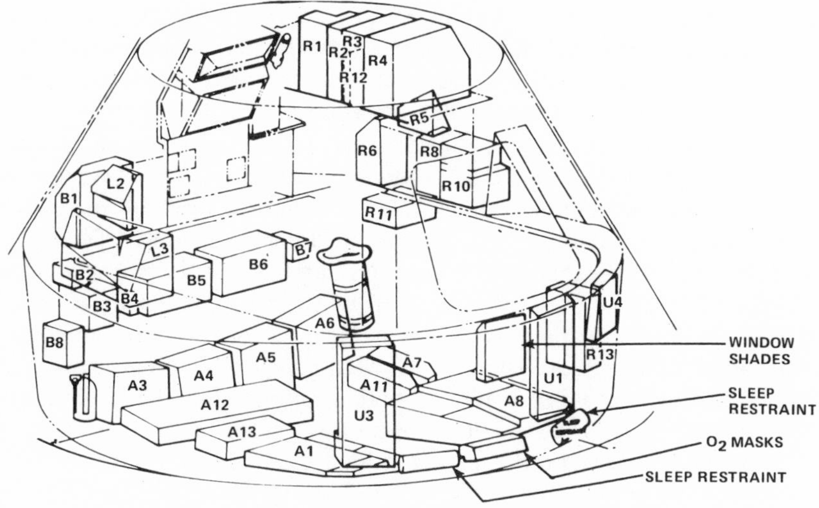

Fig. 169. Crew equipment stowage in Apollo 15 (partial).

A1 70mm camera adapter
H_2 gas separator in bag
5 tissue dispensers
2 penlights in bag
tool set
pressure garment O_2 interconnect, 3 in bag
snag line in bag
2 probe stowage straps
3 temporary stowage bags

A3 4 CO_2 absorbers
fire extinguisher
acoustic tone booster in bag
remote control cable

A5 2-speed interval timer
5 sleep restraint ropes
16mm camera sextant adapter
3 headrest pads

A6 TV monitor, monitor cable, and mounting bracket
2 CO_2 absorbers

A8 3 pilot preference kits
inflight exerciser
2 tissue dispensers
3 constant wear garments
extravehicular mobility unit maintenance kit
3 light-weight headsets
relief receptacle assembly and strap
16mm camera with magazine, powerpack, and 2 film magazines in bag

10mm lens
decontamination bags
O_2 umbilical interconnect
contingency lunar sample return container

R3 data card kit
eyepatch
2 meter covers
floodlight glare shield
fuse (16mm camera)
6 flight data file clips
flight data file books
lunar module transfer data card kit and flight data file books

R4 2 rucksack survival kits

R10 2 sanitation stowage boxes
30 fecal collection assemblies
water panel coupling assembly
waste management system water panel, quick disconnect, power cable, and quick disconnect pressure cap

R11 3 urine transfer systems
spare urine receiver assembly
roll-on cuff (red, white, blue)

U4 4 cassettes, 4 batteries for tape recorders
10 × 40 monocular
intervalometer (Hasselblad)
250mm lens

16 operations—range safety, operations safety, base support, Eastern Test Range support, Goddard's communications network support, the central instrumentation facility, technical support, and the status of the space vehicle. Despite the problem with the reaction control system fuel tank, Apollo 16 had been KSC's smoothest Apollo operation yet.[28]

One month before the scheduled liftoff, John Young, Apollo 16 commander, and Charles M. Duke, lunar module pilot, briefed KSC employees on the upcoming mission. Although 1500 attended the meeting, the crowd appeared insignificant inside the assembly building. Young and Duke discussed the problems they anticipated on landing in the high, rugged Descartes region. They outlined the goals of their extravehicular activities and explained the flight plan. After answering questions from the audience, Command Module Pilot Thomas K. Mattingly and Young circulated through the crowd, shaking hands and signing autographs. The briefing was one of the astronauts' last public appearances before the launch, as they began their three-week preflight quarantine on 26 March. This crew had special reason to appreciate the restriction; Mattingly's potential measles on Apollo 13 had prompted the quarantine and in January 1972 Duke had spent a week at the Patrick Air Force Base hospital with bacterial pneumonia.[29]

The last month of operations saw few hardware changes. The actual countdown went without a hitch. Liftoff came on a hot Sunday afternoon, 16 April, at 12:54.[30]

Apollo 17 Launch Operations

The Kennedy Space Center team saved its most spectacular liftoff for the last Apollo mission. Apollo 17, launched on a dark December night, lit up the Florida sky for miles. Despite its early hour (12:33 a.m.), the launch attracted nearly 500 000 watchers in the immediate vicinity. Where clouds did not obstruct the view, thousands more saw the ascending Apollo-Saturn from as far away as 800 kilometers. Of course there was television coverage: the Florida launch site had become familiar to millions of viewers.

Other aspects of the Apollo 17 mission reawakened the interest of the American public. It represented man's last journey to the moon for an indefinite period. Apollo 17 would carry more scientific equipment than any previous mission and would number among its crew the first scientist-astronaut, Harrison Schmitt. The mission also marked the end of a dramatic and controversial program. Appropriately for Apollo, the last mission met acclaim and success.[31]

The first launch vehicle stages for Apollo 17 arrived at KSC in late 1970 during preparations for the Apollo 14 flight. Spacecraft operations got under way in March 1972. During the next four months John Williams's directorate conducted the normal sequence of tests. Spacecraft engineers ran into some typical problems. In May Grumman engineers determined that the rendezvous radar assembly had received too much voltage during the tracking and pointing test at the boresight range. A new radar was installed on the 24th. A month later the landing radar began locking up intermittently and it was replaced. The lunar rover required several changes including replacement of forward and aft steering motors.[32]

The biggest change in command-service module operations concerned the scientific instrument module, which gained three new experiments: a lunar sounder, an infrared scanning radiometer, and a far ultraviolet spectrometer. The sounder was essentially a radar that could determine the physical properties of the lunar crust to a depth of 1.5 kilometers. This data, coupled with information gathered from cameras, the laser altimeter, and surface measurements, would allow the construction of a detailed topographical profile of the moon. The radiometer provided data from which scientists could prepare an accurate thermal map of the lunar surface. The new spectrometer measured compositional and density variations of the lunar atmosphere.[33]

The new experiments, particularly the lunar sounder, caused considerable headaches. For testing the sounder, the lunar surface had to be simulated. The sounder recorded the returning signals with an advanced optical recorder that required a special data reduction machine. After the launch team completed a lunar sounder test, the results were sent to the University of Kansas for interpretation. As the head of the Experiments Section recalled, "It would take weeks sometimes to get the results back and they might come back and say, 'You have nothing on the tapes.'" North American had trouble integrating the new experiments with the service module hardware.[34]

The stacked vehicle emerged from the assembly building on 28 August. Although another Saturn V would make the slow journey for Skylab, area residents reacted as if this were the last one. Five thousand spectators watched Apollo 17 creep toward pad A. The astronauts (Eugene Cernan, Ronald Evans, and Harrison Schmitt) joined the Bendix crew aboard the crawler for part of the trip.[35]

Launch operations during the next three months followed the routine established in earlier missions. The few changes in hardware went smoothly. There was one scare in late September, again involving the command module's reaction control system. While conducting a leak check, a technician overpressurized one of the oxidizer tanks. KSC officials feared the worst—the rupture of the bladder and the spacecraft's return to the operations and checkout building. At a press conference a few hours after the accident, NASA Administrator James Fletcher announced the possibility of a month's delay in the launch. Further tests, however, indicated that the teflon bladder was all right, and Apollo 17 stayed on schedule.[36]

In the outside world, there was an ill omen. A NASA request for 21 hours of Public Broadcasting Service network time to cover Apollo 17 stirred little excitement among the stations. Of some 70 replies, ten were favorable, ten opposed, and 50 expressed serious reservations. While this was blamed on a fear of governmental interference in programming, the commercial networks were no more enthusiastic. The prelaunch word was that they planned to cover only highlights of the flight.[37]

The morale at the spaceport remained generally high. For most companies, KSC contracts continued through Skylab and the Apollo-Soyuz flight. Apollo 17, however, marked the end of the road for the 600 members of the Grumman team. During its years at Merritt Island, Charles Kroupa's group had earned an excellent reputation with NASA counterparts and fellow contractors. The men working for test supervisor Ray Erickson wanted to assure the astronaut crew of their continued support. The result was a large poster at the lunar module working level of the mobile service structure. Signed by Grumman's employees, it read: THIS MAY BE OUR LAST BUT IT WILL BE OUR BEST. Fletcher said the slogan "should be the watchword for the entire Apollo team."[38]

The last Apollo mission was the first Saturn V launched after dark. As dusk approached, thousands of cars poured across the causeways leading onto Merritt Island. In front of the headquarters building, children threw footballs while the parents talked and listened for the progress of the countdown. The December weather did justice to Chamber of Commerce claims; in the mid-80s during the day, the temperature was 72° at launch.

The countdown proceeded. At T – 82 minutes launch control reported the cabin purge had been completed, and the booster protective cover closed. The spacecraft was pressurized and checked for leaks. Houston tested its command signals to the launch vehicle, and the first-motion signal was checked out with Houston and the Eastern Test Range; the next time, it would bring them word of liftoff. The last weather balloon was released to determine wind direction.

In the meantime the C-band and Q-ball tests were in hand. The first was used in tracking to report range velocity during the powered phase. The Q-ball, perched above the launch escape system, would warn the spacecraft commander of deviations in the first stages of flight. Cernan reported things looking good "up here." His next task was to check out the emergency hand control for the service module engine, normally operated by a computer. Far below him, little white wisps marked the topping off of the propellant loads.

At T - 1 hour, the close-out crew had secured the white room and was clearing the pad area. The elevators were set at the 96-meter level, for the astronauts' use in an emergency. At T – 50 minutes the launch control center initiated the power transfer test, switching the vehicle momentarily onto its own battery power and then restoring external power. Some five minutes later, swing arm 9—the access arm to the spacecraft—retracted 12° to a standby position. Range safety test signals were flashing to the still unarmed destruct receivers.[39]

At T – 30 minutes, reports came from around the circuit. The water system was ready to flush the pad two seconds after liftoff. Final propulsion checks were completed, the C-band tests repeated, and the reaction control systems armed on the service module. The recovery helicopters were on station, and the weather looked good—a major front remaining well to the west. The launch control center began chilling the second- and third-stage propulsion systems to condition them for the final flow of cryogenic propellants. Swing arm 9 was coming back to a fully retracted position. With the swing arm back, the launch escape system, with twice the power of a Redstone, could loft the astronauts to parachute deployment height. At T – 3 minutes and 7 seconds, the automatic sequencer took over.

This sequencer, the oldest and most reliable piece of automation on LC-39, chose this moment in the launching of the last Apollo to cause trouble. At T – 30 seconds it went into an automatic cutoff indicating that one of the essential operations leading to the launch of the space vehicle had not been properly completed. Besides halting the countdown, the cutoff started a series of "safing" procedures which included the return of swing arm 9 to a standby position.[40]

As Launch Director Walter Kapryan explained in a postlaunch press conference:

> At two minutes, 47 seconds, the countdown sequencer failed to output the proper command to pressurize the S-IVB LOX tank. The control room monitors noted it and immediately took steps to perform that pressurization manually. This was done, and at the time that we had the cutoff, we were up to pressure and everything was normal. The problem was that since the Terminal Count Sequencer did output the command, the logic circuitry said that we really didn't complete all of the

> launch preparation for the S-IVB stage. And we didn't have an interlock in our countdown circuitry that precludes the retracting of Swing Arm #1 which occurs at T – 30 seconds, and this is the reason for the cutoff. Now, it didn't take us very long to determine that we should bypass this command failure and go through the pressurization manually and go through the rest of the countdown.[41]

With the count returned to (and held at) T – 22 minutes the launch team installed jumpers that took the countdown around the faulty relays. The fix was verified on Huntsville's Saturn breadboard, the two centers making good use of the launch information exchange facility. The work took about an hour, and Marshall's confirmation took somewhat longer. Finally the launch team was satisfied that there was no problem. In Kapryan's words: "We picked up the count and went on our merry way."[42]

Apollo 17 lifted off into space at 12:33 a.m., 7 December. The flames, exploding into the darkness, made KSC momentarily as light as day. The launch was expected to be visible as far away as Montgomery. Miami observers saw a red streak crossing the northern sky, but Tampa was blacked out by a heavy ground fog and much of the Orlando area was under cloud cover.

During three days in the Taurus-Littrow valley on the moon, Cernan and Schmitt set up their multimillion-dollar array of scientific experiments, using the lunar rover to get them about the crater-pocked landscape. They took three excursions for a total of more than 32 kilometers in the rover, gathering rock samples and taking gravity measurements. Upon return to the command module, the team orbited the moon for nearly two more days of experimentation. They left the last of the Apollo lunar surface experiment packages. With four previously established nuclear-powered stations, the Apollo 17 equipment would allow scientists to monitor the moon's heat flow, volcanic activity, meteor impacts, and other phenomena. Also left behind were eight time bombs scheduled to go off after the astronauts started their return to earth. With the lunar module ascent stage, which was jettisoned into the moon, the bombs were expected to create artificial moonquakes that could be measured by seismometers and perhaps reveal more secrets of the moon's structure.[43] The Apollo program was leaving the moon with nine bangs and no whimpers.

24

Five Years After

On the morning of 16 July 1974, a large crowd gathered at the LC-39 press site to dedicate the launch complex as a national historical site. At the front of the press stands, a countdown clock ticked off the minutes. At 9:32 a.m., exactly five years after the liftoff of Apollo 11, astronauts Armstrong, Aldrin, and Collins unveiled a plaque commemorating their historic journey. The inscription read in part:

> Men began the first journeys to the moon from this complex. The success of these explorations was made possible by the united efforts of Government, and Industry, and the support of the American people.

Without question, the teamwork that joined together thousands of men and machines was Kennedy Space Center's greatest contribution to the lunar landing. Other elements undergirding KSC's success included the confidence, diligence, and technological skills of the launch team and the generous support of Congress.

A spirit of optimism marked the launch team's efforts throughout the Apollo program. Wernher von Braun exemplified this attitude in 1962 when he defended the choice of a mobile launch complex for LC-39. As von Braun noted, the fundamental question was whether NASA leaders believed "a space program is here to stay, and will continue to grow." Grumman workers typified this same outlook ten years later with their Apollo 17 slogan, "This may be our last, but it will be our best." At times during the program, the optimism wavered, most notably in 1967 in the aftermath of the AS-204 fire and with the interminable checkout of AS-501. Despite these setbacks, the launch team continued to believe that it could meet President Kennedy's challenge.

Another vital ingredient in KSC's success was the old-fashioned virtue, perseverance. Problems were the norm during most Apollo launch operations. With so much new, exotic hardware, strenuous efforts at quality control did not eliminate defective parts; equipment failures were common. The situation was complicated further by frequent last-minute modifications to the spacecraft, particularly in the hectic years of 1967–1968. From the

Debus-Davis Study to the Apollo 13 rescue, there were numerous occasions when time clocks, Sundays, and holidays were ignored. The launch team's diligence allowed KSC management to recover from many schedule slips and maintain NASA's timetable.

The launch team overcame significant technical problems on its way to the lunar launch. Although the design of the Merritt Island facilities was generally straightforward and within the state of the art, LC-39's size posed a great challenge. URSAM's assignment on the vehicle assembly building was to design one of the world's largest buildings on a marsh, in hurricane country, with openings along the sides that precluded a conventional framework. The extensible platforms, enclosing the space vehicles inside the assembly building, did not allow any appreciable sidesway. The 8000-metric-ton load intended for Marion's crawler-transporter ruled out any pre-construction tests of its design. The many changes in space-vehicle requirements and the pressing construction schedule added to the problems of size. As Col. N. A. Lore of the Corps of Engineers wrote in 1966, nearly all of LC-39 "was designed prior to [firm] definition of Apollo systems and built to support concepts rather than detailed systems." Consequently, important parts of the launch facility required extensive design changes; the swing arms and mobile service structure were prominent examples.

While visitors marvel at the size of LC-39's major facilities, the automating of launch operations represented KSC's most important technological advance. The Saturn V ground computer complex and the spacecraft's automated checkout system were in the vanguard of industrial automation. Whereas computers had been employed previously in monitoring industrial operations, KSC's electrical engineers used their computers to command lengthy processes. The automation of launch operations took nearly a decade and caused many frustrations, at times threatening the entire operation. It is unlikely, however, that KSC could have launched an Apollo–Saturn V on time, without computers.

A major reason for the launch team's success was its ability to profit from mistakes. The AS-204 fire prompted necessary changes in test procedures and safety requirements. Just as importantly, it brought the Cape's spacecraft operations completely under KSC's direction for the first time. The lightning strike on Apollo 12 caused a thorough review of LC-39's electrical protection and a tightening of weather restrictions. After the blind flange incident on the SA-5 countdown, launch officials adopted the countdown demonstration test as the final test. The launch team failed to anticipate problems in a number of areas; when difficulties appeared, however, officials profited from the experience.

Congressional support paved the way to the moon. When the launch facilities were planned in 1961–1962, Congress was willing to fund whatever was necessary to overtake the Russians. NASA's ambitious requests were largely met. With a decline of congressional support after 1962, KSC had to lower its sights—the assembly building shrank from six to four high bays and there were similar reductions in other facilities. Although congressional generosity declined, the launch operation fared well through 1969. There were ample funds for overtime, cost overruns, and special efforts such as the Boeing-TIE contract. The cutback after Apollo 11 brought a sizable reduction in KSC's workforce, but in other areas (e.g., civil service grade level and contractor overtime) there were no significant changes until the program's end.

In retrospect NASA and Congress appear to have overbuilt the launch complex. NASA engineers developed the plans for the launch facilities in 1961–1962 when other aspects of Apollo were still undecided. (The decision to employ a mobile launch preceded the selection of lunar orbital rendezvous.) The plans for LC-39 were based on predictions of high launch rates. For two decades the von Braun team had employed a building-block approach to rocket testing. It was assumed that a new launch vehicle would undergo many test flights before qualification; 16 were scheduled initially for the Saturn I. The Huntsville center also believed that lunar landing could best be achieved via earth-orbital rendezvous, which required several launches per mission. Together, the building block philosophy and earth-orbital rendezvous might require 50 launches per year, a rate justifying a mobile launch complex. However, the high launch rate never materialized, partly because of NASA's "all-up" decision (made after congressional cutbacks in 1963). After Apollo 11, a significant portion of LC-39 was not needed.

Changes during the Apollo program had a similar impact on spacecraft facilities. Several activities planned for the launch site, such as parachute packing and static test-firing, were eventually conducted elsewhere. The size of the facilities also anticipated a higher launch rate. In some cases the vacant space was used for other purposes; thus the parachute-packing facility became a news center in 1968. KSC had few white elephants at the peak of Apollo operations, but much of the spacecraft facilities went unused during the program's last three years. Viewed from the perspective of the mid-1970s, midway between the eras of Apollo and the Space Shuttle, the manned launch complex appears grossly overbuilt. It can be argued, however, that the Apollo-Saturn launch facilities provided a margin for error in the hectic months of 1968–1969 when KSC had three vehicles in the operational flow. It can also be argued that those facilities may be used in the future.

Apollo placed a severe strain on the larger Cape community. Brevard County had grown with amazing rapidity during the 1950s. The increase brought about by the Apollo program further taxed the social and economic resources of the area and took a heavy toll of family life, as the divorce rates of the time indicate. Race relations at Kennedy Space Center seemed harmonious, but the limited numbers of black engineers and trained technicians kept most blacks in service or maintenance areas.

Labor disputes were among the most distressing aspects of the launch facility construction. Unions quarreled with NASA over tasks performed by civil servants; union members refused to work alongside nonunion labor; and most frequently they fought each other over jurisdictional rights to jobs. While not noticeably greater than in most industrial areas, work stoppages seemed as totally out of place in the space race as they had during World War II. The contrast between the total dedication of some workers intent on getting men to the moon, and others arguing about jurisdiction in areas of employment, tended to shock the nation.

The conflict at the Cape was not limited to the labor unions. Some members of the Air Force viewed the civilian agency's program as an infringement on their preserve. The manned spaceflight centers questioned each other's performance and objectives. Houston's and Huntsville's mistakes were magnified at Merritt Island, where the launch team corrected space-vehicle errors. It was easy to forget the thousands of parts that worked when the failure of one piece delayed a launch. The subordination of Houston and Goddard launch teams to KSC also caused hard feelings. Finally there were differing opinions as to the relative contributions of contractor and civil servant at KSC.

While many disagreements sprang up during the launch operations, the Apollo team subordinated its differences to the goal of a lunar landing. At the fifth anniversary of the Apollo 11 launch, James Webb noted:

> The successes achieved here resulted not only from teamwork between individuals, not only from effective interfaces between men and machines, but also because Dr. Kurt Debus and his associates in NASA, in the Air Force and other government agencies, in industry and in universities have created a team of organizations which is a much more difficult undertaking than to create a team of individuals.

The leadership was the more remarkable, coming in large part from engineers with little previous schooling in management. The demonstration of this teamwork of organizations—from the planning for LC-39 through the successful launch of Apollo 17—is the most impressive legacy of the Apollo launch program.

Appendix A

LAUNCHES OF SATURN IB AND SATURN V

Mission	AS-201	AS-202	AS-203	Apollo 4
Objective	Launch vehicle develop-ment; sub-orbital	Repeat AS-201; suborbital; test heat shield	Repeat AS-201; test liquid hydrogen systems	First Saturn V launch
Vehicles				
launch	SA-201	SA-202	SA-203	AS-501
command-service	009	011	—	017
lunar	—	—	—	10R
Facilities				
high bay	—	—	—	1
mobile launcher	—	—	—	1
complex, pad	34	34	37B	39A
firing room	—	—	—	1
Erection				
launch veh.	25 Oct 65	11 Mar 66	21 Apr 66	19 Jun 67
spacecraft	26 Dec 65	2 Jul 66	—	20 Jun 67
CDDT	9 Feb 66	5 Aug 66	1 Jul 66	13 Oct 67
Crew	—	—	—	—
Lunar landing site	—	—	—	—
Test supervisor	Donnelly	Phillips	Donnelly	Henschel
Lunar window	—	—	—	—
Launched	26 Feb 66	25 Aug 66	5 Jul 66	9 Nov 67

Apollo 5	Apollo 6	Apollo 7	Apollo 8	Apollo 9
Lunar module develop- ment	Launch vehicle develop- ment	Manned test of command- service module	Manned flight of Saturn V and lunar orbit	Manned test of lunar hardware in earth orbit
AS-204	AS-502	SA-205	AS-503	AS-504
—	020	101	103	104
1	2R	—	B	3
—	3	—	1	3
—	2	—	1	2
37B	39A	34	39A	39A
—	2	—	1	2
11 Apr 67	14 Jul 67	11 May 68	14 Aug 68	7 Oct 68
19 Nov 67	10 Dec 67	9 Aug 68	7 Oct 68	3 Dec 68
19 Jan 68	31 Mar 68	17 Sep 68	11 Dec 68	19 Feb 69
—	—	Schirra Eisele Cunning- ham	Borman Lovell Anders	McDivitt Scott Schweick- art
—	—	—	—	—
Phillips	Harrington	Phillips	Schick	Harrington
—	—	—	—	—
22 Jan 68	4 Apr 68	11 Oct 68	21 Dec 68	3 Mar 69

Apollo 10	Apollo 11	Apollo 12	Apollo 13	Apollo 14
Manned test of lunar hardware near moon	Lunar landing	Deploy lunar experiments; investigate Surveyor III	Deploy lunar experiments; photograph later sites	Repeat 13
AS-505	AS-506	AS-507	AS-508	AS-509
106	107	108	109	110
4	5	6	7	8
2	1	3	1	3
3	1	2	3	2
39B	39A	39A	39A	39A
3	1	2	1	2
30 Dec 68	5 Mar 69	22 May 69	1 Aug 69	14 May 70
6 Feb 69	14 Apr 69	30 Jun 69	10 Dec 69	5 Nov 70
6 May 69	3 Jul 69	29 Oct 69	26 Mar 70	19 Jan 71
Stafford Young Cernan	Armstrong Collins Aldrin	Conrad Gordon Bean	Lovell Swigert Haise	Shepard Roosa Mitchell
—	Sea of Tran- quillity	Ocean of Storms	(Fra Mauro intended)	Fra Mauro
Phillips	Schick	Harrington	Grenville	Henschel
—	16–22 Jul	10–17 Nov	7–14 Apr	29 Jan– 3 Feb
18 May 69	16 Jul 69	14 Nov 69	11 Apr 70	31 Jan 71

Apollo 15	Apollo 16	Apollo 17
Extended investigations with lunar rover	Repeat 15	Repeat 15; return largest load of lunar samples
AS-510	AS-511	AS-512
112 10	113 11	114 12
3	3	3
3	3	3
39A 1	39A 1	39A 1
17 Sep 70 8 May 71	6 Oct 71 8 Dec 71, 5 Feb 72	27 Jun 72 23 Aug 72
14 Jul 71	31 Mar 72	21 Nov 72
Scott Worden Irwin	Young Mattingly Duke	Cernan Evans Schmitt
Hadley Apennine	Descartes	Taurus Littrow
Harrington	Turner	Schick
24–30 Jul	16–23 Apr	5–17 Dec
26 Jul 71	16 Apr 72	7 Dec 72

Appendix B

Launch Complex 39

1. Vehicle (Vertical) Assembly Building (VAB)

Area: 32 500 sq m (8 acres).

Dimensions: 218 × 158 m, 160 m tall. Compare to Statue of Liberty, 93 m tall.

Volume: 3 665 000 cu m. Compare to Pentagon, 2 181 000 cu m.

Features: 4 high bays for assembly and checkout of launch vehicles with spacecraft, low bays for checkout of individual stages. 4 high bay doors, opening height 139 m. 71 lifting devices. 2 bridge cranes of 227-metric-ton capacity. 9070 metric tons of air conditioning, 125 ventilators.

Construction: 89 400 metric tons of steel, 49 700 cu m of concrete. 4225 open-end steel pipe piles, 0.4 m diameter, driven to depth of 49 m. Siding of 100 800 sq m of insulated aluminum panels and 6500 sq m of plastic panels.

Cost of construction: $117 000 000.

2. Launch Control Center (LCC)

The 4-story, electronic brain of LC-39, the LCC was built adjacent to the VAB and 5.6 km from pad A. During launch, 62 TV cameras provided closed-circuit pictures to 100 monitor screens in the LCC. The LCC was connected to the mobile launchers by a high-speed data link.

1st floor: offices, cafeteria, shops.

2d floor: telemetry, RF and tracking equipment, instrumentation, data reduction and evaluation equipment.

3d floor: 4 firing rooms, one for each of the high bays in the VAB. Each active room had 470 sets of control and monitoring equipment.

4th floor: conference and display rooms, offices, mechanical equipment.

Cost of construction: $10 000 000.

3. Mobile Launchers (3)

Weight: 5715 metric tons, with unfueled vehicle.

Height (on pedestals): 136 m to top of crane.

Launch platform: 2-story steel structure 49 × 41 m, 7.6 m high. Exhaust hole 14 m square. 4 hold-down arms, each 18 100 kg, held rocket vertical during thrust buildup, approximately 8.9 seconds to reach 95% of total thrust. Platform supported by 6 steel pedestals 7 meters high when in VAB or on pad. 4 additional extensible columns used at pad, to stiffen platform during firing.

Umbilical tower, mounted on platform, 18 levels, 2 elevators. 9 swing (service) arms for personnel access, propellant, electrical, pneumatic, and instrumentation lines. Arms weighed 15 900–23 600 kg, length 13.7–18.3 m. Top arm (9) used by astronauts to enter spacecraft. 4 arms retracted before liftoff, 5 at T – 0.

Cost of construction: $33 963 000.

4. *Transporters (2)*

Used to move mobile launcher, with assembled space vehicle, from VAB to pad, also to move mobile service structure to and from pad.

Weight: 2720 metric tons, largest tracked vehicle known.

Dimensions: 35 × 40 m, with top deck about size of baseball infield; height, 6–8 m.

4 double-tracked crawlers each 3 m high, 12 m long. 8 tracks per transporter, 57 shoes per track. Each tread shoe (or link in the track) weighed 0.9 metric ton.

Power: 16 traction motors powered by four 1000-kw generators, driven by 2 diesel engines; two 750-kw generators, driven by 2 diesel engines for jacking, steering, lighting, ventilating; two 150-kw generators for power to the mobile launcher.

Maximum speed: 1.6 km/hr loaded, 3.2 unloaded. Pad-to-VAB trip time, loaded, 7 hrs.

Levelling: top of space vehicle kept vertical within ± 10′ of arc, including negotiation of 5% grade leading up to pad.

Cost of construction: $13 600 000.

5. *Mobile Service Structure (1)*

Weight: 4760 metric tons.

Height: 125 m.

Elevators: 2 for personnel and equipment in tower, 1 from ground to base work area.

Work platforms: 2 self-propelled, 3 fixed. Top 3 platforms served spacecraft, bottom 2 served Saturn V.

Parking position during launch: 2100 m from pad A.

Cost of construction: $11 600 000.

6. *Crawlerway*

Length: VAB to pad A, 5500 m; VAB to pad B, 6800 m.
Width: 2 lanes, each 12 m, separated by 15 m median.
Depth: average 2 m.
Cost of construction: $7 500 000.

7. *Launch Pads (2)*

Construction: 52 000 cu m of concrete, roughly octagonal in shape. 2 pads are virtually identical, 2660 m apart.
Flame trench: 13 m deep, 18 m wide, 137 m long.
Flame deflector: 635 metric tons, 12 m high, 15 m wide, 23 m long.
Lighting: 40 xenon high-intensity searchlights in 5 clusters around perimeter.
Emergency Egress System: 61-m escape tube from mobile launcher platform to blast-resistant room 12 m below pad, which contained survival supplies for 20 persons for 24 hours. Also, cab on slidewire from 98-m level to revetment 763 m away.

Appendix C

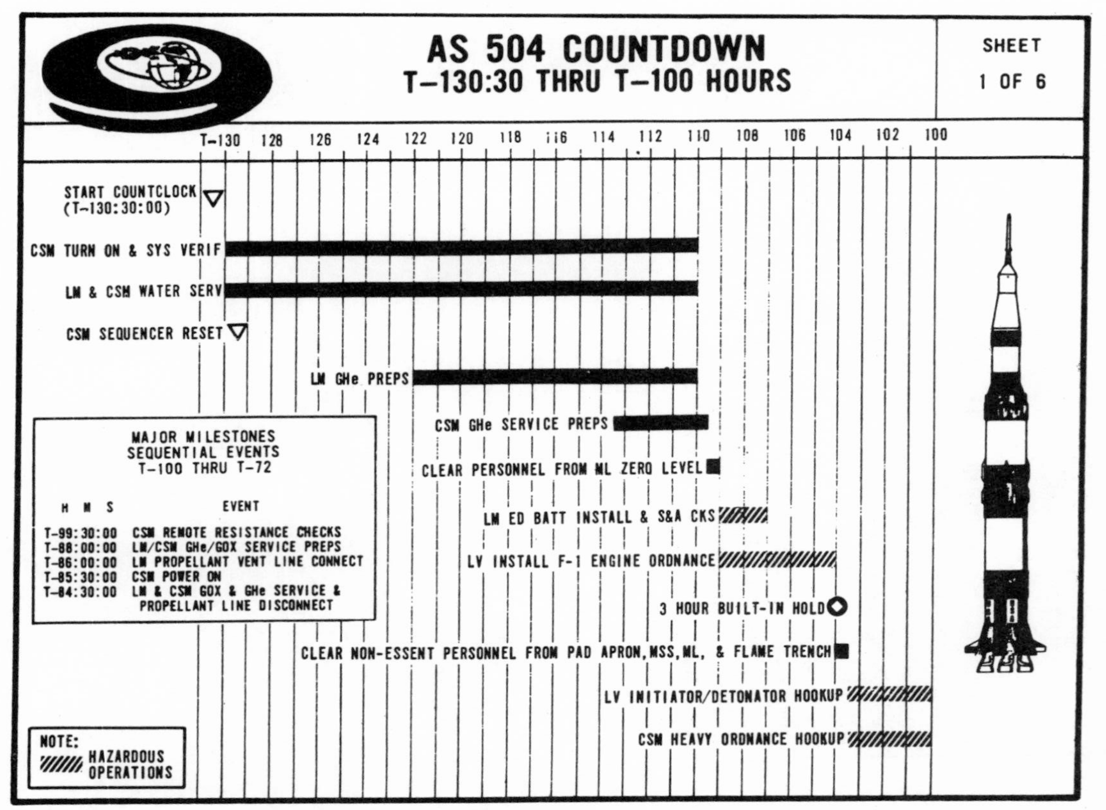

AS 504 COUNTDOWN
T–130:30 THRU T–100 HOURS
SHEET
1 OF 6
T–130 128 126 124 122 120 118 116 114 112 110 108 106 104 102 100
START COUNTCLOCK
(T–130:30:00)
CSM TURN ON & SYS VERIF
LM & CSM WATER SERV
CSM SEQUENCER RESET
LM GHe PREPS
CSM GHe SERVICE PREPS
CLEAR PERSONNEL FROM ML ZERO LEVEL
LM ED BATT INSTALL & S&A CKS
LV INSTALL F-1 ENGINE ORDNANCE
3 HOUR BUILT-IN HOLD
CLEAR NON-ESSENT PERSONNEL FROM PAD APRON, MSS, ML, & FLAME TRENCH
LV INITIATOR/DETONATOR HOOKUP
CSM HEAVY ORDNANCE HOOKUP
MAJOR MILESTONES
SEQUENTIAL EVENTS
T–100 THRU T–72
H M S
EVENT
T-99:30:00 CSM REMOTE RESISTANCE CHECKS
T-88:00:00 LM/CSM GHe/GOX SERVICE PREPS
T-86:00:00 LM PROPELLANT VENT LINE CONNECT
T-85:30:00 CSM POWER ON
T-84:30:00 LM & CSM GOX & GHe SERVICE &
PROPELLANT LINE DISCONNECT
NOTE:
HAZARDOUS
OPERATIONS

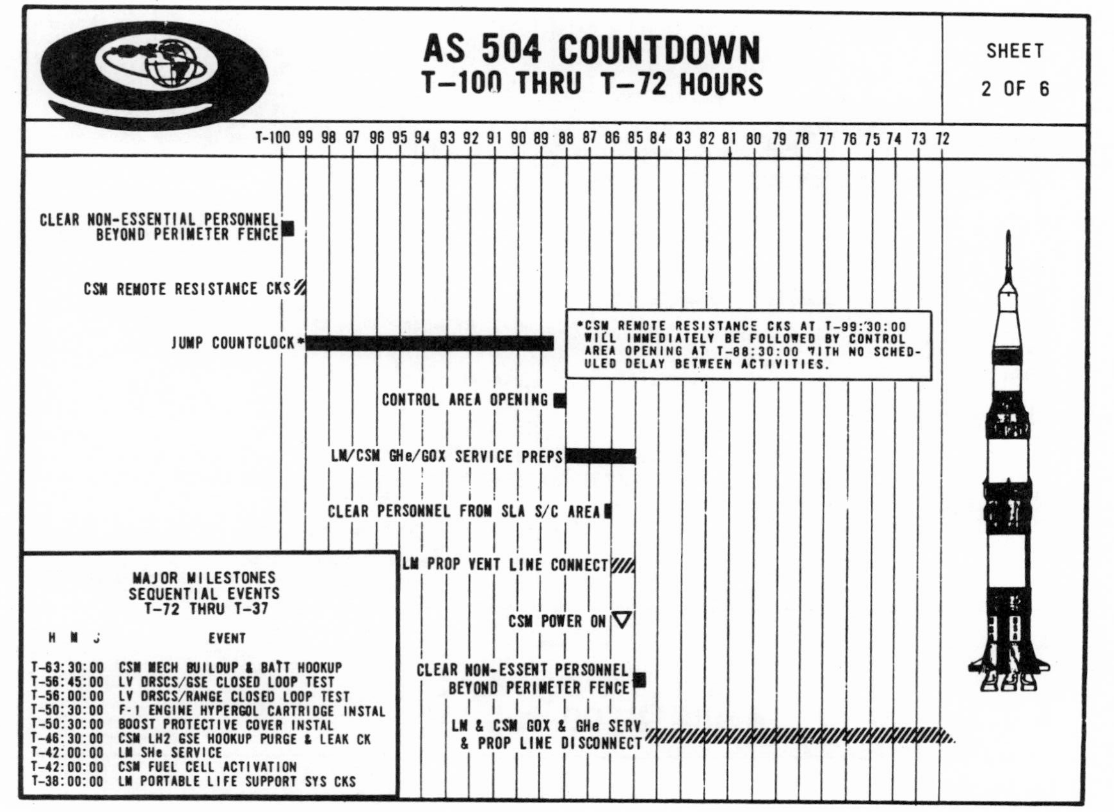
AS 504 COUNTDOWN
T–100 THRU T–72 HOURS
SHEET 2 OF 6
T-100 99 98 97 96 95 94 93 92 91 90 89 88 87 86 85 84 83 82 81 80 79 78 77 76 75 74 73 72
CLEAR NON-ESSENTIAL PERSONNEL BEYOND PERIMETER FENCE
CSM REMOTE RESISTANCE CKS
JUMP COUNTCLOCK*
*CSM REMOTE RESISTANCE CKS AT T-99:30:00 WILL IMMEDIATELY BE FOLLOWED BY CONTROL AREA OPENING AT T-88:30:00 WITH NO SCHEDULED DELAY BETWEEN ACTIVITIES.
CONTROL AREA OPENING
LM/CSM GHe/GOX SERVICE PREPS
CLEAR PERSONNEL FROM SLA S/C AREA
LM PROP VENT LINE CONNECT
CSM POWER ON
CLEAR NON-ESSENT PERSONNEL BEYOND PERIMETER FENCE
LM & CSM GOX & GHe SERV & PROP LINE DISCONNECT
MAJOR MILESTONES SEQUENTIAL EVENTS T–72 THRU T–37
H M S EVENT
T-63:30:00 CSM MECH BUILDUP & BATT HOOKUP
T-56:45:00 LV DRSCS/GSE CLOSED LOOP TEST
T-56:00:00 LV DRSCS/RANGE CLOSED LOOP TEST
T-50:30:00 F-1 ENGINE HYPERGOL CARTRIDGE INSTAL
T-50:30:00 BOOST PROTECTIVE COVER INSTAL
T-46:30:00 CSM LH2 GSE HOOKUP PURGE & LEAK CK
T-42:00:00 LM SHe SERVICE
T-42:00:00 CSM FUEL CELL ACTIVATION
T-38:00:00 LM PORTABLE LIFE SUPPORT SYS CKS

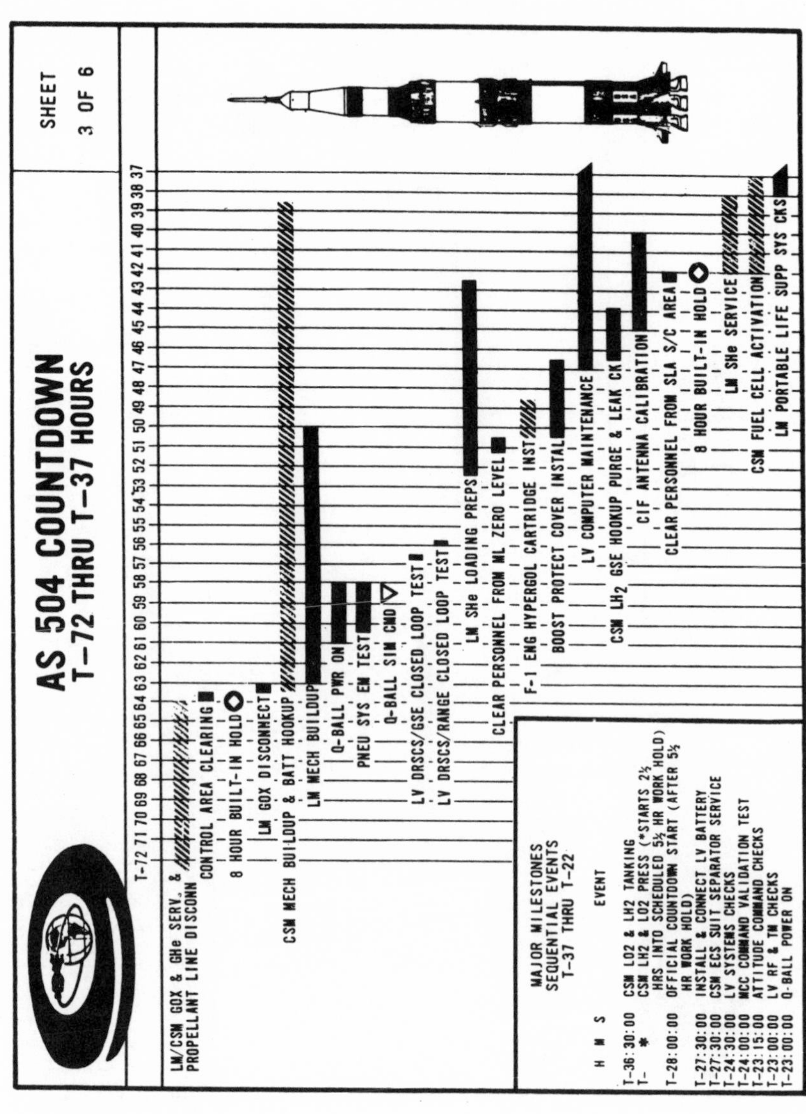
AS 504 COUNTDOWN
T–72 THRU T–37 HOURS
SHEET
3 OF 6
T-72 71 70 69 68 67 66 65 64 63 62 61 60 59 58 57 56 55 54 53 52 51 50 49 48 47 46 45 44 43 42 41 40 39 38 37
LM/CSM GOX & GHe SERV. &
PROPELLANT LINE DISCONN
CONTROL AREA CLEARING
8 HOUR BUILT-IN HOLD
LM GOX DISCONNECT
CSM MECH BUILDUP & BATT HOOKUP
LM MECH BUILDUP
Q-BALL PWR ON
PNEU SYS EM TEST
Q-BALL SIM CMD
LV DRSCS/GSE CLOSED LOOP TEST
LV DRSCS/RANGE CLOSED LOOP TEST
LM SHe LOADING PREPS
CLEAR PERSONNEL FROM ML ZERO LEVEL
F-1 ENG HYPERGOL CARTRIDGE INST
BOOST PROTECT COVER INSTAL
LV COMPUTER MAINTENANCE
CSM LH2 GSE HOOKUP PURGE & LEAK CK
CIF ANTENNA CALIBRATION
CLEAR PERSONNEL FROM SLA S/C AREA
8 HOUR BUILT-IN HOLD
LM SHe SERVICE
CSM FUEL CELL ACTIVATION
LM PORTABLE LIFE SUPP SYS CKS
MAJOR MILESTONES
SEQUENTIAL EVENTS
T-37 THRU T-22
H M S
EVENT
T-36:30:00 CSM LO2 & LH2 TANKING
T- * CSM LH2 & LO2 PRESS (*STARTS 2½
HRS INTO SCHEDULED 5½ HR WORK HOLD)
T-28:00:00 OFFICIAL COUNTDOWN START (AFTER 5½
HR WORK HOLD)
T-27:30:00 INSTALL & CONNECT LV BATTERY
T-27:30:00 CSM ECS SUIT SEPARATOR SERVICE
T-24:30:00 LV SYSTEMS CHECKS
T-24:00:00 MCC COMMAND VALIDATION TEST
T-23:15:00 ATTITUDE COMMAND CHECKS
T-23:00:00 LV RF & TM CHECKS
T-23:00:00 Q-BALL POWER ON

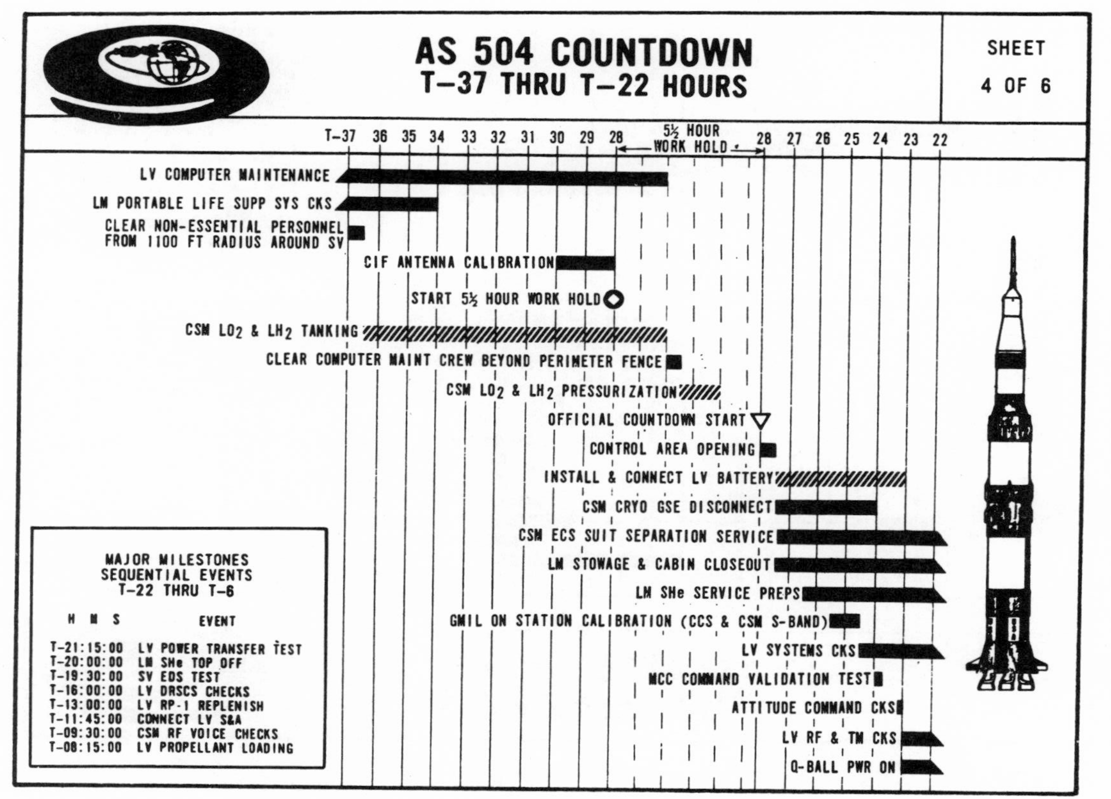
AS 504 COUNTDOWN
T–37 THRU T–22 HOURS
SHEET
4 OF 6
T–37 36 35 34 33 32 31 30 29 28
5½ HOUR
WORK HOLD
28 27 26 25 24 23 22
LV COMPUTER MAINTENANCE
LM PORTABLE LIFE SUPP SYS CKS
CLEAR NON-ESSENTIAL PERSONNEL FROM 1100 FT RADIUS AROUND SV
CIF ANTENNA CALIBRATION
START 5½ HOUR WORK HOLD
CSM LO2 & LH2 TANKING
CLEAR COMPUTER MAINT CREW BEYOND PERIMETER FENCE
CSM LO2 & LH2 PRESSURIZATION
OFFICIAL COUNTDOWN START
CONTROL AREA OPENING
INSTALL & CONNECT LV BATTERY
CSM CRYO GSE DISCONNECT
CSM ECS SUIT SEPARATION SERVICE
LM STOWAGE & CABIN CLOSEOUT
LM SHe SERVICE PREPS
GMIL ON STATION CALIBRATION (CCS & CSM S-BAND)
LV SYSTEMS CKS
MCC COMMAND VALIDATION TEST
ATTITUDE COMMAND CKS
LV RF & TM CKS
Q-BALL PWR ON
MAJOR MILESTONES
SEQUENTIAL EVENTS
T–22 THRU T–6
H M S EVENT
T–21:15:00 LV POWER TRANSFER TEST
T–20:00:00 LM SHe TOP OFF
T–19:30:00 SV EDS TEST
T–16:00:00 LV DRSCS CHECKS
T–13:00:00 LV RP-1 REPLENISH
T–11:45:00 CONNECT LV S&A
T–09:30:00 CSM RF VOICE CHECKS
T–08:15:00 LV PROPELLANT LOADING

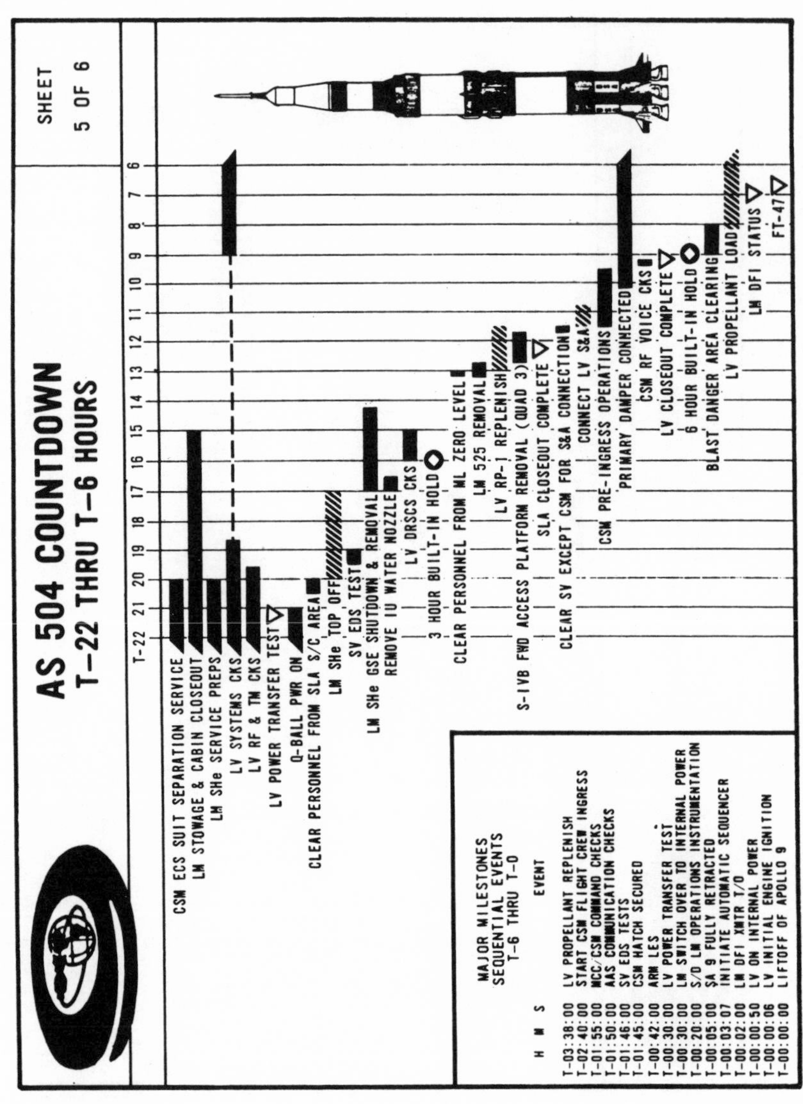
AS 504 COUNTDOWN
T–22 THRU T–6 HOURS
SHEET
5 OF 6
T-22 21 20 19 18 17 16 15 14 13 12 11 10 9 8 7 6
CSM ECS SUIT SEPARATION SERVICE
LM STOWAGE & CABIN CLOSEOUT
LM SHe SERVICE PREPS
LV SYSTEMS CKS
LV RF & TM CKS
LV POWER TRANSFER TEST
Q-BALL PWR ON
CLEAR PERSONNEL FROM SLA S/C AREA
LM SHe TOP OFF
SV EDS TEST
LM SHe GSE SHUTDOWN & REMOVAL
REMOVE IU WATER NOZZLE
LV DRSCS CKS
3 HOUR BUILT-IN HOLD
CLEAR PERSONNEL FROM ML ZERO LEVEL
LM 525 REMOVAL
LV RP-1 REPLENISH
S-IVB FWD ACCESS PLATFORM REMOVAL (QUAD 3)
SLA CLOSEOUT COMPLETE
CLEAR SV EXCEPT CSM FOR S&A CONNECTION
CONNECT LV S&A
CSM PRE-INGRESS OPERATIONS
PRIMARY DAMPER CONNECTED
CSM RF VOICE CKS
LV CLOSEOUT COMPLETE
6 HOUR BUILT-IN HOLD
BLAST DANGER AREA CLEARING
LV PROPELLANT LOAD
LM DFI STATUS
FT-47
MAJOR MILESTONES
SEQUENTIAL EVENTS
T-6 THRU T-0
H M S EVENT
T-03:38:00 LV PROPELLANT REPLENISH
T-02:40:00 START CSM FLIGHT CREW INGRESS
T-01:55:00 MCC/CSM COMMAND CHECKS
T-01:50:00 AAS COMMUNICATION CHECKS
T-01:46:00 SV EDS TESTS
T-01:45:00 CSM HATCH SECURED
T-00:42:00 ARM LES
T-00:30:00 LV POWER TRANSFER TEST
T-00:30:00 LM SWITCH OVER TO INTERNAL POWER
T-00:20:00 S/D LM OPERATIONS INSTRUMENTATION
T-00:05:00 SA 9 FULLY RETRACTED
T-00:03:07 INITIATE AUTOMATIC SEQUENCER
T-00:02:00 LM DFI XMTR T/O
T-00:00:50 LV ON INTERNAL POWER
T-00:00:06 LV INITIAL ENGINE IGNITION
T-00:00:00 LIFTOFF OF APOLLO 9

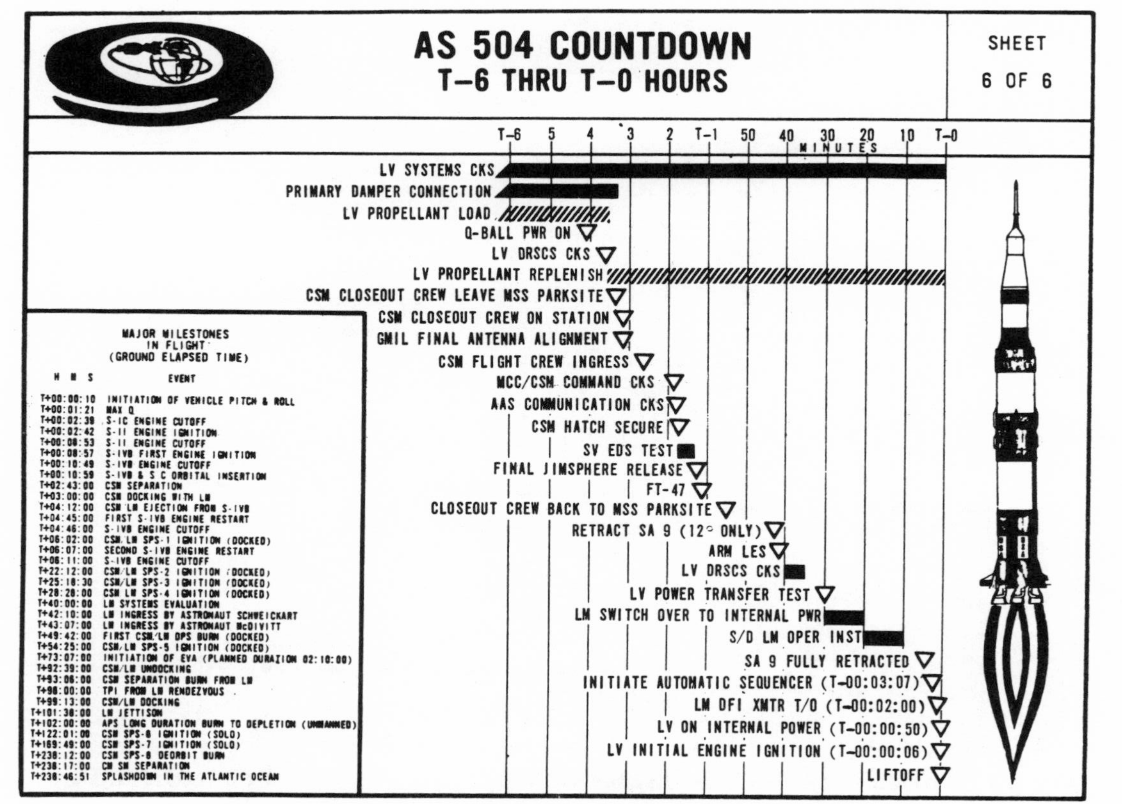
AS 504 COUNTDOWN
T–6 THRU T–0 HOURS
SHEET
6 OF 6
T–6 5 4 3 2 T–1 50 40 30 20 10 T–0
MINUTES
LV SYSTEMS CKS
PRIMARY DAMPER CONNECTION
LV PROPELLANT LOAD
Q-BALL PWR ON
LV DRSCS CKS
LV PROPELLANT REPLENISH
CSM CLOSEOUT CREW LEAVE MSS PARKSITE
CSM CLOSEOUT CREW ON STATION
GMIL FINAL ANTENNA ALIGNMENT
CSM FLIGHT CREW INGRESS
MCC/CSM COMMAND CKS
AAS COMMUNICATION CKS
CSM HATCH SECURE
SV EDS TEST
FINAL JIMSPHERE RELEASE
FT-47
CLOSEOUT CREW BACK TO MSS PARKSITE
RETRACT SA 9 (12° ONLY)
ARM LES
LV DRSCS CKS
LV POWER TRANSFER TEST
LM SWITCH OVER TO INTERNAL PWR
S/D LM OPER INST
SA 9 FULLY RETRACTED
INITIATE AUTOMATIC SEQUENCER (T–00:03:07)
LM DFI XMTR T/O (T–00:02:00)
LV ON INTERNAL POWER (T–00:00:50)
LV INITIAL ENGINE IGNITION (T–00:00:06)
LIFTOFF
MAJOR MILESTONES
IN FLIGHT
(GROUND ELAPSED TIME)
H M S EVENT
T+00:00:10 INITIATION OF VEHICLE PITCH & ROLL
T+00:01:21 MAX Q
T+00:02:39 S-IC ENGINE CUTOFF
T+00:02:42 S-II ENGINE IGNITION
T+00:08:53 S-II ENGINE CUTOFF
T+00:08:57 S-IVB FIRST ENGINE IGNITION
T+00:10:49 S-IVB ENGINE CUTOFF
T+00:10:59 S-IVB & S C ORBITAL INSERTION
T+02:43:00 CSM SEPARATION
T+03:00:00 CSM DOCKING WITH LM
T+04:12:00 CSM LM EJECTION FROM S-IVB
T+04:45:00 FIRST S-IVB ENGINE RESTART
T+04:46:00 S-IVB ENGINE CUTOFF
T+06:02:00 CSM/LM SPS-1 IGNITION (DOCKED)
T+06:07:00 SECOND S-IVB ENGINE RESTART
T+06:11:00 S-IVB ENGINE CUTOFF
T+22:12:00 CSM/LM SPS-2 IGNITION (DOCKED)
T+25:18:30 CSM/LM SPS-3 IGNITION (DOCKED)
T+28:28:00 CSM LM SPS-4 IGNITION (DOCKED)
T+40:00:00 LM SYSTEMS EVALUATION
T+42:10:00 LM INGRESS BY ASTRONAUT SCHWEICKART
T+43:07:00 LM INGRESS BY ASTRONAUT McDIVITT
T+49:42:00 FIRST CSM/LM DPS BURN (DOCKED)
T+54:25:00 CSM/LM SPS-5 IGNITION (DOCKED)
T+73:07:00 INITIATION OF EVA (PLANNED DURATION 02:10:00)
T+92:39:00 CSM/LM UNDOCKING
T+93:06:00 CSM SEPARATION BURN FROM LM
T+98:00:00 TPI FROM LM RENDEZVOUS
T+99:13:00 CSM/LM DOCKING
T+101:38:00 LM JETTISON
T+102:00:00 APS LONG DURATION BURN TO DEPLETION (UNMANNED)
T+122:01:00 CSM SPS-6 IGNITION (SOLO)
T+169:49:00 CSM SPS-7 IGNITION (SOLO)
T+238:12:00 CSM SPS-8 DEORBIT BURN
T+238:17:00 CM SM SEPARATION
T+238:46:51 SPLASHDOWN IN THE ATLANTIC OCEAN

Appendix D

Processing Apollo 11 Through LC-39

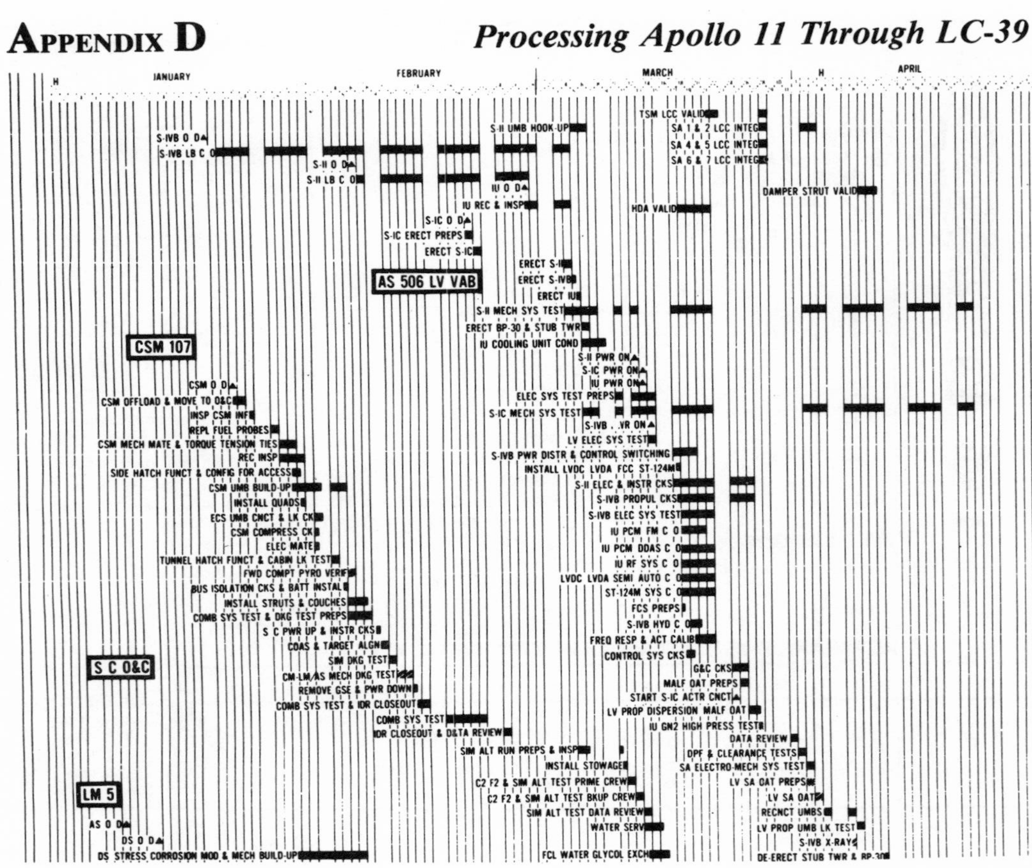

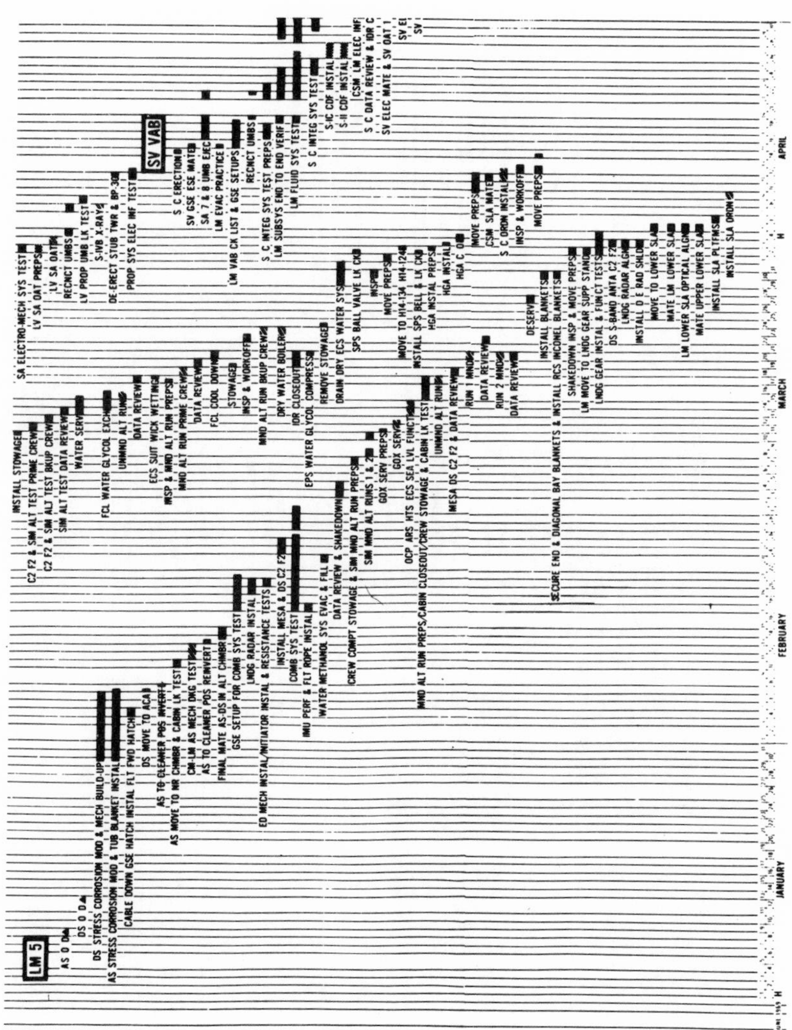

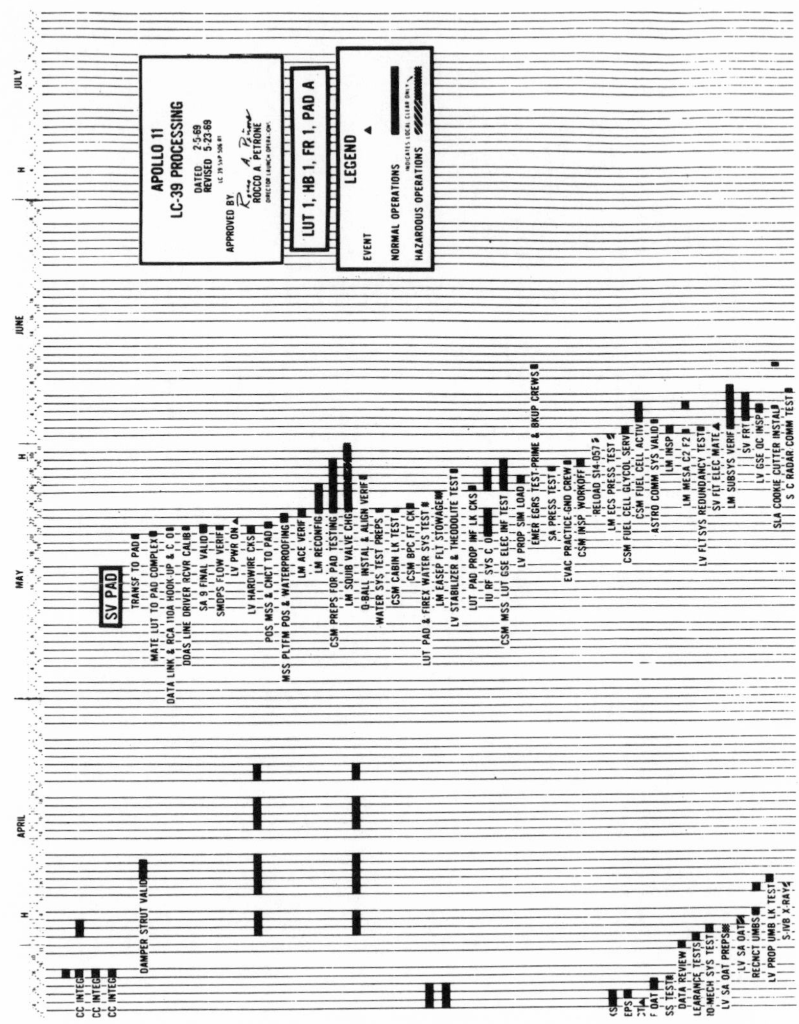
APOLLO 11
LC-39 PROCESSING
DATED 2-5-69
REVISED 5-23-69
APPROVED BY
ROCCO A PETRONE
LUT 1, HB 1, FR 1, PAD A
LEGEND
EVENT
NORMAL OPERATIONS
HAZARDOUS OPERATIONS
APRIL
MAY
JUNE
JULY
SV PAD
CC INTEG
DAMPER STRUT VALID
TRANSF TO PAD
MATE LUT TO PAD COMPLEX
DATA LINK & RCA 110A HOOK-UP & C O
DDAS LINE DRIVER RCVR CALIB
SA 9 FINAL VALID
SMDPS FLOW VERIF
LV PWR ON
LV HARDWIRE CKS
POS MSS & CNCT TO PAD
MSS PLTFM POS & WATERPROOFING
LM ACE VERIF
LM RECONFIG
CSM PREPS FOR PAD TESTING
LM SQUIB VALVE CHG
Q-BALL INSTAL & ALIGN VERIF
WATER SYS TEST PREPS
CSM CABIN LK TEST
CSM BPC FIT CK
LUT PAD & FIREX WATER SYS TEST
LM EASEP FLT STOWAGE
LV STABILIZER & THEODOLITE TEST
LUT PAD PROP INF LK CKS
IU RF SYS C O
CSM MSS LUT GSE ELEC INF TEST
LV PROP SIM LOAD
EMER EGRS TEST-PRIME & BKUP CREWS
SA PRESS TEST
EVAC PRACTICE-GND CREW
CSM INSP WORKOFF
RELOAD S14-057
LM ECS PRESS TEST
CSM FUEL CELL GLYCOL SERV
CSM FUEL CELL ACTIV
ASTRO COMM SYS VALID
LM INSP
LM MESA C2 F2
LV FLT SYS REDUNDANCY TEST
SV FLT ELEC MATE
LM SUBSYS VERIF
SV FRT
LV GSE QC INSP
SLA COOKIE CUTTER INSTAL
S C RADAR COMM TEST
DATA REVIEW
LV SA OAT
RECNCT UMBS
LV PROP UMB LK TEST
S-IVB X-RAY
LV SA OAT PREPS

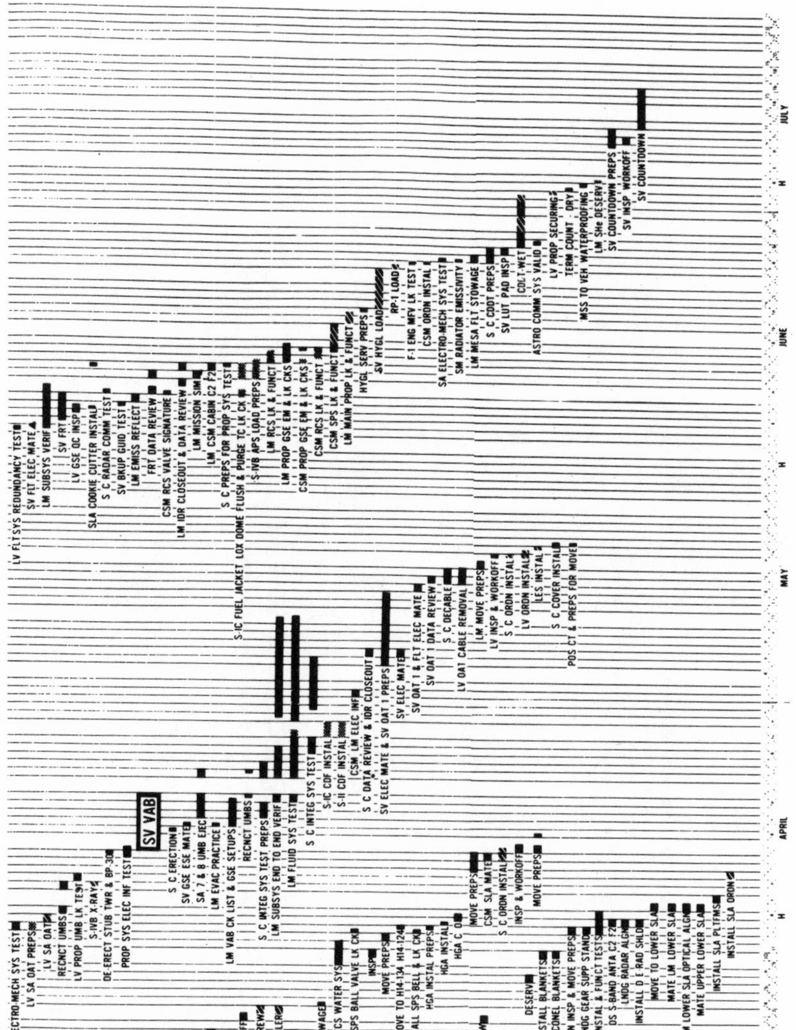

Appendix E

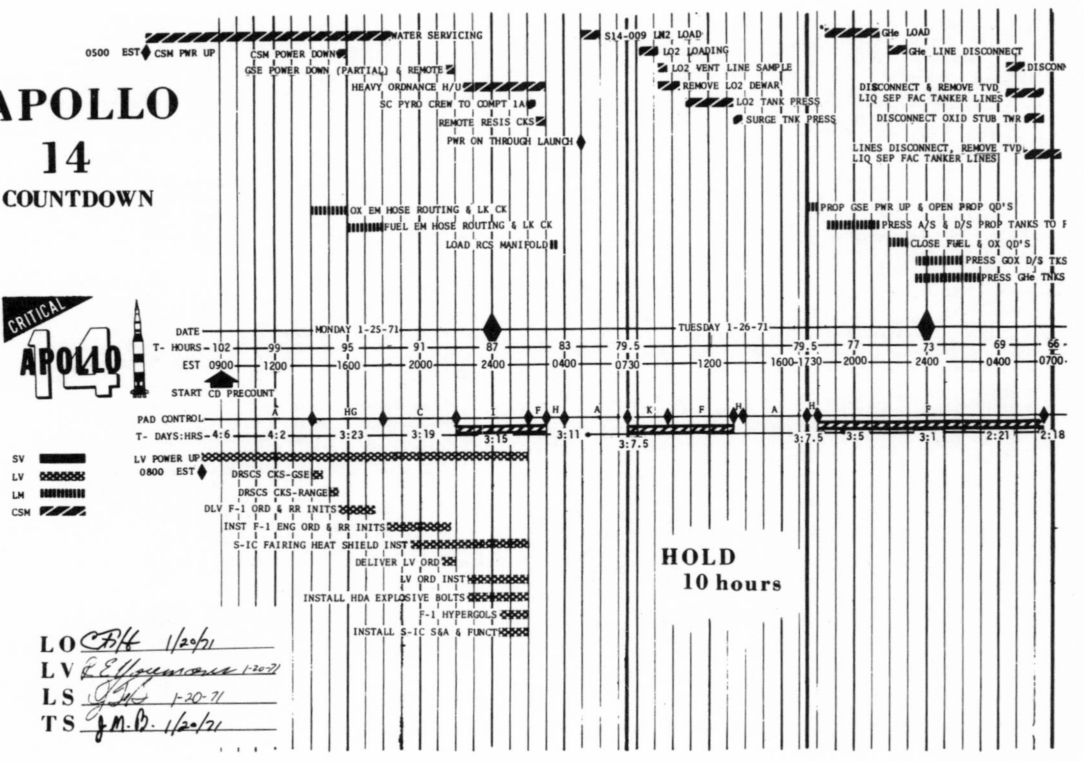
APOLLO
14
COUNTDOWN
CRITICAL
APOLLO 14
0500 EST CSM PWR UP
WATER SERVICING
CSM POWER DOWN
GSE POWER DOWN (PARTIAL) & REMOTE
HEAVY ORDNANCE H/U
SC PYRO CREW TO COMPT 1A
REMOTE RESIS CKS
PWR ON THROUGH LAUNCH
S14-009 LN2 LOAD
LO2 LOADING
LO2 VENT LINE SAMPLE
REMOVE LO2 DEWAR
LO2 TANK PRESS
SURGE TNK PRESS
GHe LOAD
GHe LINE DISCONNECT
DISCONNECT & REMOVE TVD
LIQ SEP FAC TANKER LINES
DISCONNECT OXID STUB TWR
LINES DISCONNECT, REMOVE TVD,
LIQ SEP FAC TANKER LINES
OX EM HOSE ROUTING & LK CK
FUEL EM HOSE ROUTING & LK CK
LOAD RCS MANIFOLD
PROP GSE PWR UP & OPEN PROP QD'S
PRESS A/S & D/S PROP TANKS TO F
CLOSE FUEL & OX QD'S
PRESS GOX D/S TKS
PRESS GHe TNKS
DATE
MONDAY 1-25-71
TUESDAY 1-26-71
T- HOURS - 102 99 95 91 87 83 79.5 79.5 77 73 69 66
EST 0900 1200 1600 2000 2400 0400 0730 1200 1600-1730 2000 2400 0400 0700
START CD PRECOUNT
PAD CONTROL
L A HG C I F H A K F H A H F
T- DAYS:HRS - 4:6 4:2 3:23 3:19 3:15 3:11 3:7.5 3:7.5 3:5 3:1 2:21 2:18
SV
LV
LM
CSM
LV POWER UP
0800 EST
DRSCS CKS-GSE
DRSCS CKS-RANGE
DLV F-1 ORD & RR INITS
INST F-1 ENG ORD & RR INITS
S-IC FAIRING HEAT SHIELD INST
DELIVER LV ORD
LV ORD INST
INSTALL HDA EXPLOSIVE BOLTS
F-1 HYPERGOLS
INSTALL S-IC S&A & FUNCT
HOLD
10 hours
LO
LV
LS
TS

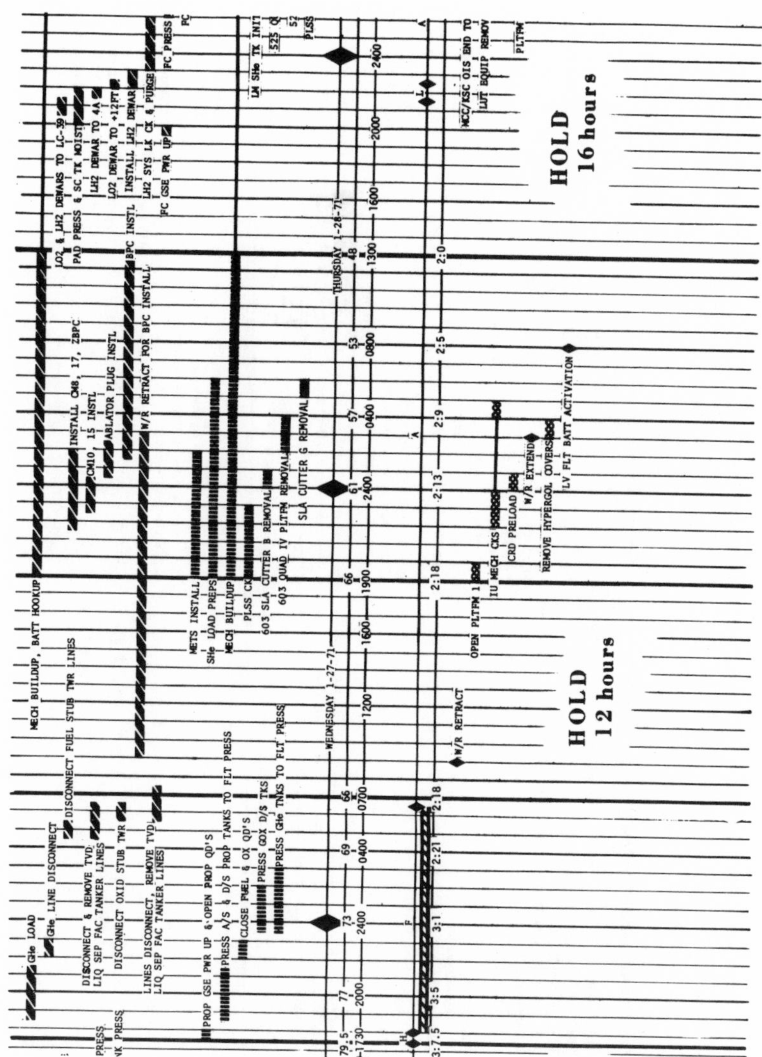
HOLD
16 hours
HOLD
12 hours

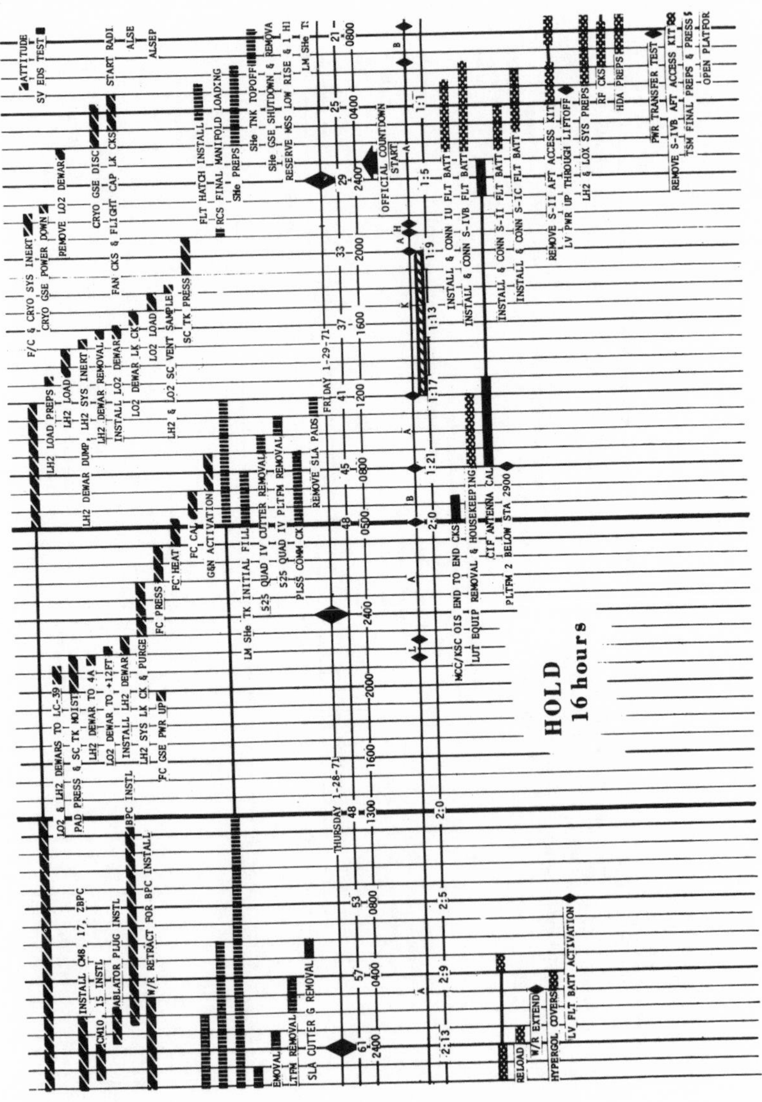
HOLD
16 hours
THURSDAY 1-28-71
FRIDAY 1-29-71
OFFICIAL COUNTDOWN START
LV FLT BATT ACTIVATION
HYPERGOL COVERS
W/R EXTEND
RELOAD
SLA CUTTER G REMOVAL
LTPM REMOVAL
INSTALL CM8, 17, ZBPC
CM10, 15 INSTL
ABLATOR PLUG INSTL
W/R RETRACT FOR BPC INSTALL
BPC INSTL
LO2 & LH2 DEWARS TO LC-39
PAD PRESS & SC TK MOIST
LH2 DEWAR TO 4A
LO2 DEWAR TO +12FT
INSTALL LH2 DEWAR
LH2 SYS LK CK & PURGE
FC GSE PWR UP
LM SHe TK INITIAL FILL
S25 QUAD IV CUTTER REMOVAL
S25 QUAD IV PLTFM REMOVAL
PLSS COMM CK
REMOVE SLA PADS
FC PRESS
FC HEAT
FC CAL
G&N ACTIVATION
LH2 DEWAR DUMP
LH2 LOAD PREPS
LH2 LOAD
LH2 SYS INERT
LH2 DEWAR REMOVAL
INSTALL LO2 DEWAR
LO2 DEWAR LK CK
LO2 LOAD
LH2 & LO2 SC VENT SAMPLE
SC TK PRESS
F/C & CRYO SYS INERT
CRYO GSE POWER DOWN
REMOVE LO2 DEWAR
FAN CKS & FLIGHT CAP LK CKS
CRYO GSE DISC
START RADI
ALSE
ALSEP
ATTITUDE
SV EDS TEST
FLT HATCH INSTALL
RCS FINAL MANIFOLD LOADING
SHe PREPS
SHe TNK TOPOFF
SHe GSE SHUTDOWN & REMOVA
RESERVE MSS LOW RISE & 1 HI
MCC/KSC OIS END TO END CKS
LUT EQUIP REMOVAL & HOUSEKEEPING
CIF ANTENNA CAL
PLTFM 2 BELOW STA 2900
INSTALL & CONN IU FLT BATT
INSTALL & CONN S-IVB FLT BATT
INSTALL & CONN S-II FLT BATT
INSTALL & CONN S-IC FLT BATT
REMOVE S-II AFT ACCESS KIT
LV PWR UP THROUGH LIFTOFF
LH2 & LOX SYS PREPS
RF CKS
HDA PREPS
PWR TRANSFER TEST
REMOVE S-IVB AFT ACCESS KIT
TSM FINAL PREPS & PRESS
OPEN PLATFOR

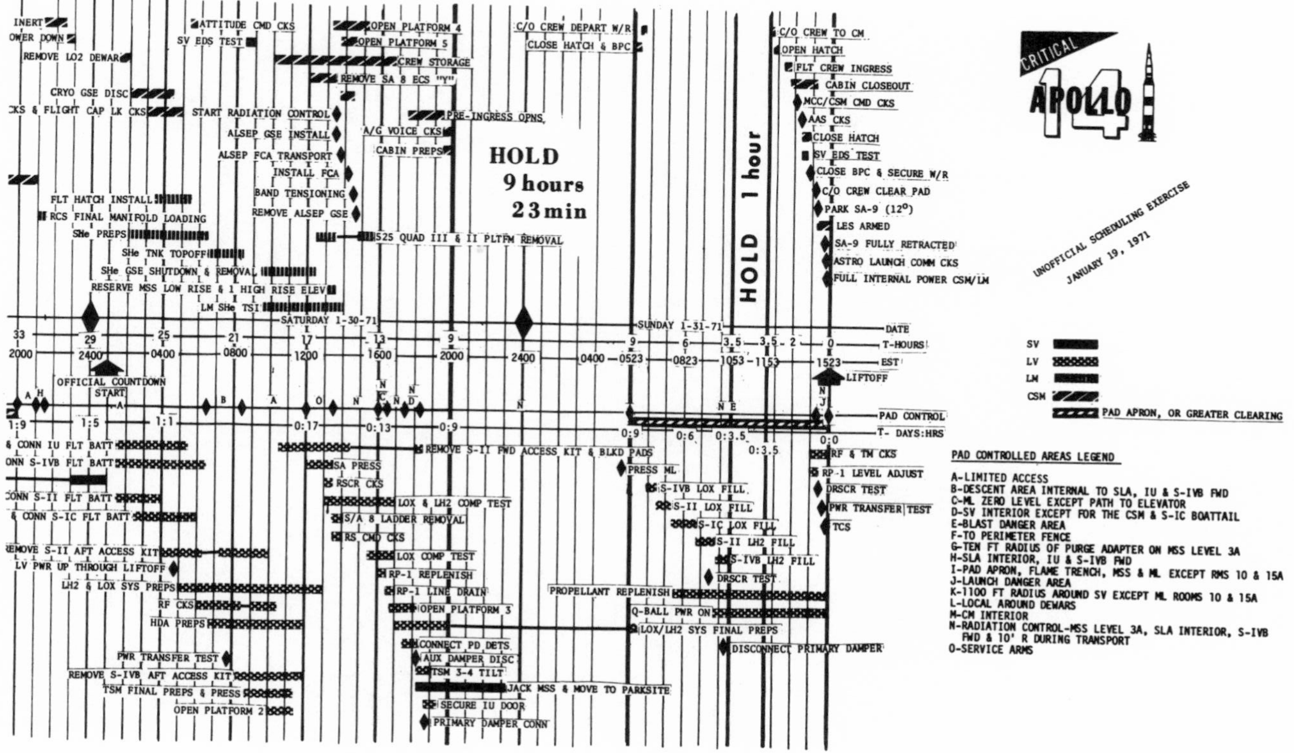
INERT
OWER DOWN
REMOVE LO2 DEWAR
CRYO GSE DISC
CKS & FLIGHT CAP LK CKS
START RADIATION CONTROL
ALSEP GSE INSTALL
ALSEP FCA TRANSPORT
INSTALL FCA
BAND TENSIONING
REMOVE ALSEP GSE
FLT HATCH INSTALL
RCS FINAL MANIFOLD LOADING
SHe PREPS
SHe TNK TOPOFF
SHe GSE SHUTDOWN & REMOVAL
RESERVE MSS LOW RISE & 1 HIGH RISE ELEV
LM SHe TSI
ATTITUDE CMD CKS
SV EDS TEST
OPEN PLATFORM 4
OPEN PLATFORM 5
CREW STORAGE
REMOVE SA 8 ECS "Y"
PRE-INGRESS OPNS
A/G VOICE CKS
CABIN PREPS
525 QUAD III & II PLTFM REMOVAL
C/O CREW DEPART W/R
CLOSE HATCH & BPC
HOLD 9 hours 23 min
HOLD 1 hour
C/O CREW TO CM
OPEN HATCH
FLT CREW INGRESS
CABIN CLOSEOUT
MCC/CSM CMD CKS
AAS CKS
CLOSE HATCH
SV EDS TEST
CLOSE BPC & SECURE W/R
C/O CREW CLEAR PAD
PARK SA-9 (12°)
LES ARMED
SA-9 FULLY RETRACTED
ASTRO LAUNCH COMM CKS
FULL INTERNAL POWER CSM/LM
CRITICAL
APOLLO 14
UNOFFICIAL SCHEDULING EXERCISE
JANUARY 19, 1971
SATURDAY 1-30-71
SUNDAY 1-31-71
DATE
33 29 25 21 17 13 9 9 6 3.5 3.5 2 0
T-HOURS
2000 2400 0400 0800 1200 1600 2000 2400 0400 0523 0823 1053 1153 1523
EST
OFFICIAL COUNTDOWN START
LIFTOFF
PAD CONTROL
1:9 1:5 1:1 0:17 0:13 0:9 0:9 0:6 0:3.5 0:3.5 0:0
T- DAYS:HRS
& CONN IU FLT BATT
ONN S-IVB FLT BATT
CONN S-II FLT BATT
& CONN S-IC FLT BATT
REMOVE S-II AFT ACCESS KIT
LV PWR UP THROUGH LIFTOFF
LH2 & LOX SYS PREPS
RF CKS
HDA PREPS
PWR TRANSFER TEST
REMOVE S-IVB AFT ACCESS KIT
TSM FINAL PREPS & PRESS
OPEN PLATFORM 2
REMOVE S-II FWD ACCESS KIT & BLKD PADS
SA PRESS
RSCR CKS
LOX & LH2 COMP TEST
S/A 8 LADDER REMOVAL
RS CMD CKS
LOX COMP TEST
RP-1 REPLENISH
RP-1 LINE DRAIN
OPEN PLATFORM 3
CONNECT PD DETS
AUX DAMPER DISC
TSM 3-4 TILT
JACK MSS & MOVE TO PARKSITE
SECURE IU DOOR
PRIMARY DAMPER CONN
PRESS ML
S-IVB LOX FILL
S-II LOX FILL
S-IC LOX FILL
S-II LH2 FILL
S-IVB LH2 FILL
DRSCR TEST
PROPELLANT REPLENISH
Q-BALL PWR ON
LOX/LH2 SYS FINAL PREPS
DISCONNECT PRIMARY DAMPER
RF & TM CKS
RP-1 LEVEL ADJUST
DRSCR TEST
PWR TRANSFER TEST
TCS
SV
LV
LM
CSM
PAD APRON, OR GREATER CLEARING
PAD CONTROLLED AREAS LEGEND
A-LIMITED ACCESS
B-DESCENT AREA INTERNAL TO SLA, IU & S-IVB FWD
C-ML ZERO LEVEL EXCEPT PATH TO ELEVATOR
D-SV INTERIOR EXCEPT FOR THE CSM & S-IC BOATTAIL
E-BLAST DANGER AREA
F-TO PERIMETER FENCE
G-TEN FT RADIUS OF PURGE ADAPTER ON MSS LEVEL 3A
H-SLA INTERIOR, IU & S-IVB FWD
I-PAD APRON, FLAME TRENCH, MSS & ML EXCEPT RMS 10 & 15A
J-LAUNCH DANGER AREA
K-1100 FT RADIUS AROUND SV EXCEPT ML ROOMS 10 & 15A
L-LOCAL AROUND DEWARS
M-CM INTERIOR
N-RADIATION CONTROL-MSS LEVEL 3A, SLA INTERIOR, S-IVB FWD & 10' R DURING TRANSPORT
O-SERVICE ARMS

Source Notes

Notes, bibliography, and index encompass both *Gateway to the Moon* and its companion volume, *Moon Launch!*

Chapter 1: The First Steps

1. Army Ballistic Missile Agency (hereafter ABMA), *Juno V Space Vehicle Development Program (Phase I), Booster Feasibility Demonstration*, by H. H. Koelle et al., report DSP-TM-10-58 (Redstone Arsenal, AL, 13 Oct. 1958), p. 1; Oswald Lange, "Development of the Saturn Space Vehicle," in *From Peenemünde to Outer Space*, ed. Ernst Stuhlinger et al. (Huntsville, AL: Marshall Space Flight Center, 1962), p. 6. Probably the best source for an understanding of the complex developments of the American space program during the late 1950s is *The History of Rocket Technology*, ed. Eugene Emme (Detroit: Wayne State Press, 1964). Maj. Gen. John B. Medaris gives an interesting, albeit one-sided, account of ABMA's activities during this period in *Countdown for Decision* (New York: G. P. Putnam's Sons, 1960).
2. R. Cargill Hall, *Project Ranger: A Chronology* (Pasadena: Jet Propulsion Laboratory, California Institute of Technology, 1971), pp. 48–52; ABMA, *Juno V Development*, pp. 1–2.
3. ABMA, *Juno V Development*, pp. 1–2; Lange, "Saturn Space Vehicle," p. 6; Medaris, *Countdown*, pp. 151–241, passim.
4. Memo of agreement, Advanced Research Projects Agency and Army Ordnance Missile Command, "High Thrust Booster Program Using Clustered Engines," 23 Sept. 1958, printed in ABMA, *Juno V Development*, Appendix A; NASA, *Historical Pocket Statistics*, July 1972 (Washington, 1972), p. E-4. The tenfold increase in the cost of the Saturn I program can be explained in large part by the changing purposes of the program. Initially the Defense Department viewed it as a four-vehicle test series relying extensively on available engines, fuel tanks, and tooling machinery. The program evolved into something quite different, requiring much unanticipated construction for launch vehicles and facilities. Warren G. Hunter, ARPA Coordinator, SSEL, to Hans Hueter, Dir., SSEL, "Juno V (Saturn) Program," 3 Oct. 1958. Unless specified otherwise, manuscript sources are in KSC Archives.
5. ABMA, *Juno V Development*, pp. 7–11, 19–20, 25–27, 47–51; ABMA, *Juno V Transportation Feasibility Study*, by J. S. Hamilton, J. L. Fuller, and P. F. Keyes, report DLMT-TM-58-58 (Redstone Arsenal, AL, 5 Jan. 1959), pp. 1–4; ABMA, *Juno V Space Vehicle Development Program* (Status Report—15 Nov. 1958), by H. H. Koelle et al., report DSP-TM-11-58 (Redstone Arsenal, AL, 15 Nov. 1958), pp. 2–3, 19–20.
6. NASA Special Committee on Space Technology, *Recommendations Regarding a National Civil Space Program* (Stever Committee Report), Washington, 28 Oct. 1958; ABMA, *Juno V Development*, pp. 19–20, 65; Army Ordnance Missile Command (hereafter cited as AOMC), *Saturn Systems Study*, by H. H. Koelle, F. L. Williams, and W. C. Huber, report DSP-TM-1-59 (Redstone Arsenal, AL, 13 Mar. 1959), pp. 16–19, 61–63, 183–89; House Committee on Science and Astronautics, *Equatorial Launch Sites—Mobile Sea Launch Capability*, report 710, 87th Cong., 1st sess., 12 July 1961, pp. 1–5 (see hearings of same committee and topic, 15–16 May 1961, for fuller discussion); Mrazek interview. The debate over the merits of an equatorial launch site or a mobile sea launch capability continued for several years with congressional hearings in the spring of 1961. Vice Adm. John T. Hayward was a leading advocate of shipboard launches.
7. Missile Firing Laboratory, "Project Saturn, Facilities for Launch Site," n.d.

8. "Champagne Flight," *Spaceport News* 2 (18 July 1963): 3. For other details of this first attempt, see L. B. Taylor, *Liftoff: The Story of America's Spaceport* (New York: E. P. Dutton & Co., 1968), pp. 42–44.
9. House Committee on Science and Astronautics, *Management and Operation of the Atlantic Missile Range*, 86th Cong., 2d sess., 5 July 1960, pp. 1–2.
10. Zeiler interview, 24 Aug. 1972.
11. H. H. Koelle, ed., *Handbook of Astronautical Engineering* (New York: McGraw-Hill Book Co., 1961), pp. 28-8 through 28-10.
12. Deese interview, 16 Mar. 1973.
13. E. R. Bramlitt, *History of Canaveral District, 1950–1971* (So. Atlantic Dist. U.S. Corps of Engineers, 1971), pp. 17–21.
14. AOMC, *Saturn System Study*, pp. 4–5, 21; AOMC, *Saturn System Study II*, report DSP-TM-13-59 (Redstone Arsenal, AL, 13 Nov. 1959), pp. 1–2.
15. Dept. of the Army, *Project Horizon, A U.S. Army Study for the Establishment of a Lunar Military Outpost*, I, *Summary* (Redstone Arsenal, AL, 8 June 1959).
16. Minutes, NASA Research Steering Committee on Manned Space Flight (the Goett Committee), 25–26 May 1959, pp. 2–10, NASA Hq. History Office. The authors wish to thank historian Thomas Ray of NASA Hq. for assistance on this subject.
17. NASA Hq. working draft, "Long Range Objectives," 1 June 1959, NASA Hq. History Office.
18. Medaris, *Countdown*, pp. 241–47; Eugene Emme, "Historical Perspectives on Apollo," NASA Historical Note 75 (Oct. 1967), pp. 14–17.
19. Medaris, *Countdown*, pp. 247–69.
20. Emme, "Historical Perspectives," p. 17.
21. Medaris, *Countdown*, pp. 262–66; ABMA, *Saturn System Study II*, pp. 1–2.
22. AOMC, *Saturn System Study II*, pp. 5–10; *Report to the Administrator, NASA, on Saturn Development Plan by Saturn Vehicle Team*, 15 Dec. 1959, p. 1.
23. *Report on Saturn*, pp. 4, 7, 8, and table III.
24. Emme, "Historical Perspectives," p. 18; Robert L. Rosholt, *An Administrative History of NASA, 1958-1963*, NASA SP-4101 (Washington, 1966), p. 114.

Chapter 2: Launch Complex 34

1. Chief, MFL, to Chief, Ops. Off., Guided Missile Development Div. (GMDD), "Manning Charts," 5 Jan. 1953; Chief, MFL, to Dep. Chief, GMDD, "Official List of Operations Personnel for Missile #1"; Launch Operations Directorate (hereafter cited as LOD), "Special Report on Support Operations at the AMR by LOD," 21 Dec. 1960, part 5.
2. ABMA, *Juno V Development*, p. 47; Georg von Tiesenhausen, "Saturn Ground Support and Operations," *Astronautics* 5 (Dec. 1960): 30.
3. Debus to C.O., Atlantic Missile Range (hereafter cited as AMR), "Juno V Program," 1 Oct. 1958; Robert F. Heiser, Technical Asst., Off. of the Dir., MFL, memo for record, "Juno V," 26 Sept. 1958.
4. Deese to Debus, priority TWX, "Feasibility Study and/or Criteria for a Launch Site at AMR for a Clustered First Stage of Juno V Project," 8 Oct. 1958; Koelle, ed., *Handbook of Astronautical Engineering*, pp. 28-1 to 28-10; MFL, *Juno V (Saturn) Heavy Missile Launch Facility, 1st Phase Request, 2d Phase Estimate*, by R. P. Dodd and J. H. Deese, 14 Feb. 1959, pp. 2–3.
5. Deese to Debus, "Feasibility Study for a Launch Site"; Warren G. Hunter, ARPA Coordinator, SSEL, memo, "Meeting at MFL, CCMTA on Juno V Launch Complex," 10 Nov. 1958; Pan American Aviation, "Juno V Program Siting Study," 24 Oct. 1958, pp. 1–3.

6. Glen W. Stover, Chief, Facilities Br., AMR, Army Field Off., memo for record, "Criteria Contract, Juno V Facilities," 10 Nov. 1958; Maurice H. Connell and Assoc., *Heavy Missile Launch Facility Criteria* (Miami, FL, 15 Mar. 1959).
7. ABMA, *Juno V Development*, p. 55; LOD, "Complex 34 Safety Plan for SA-1 Launch," 24 Oct. 1961, p. 2; Porcher interview.
8. MFL, *Juno V (Saturn) Facility*; Connell and Assoc., *Launch Facility Criteria*; Sparkman interview, 13 June 1974. For detailed descriptions of the Saturn C-1 Launch Complex with its ground support equipment, see Marshall Space Flight Center (hereafter cited as MSFC), *Saturn SA-1 Vehicle Data Book*, report MTP-M-S&M-E-61-3 (Huntsville, AL, 26 June 1961), pp. 133–65, and MSFC, *Project Saturn C-1, C-2 Comparison*, report M-MS-G-113-60 (Huntsville, AL, 16 Nov. 1960), pp. 33–47, 123–290.
9. Davis interview; Walter interview, 21 Sept. 1973.
10. Von Tiesenhausen interview, 20 July 1973; Buchanan interview, 22 Sept. 1972.
11. Davis interview; Koelle, ed., *Handbook of Astronautical Engineering*, p. 28-44.
12. MSFC, *C-1, C-2 Comparison*, pp. 167–83; Wasileski interview, 14 Dec. 1972.
13. Zeiler interview, 24 Aug. 1972; Connell and Assoc., *Launch Facility Criteria*, p. 2-9.
14. Zeiler, Chief of Mechanical Br., MFL, to Debus, "Servicing Equipment for Juno V on Launch Site," 24 Nov. 1958.
15. Connell and Assoc., *Heavy Missile Service Structure Criteria*; Army Engineer Dist., Jacksonville Corps of Engineers, "Minutes of Conference on Review of Criteria for Saturn Facilities," 7 Apr. 1959.
16. Robert E. Linstrom, DOD Saturn Project Engineer, memo for record, "Summary of Fifth Saturn Meeting," 1 Apr. 1959; Debus's Daily Journal (hereafter cited as DDJ), 13 Apr. 1959, KSC Director's Office; Debus, memo for record, "Meeting on Saturn Service Structure," 9 Apr. 1959.
17. Debus, memo for record, "Meeting on Saturn Service Structure," 9 Apr. 1959; Connell and Assoc., *Alternate "I" Heavy Missile Service Structure*, 24 Apr. 1959, R. P. Dodd's personal papers.
18. MSFC, *SA-1 Data Book*, pp. 131–33.
19. DDJ, 13 Apr., 22 June, 11 July 1959; "Corps of Engineers Contract Tabulations for LC-34, LC-37, and MILA Facilities," p. 3.
20. "Vibroflotation," Vibroflotation Foundation Co., Div. of Litton Industries, Pittsburgh, PA, undated pamphlet.
21. J. P. Claybourne to Robert Heiser, priority TWX, "Saturn Launch Facility Costs," 7 Sept. 1960.
22. J. P. Claybourne, memo for record, "Cost of Saturn Launch Facilities and Ground Support Equipment," 13 Sept. 1960.
23. C. C. Parker, Technical Program Dir., memo for record, "Saturn Launch Facilities at AMR," 3 Dec. 1959; Connell and Assoc., *Siting Study and Recommendation, Saturn Staging Building and Service Structure, Complex 34 AFMTC*, Jan. 1960, pp. 2–3; MSFC, *C-1, C-2 Comparison*, p. 35; Debus to Rees, "Additional Saturn Launch Complex," 29 Jan. 1960; ABMA, *A Committee Study of Blast Potentials at the Saturn Launch Site*, by Charles J. Hall, report DHM-TM-9-60 (Redstone Arsenal, AL, Feb. 1960).
24. MSFC, *C-1, C-2 Comparison*, pp. 100–102; Poppel interview, 12 Feb. 1973.
25. MSFC, *C-1, C-2 Comparison*, pp. 123–63, 206–20, 244–67; Poppel, memo, "Launch Equipment Installation at Complex 34, AMR," 27 May 1960; Sparkman interview, 15 Dec. 1972; Wasileski interview, 14 Dec. 1972.
26. Davis interview, 2 Feb. 1972; Poppel, memo for record, "Contract NAS8-46 Extension," 29 Dec. 1960; Poppel to Petrone, "Documentation of Major Facility and/or GSE Changes for Project Saturn," 23 Feb. 1961, p. 5.
27. C. C. Parker, Chief, Ops. Off., MSFC, "Saturn FY-61 LOD Budget," 29 July 1960, encl. 3.
28. Debus, memo, "Priority of Effort on Saturn Launch Facilities," 8 July 1960.

29. R. P. Dodd to C. C. Parker, 8 Aug. 1960.
30. Debus to Wernher von Braun, "Labor Situation," 21 Sept. 1960.
31. George V. Hanna, "Chronology of Work Stoppages and Related Events, KSC/NASA and AFETR through July 1965," KSC historical report (KSC, FL, Oct. 1965), pp. 26–27.
32. Ibid., pp. 30–35; DDJ, 15 and 28 Nov., 5 and 22 Dec. 1960.

Chapter 3: Launching the First Saturn I Booster

1. W. R. McMurran, ed., "The Evolution of Electronic Tracking, Optical, Telemetry, and Command Systems at the Kennedy Space Center," mimeographed paper (KSC, 17 Apr. 1973), fig. 2; MSFC, *Saturn SA-1 Flight Evaluation*, report MPR-SAT-WF-61-8 (Huntsville, AL, 14 Dec. 1961), p. 235. The Saturn Flight Evaluation Working Group at MSFC published reports on all the Saturn C-1 launches. See also MSFC, *Results of the First Saturn I Launch Vehicle Test Flight, SA-1*, report MPR-SAT-64-14 (Huntsville, AL, 27 Apr. 1964) which superseded the above report, and MSFC, *Results of the Saturn I Launch Vehicle Test Flights*, report MPR-SAT-FE-66-9 (Huntsville, AL, 9 Dec. 1966).
2. MSFC, *C-1, C-2 Comparison*, pp. 3–7. See pp. 68, 78–82 for Long-Range Program (Sloop Committee Report) of Sept. 1960.
3. Oswald H. Lange, "Saturn Program Review," 27 Jan. 1961; Akens, *Saturn Illustrated Chronology*, pp. 13, 19.
4. F. A. Speer, "Saturn I Flight Test Evaluation," American Institute of Aeronautics and Astronautics paper 64-322 given at Washington, D.C., 29 June–2 July 1964, p. 2; ARINC Research Corp., *Reliability Study of Saturn SA-3 Pre-Launch Operations*, by Arthur W. Green et al., publication 247-1-399 (Washington, 3 Jan. 1963), pp. 2-21 through 2-23.
5. ABMA, *Organizational Manual* (Redstone Arsenal, AL, 5 Mar. 1959), Sec. 530, pp. 13–15, copy available in Historical Div., Sec. of General Staff, Army Missile Div., Redstone Arsenal, AL; Russell interview.
6. Moser interview, 30 Mar. 1973.
7. Grady Williams interview.
8. Debus to Joseph Shea, "Principles of Operations of the MSFC Firing Team at Cape Canaveral," 22 Mar. 1962, Debus papers, KSC Archives.
9. Ibid.
10. Ibid.
11. Debus to Committee for LOD Scheduling and Test Procedures, "Day-by-Day Test Schedule for Saturn SA-1," 16 Mar. 1961, ibid.; DDJ, 14 Mar. 1961.
12. Moser interview, 30 Mar. 1973; DDJ, 17 Apr. 1961.
13. MSFC, *Catalog of Systems Tests for Saturn S-1 Stage*, pp. III-46–III-54 (Huntsville, AL, 1 Sept. 1961), Moser papers, Federal Archives and Records Center, East Point, GA, accession 68A1230, boxes 436257, 436259.
14. "Launch Facilities and Support Equipment Office [hereafter cited as LFSEO] Monthly Progress Report," 12 June 1961, p. 2; "LFSEO Monthly Progress Report," 13 July 1961, pp. 2–3.
15. DDJ, 11, 12 May 1961; "LFSEO Monthly Progress Report," 12 June 1961, p. 3.
16. Zeiler to Debus, "Work Statement," 14 July 1961.
17. "Saturn SA-1 Schedule," 15 Aug. 1961, Moser papers, Federal Archives and Records Center, East Point, GA, accession 68A1230, boxes 436257, 436259.
18. DDJ, 24 Mar., 26 Apr. 1961; Akens, *Saturn Illustrated Chronology*, p. 18; Georg von Tiesenhausen, "Ground Equipment to Support the Saturn Vehicle," paper 1425-60 presented at the 15th annual meeting of the American Rocket Society, Washington, D.C., 5–8 Dec. 1960, pp. 2–3.

19. "LFSEO Monthly Progress Report," 13 July 1961, p. 4; Akens, *Saturn Illustrated Chronology*, pp. 21, 26–27.
20. MSFC, *SA-1 Vehicle Data Book*, pp. 123–30.
21. Karl L. Heimburg to MSFC Dep. Dir. for R & D, "Water Route for NASA Vessels to Cape Canaveral," 9 Feb. 1962, attached to a response from Debus, 14 Feb. 1962, in Debus papers.
22. MSFC, *SA-1 Flight Evaluation*, p. 7; Akens, *Saturn Illustrated Chronology*, p. 26; Crunk interview; Zeiler interview, 23 July 1973; von Tiesenhausen, "Equipment to Support the Saturn," pp. 2–3.
23. MSFC, *SA-1 Flight Evaluation*, p. 7; "LOD Daily Journal," 27 July 1961; "Saturn SA-1 Schedule," 15 Aug. 1961.
24. MSFC, *SA-1 Flight Evaluation*, p. 8; MSFC, "Saturn Quarterly Progress Report," July–Sept. 1961, p. 1; "Saturn Schedule," 15 Aug. 1961; interviews with Newall, Marsh, and Humphrey.
25. Grady Williams interview.
26. MSFC, *SA-1 Vehicle Data Book*, pp. 74–81.
27. Interviews with Edwards and Glaser.
28. White interview; MSFC, *Consolidated Instrumentation Plan for Saturn Vehicle SA-1*, by Ralph T. Gwinn and Kenneth J. Dean, report MTP-LOD-61-36.2a (Huntsville, AL, 25 Oct. 1961).
29. White interview.
30. MSFC, *SA-1 Flight Evaluation*, p. 7.
31. Moser interview, 30 Mar. 1973.
32. Ibid.; "Saturn SA-1 Schedule," 15 Aug. 1961; MSFC, *SA-1 Flight Evaluation*, pp. 8, 200–202.
33. Moser interview, 30 Mar. 1973; "Saturn SA-1 Schedule"; MSFC, *SA-1 Flight Evaluation*, pp. 7–9; LOD, "Saturn Test Procedures, SA-1 G & C Overall Test #3," Moser papers, Federal Archives and Records Center, East Point, GA, accession 68A1230, boxes 436257, 436259.
34. Moser interview, 30 Mar. 1973; "Saturn SA-1 Schedule"; MSFC, "SA-1 Flight Evaluation," pp. 8–9; George Alexander, "Telemetry Data Confirms Saturn Success," *Aviation Week and Space Technology*, 6 Nov. 1961, pp. 30–32.
35. "Saturn Test Procedures: SA-1 Mechanical Office L – 1 Day Prelaunch Preparations," Moser papers; MSFC, *SA-1 Flight Evaluation*, pp. 9–10; interview with Chester Wasileski by Benson, 14 Dec. 1972; Pantoliano interview.
36. MSFC, *Launch Countdown Saturn Vehicle SA-1*, report MIP-LOD-61-35-2 (Huntsville, AL, 3 Oct. 1961), pp. 9–15, Moser papers.
37. MSFC, *SA-1 Flight Evaluation*, p. 10; MSFC, *Countdown SA-1*.
38. MSFC, *Countdown SA-1*, pp. 20–23.
39. MSFC, *SA-1 Flight Evaluation*, pp. 11–12; MSFC, *Countdown SA-1*, pp. 28–30; LOD, *Saturn Test Procedures: Set Up LO_2 Facility for Fast Fill (T – 100)*, procedure LOD-M-703; LOD, *Saturn Test Procedures: Fast Fill LO_2 Loading (T – 60)*, procedure LOD-M-704.
40. MSFC, *Countdown SA-1*, pp. 36–39; Alexander, "Telemetry Confirms Success," p. 31; "Emergency Procedures SA-1," LOD Networks Group, pp. 2–3, Moser papers.
41. Richard Austin Smith, "Canaveral, Industry's Trial by Fire," *Fortune*, June 1962, pp. 204, 206; "Saturnalia at Canaveral," *Newsweek*, 6 Nov. 1961, p. 64; *Miami Herald*, 28 Oct. 1961, p. 1; "Saturn's Success," *Time*, 3 Nov. 1961, p. 15.
42. *New York Times*, 28 Oct. 1961, pp. 1, 9; *Miami Herald*, 28 Oct. 1961, p. 1. The MSFC news release on the SA-1 launch, dated 1 Nov. 1961, included a paragraph on the sound effect.
43. *Miami Herald*, 28 Oct. 1961, p. 1 (UPI release).
44. *New York Times*, 27 Oct. 1961, p. 1; 28 Oct. 1961, pp. 1, 9.

Chapter 4: Origins of the Mobile Moonport

1. House Committee on Science and Astronautics, *Report on Cape Canaveral Inspection*, 86th Cong., 2d sess., 27 June 1960, p. 1.
2. House Committee on Science and Astronautics, *Management and Operation of the Atlantic Missile Range*, 86th Cong., 2d sess., 5 July 1960, p. 4.
3. Francis L. Williams interview.
4. David S. Akens, *Saturn Illustrated Chronology* (MSFC, Jan. 1971), pp. 7–8; J. P. Claybourne, Saturn Project Office, memo, "Saturn C-2 Configurations," 6 July 1960; NASA, "A Plan for Manned Lunar Landing" (Low Committee report), 7 Feb. 1961, pp. 7–13, figs. 4, 7, NASA Hq. History Office.
5. Interview with Debus by Benson, 16 May 1972; H. H. Koelle, "Missiles and Space Systems," *Astronautics* 7 (Nov. 1962): 29–37.
6. Claybourne, "Saturn C-2 Configurations," 6 July 1960.
7. DDJ, 24 Apr. 1961.
8. Livingston Wever, Support Instrumentation Div., to Porcher, Facilities Br., Army Test Off., AFMTC, "Addendum to Scheme for Offshore Launching Platform for Space Vehicles," Mar. 1960; Wyle Laboratories, *Sonic and Vibration Environments for Ground Facilities—A Design Manual*, by L. C. Sutherland, report WR68-2, 1968, pp. 5-21, 10-2.
9. Nelson M. Parry, Army Test Off., AFMTC, "Land Developments for Missile Range Installations (Preliminary Notes)," 30 Dec. 1958, p. 3; Nelson Parry to Porcher, "Offshore Launch Platform for Heavy Space Vehicles," 6 Apr. 1960.
10. Porcher interview.
11. Von Tiesenhausen interview, 29 Mar. 1972; Sparks interview; von Tiesenhausen, "Vorversuche für Project Schwimmiweste," Electromechanische Werke Peenemünde, 11 Sept. 1944, typescript, von Tiesenhausen's private papers.
12. Poppel to Debus, "Offshore Complex," 6 May 1960.
13. MSFC, *Preliminary Feasibility Study on Offshore Launch Facilities for Space Vehicles*, by O. L. Sparks, report IN-LOD-DL-1-60 interim (Huntsville, AL, 29 July 1960).
14. DDJ, 28 Feb., 2 Mar. 1961.
15. Debus to Col. Asa Gibbs, Chief, NASA Test Support Off., AFMTC, "Future Saturn Launch Sites at Cape Canaveral (SR 2953)," 14 Feb. 1961.
16. Parry to Charles J. Hall, "Future Launch Sites at Cape Canaveral," 9 Mar. 1961.
17. Debus to Gibbs, "Siting for Third Saturn Launch Complex at AMR (SR 2953)," n.d.; Debus to Gibbs, "Siting of Fourth and Fifth Saturn Launch Complexes at AMR," 6 Apr. 1961. The siting requests were canceled 22 Sept. 1961, after the MILA purchase had changed the situation.
18. Future Launch Systems Study Off., LOD, "Progress Report," Jan. 1961; DDJ, 6 Feb. 1961.
19. Debus to Poppel, "Offshore Launch Facility Study," 4 Apr. 1961.
20. Nelson M. Parry, "Land Development (Offshore and Semi-Offshore Launch Sites)," 14 Apr. 1961; Deese interview, 10 May 1972; DDJ, 12 May 1961.
21. Petrone interview.
22. Joint Air Force-NASA Hazards Analysis Board, AFMTC, *Safety and Design Considerations for Static Test and Launch of Large Space Vehicles*, 1 June 1961, p. I-B-1.
23. Debus to von Braun, "Offshore Facilities Studies," 24 May 1961.
24. "Death on Old Shaky," *Time*, 27 Jan. 1961, pp. 15–16; von Tiesenhausen interview, 29 Mar. 1972; Debus interview, 16 May 1972.
25. Zeiler interview, 11 Aug. 1970.
26. Memo for record, "Phoenix Study Program," 3 July 1961, p. 2; Aerospace Corp. News Release, "Titan III Management and Technology to Be Model for Future Systems" (Los Angeles, June 1965), p. 2.

27. KSC Public Affairs Office, *Kennedy Space Center Story* (Kennedy Space Center, FL, Dec. 1972), p. 5. No document is cited for the statement that the three men met 16 days after JFK's inaugural. This would place the meeting on Sunday, 5 Feb. The Daily Journal for 6 Feb. mentions a Saturday meeting of the three men.
28. Duren interview, 16 May 1972.
29. DDJ, 22, 30, 31 Mar., 10 Apr. 1961.
30. MSFC, *Interim Report on Future Saturn Launch Facility Study*, by Olin K. Duren, report MIN-LOD-DL-1-61 (Huntsville, AL, 10 May 1961).
31. DDJ, 17, 20, 26 Apr. 1961.
32. Douglas Aircraft Co., *Saturn C-2 Operational Requirement Study*, prepared by J. Simmons, report SM-38771 (Santa Monica, CA, July 1961), p. 188.
33. Ibid., pp. 171–205; The Martin Co., *Saturn C-2 Operational Modes Study, Summary Report*, report ER-11816 (Baltimore, MD, June 1961), pp. 19–25, 46–47, 66–68.
34. John M. Logsdon, "NASA's Implementation of the Lunar Landing Decision," NASA HHN-81, Aug. 1969, typescript, pp. 1–6; Lunar Landing Working Group (Low Committee), "A Plan for Manned Lunar Landing," 7 Feb. 1961.
35. House Committee on Science and Astronautics, *Hearings, 1962 NASA Authorization*, 87th Cong., 1st sess., 23 Mar. 1961, pt. 1, p. 177.
36. Ibid.; Akens, *Saturn Illustrated Chronology*, p. 4.
37. Akens, *Saturn Illustrated Chronology*, pp. 17, 19, 22; Logsdon, "NASA's Implementation," p. 6; House, *1962 NASA Authorization*, pp. 170–77.
38. DDJ; Debus to von Braun, "Offshore Launch Facilities," 24 May 1961; MSFC, Future Projects Off., "Procurement Request," 26 Apr. 1961.
39. Logsdon, "NASA's Implementation," pp. 8–18.
40. DDJ, 6, 26 June 1961. The Fleming Master Flight Plan called for 167 flights prior to the first lunar landing, but this included launchings of Atlas, Agena, Centaur, Saturn C-1, Saturn C-3, and Nova rockets. Fourteen C-1s and 24 C-3s were to be launched in 1965–1966.
41. DDJ, 6, 20 June 1961.
42. NASA, *A Feasible Approach for an Early Manned Lunar Landing* (Fleming Committee Report), 16 June 1961, p. 26.
43. Robert C. Seamans, Jr., to Maj. Gen. Leighton I. Davis and Debus, "National Space Program Range Facilities and Resources Planning," 23 June 1961.
44. Seamans to Davis and Debus, "National Space Program Range Facilities and Resources Planning," 30 June 1961.
45. MSFC, LFSEO, *Preliminary Concepts of Launch Facilities for Manned Lunar Landing Program*, report MIN-LOD-DL-3-61, 1 Aug. 1961, pp. 4–6.
46. NASA-DOD, *Joint Report on Facilities and Resources Required at Launch Site to Support NASA Manned Lunar Landing Program* (hereafter cited as Debus-Davis Report), 31 July 1961, p. 3; Owens interview, 12 Apr. 1972; Petrone interview, 25 May 1972; Clark interview.
47. Petrone interview; Clark interview.
48. Petrone and Leonard Shapiro, "Guideline for Preparation of NASA Manned Lunar Landing Project Report," 7 July 1961; KSC Biographies, in KSC Archives.
49. Debus-Davis Report, passim; Owens interview, 12 Apr. 1972.
50. MSFC, *Interim Report on Future Saturn Launch*, p. 16.
51. Zeiler interview, 11 July 1972; von Tiesenhausen interview, 29 Mar. 1972.
52. Debus-Davis Report, pp. B-1 through B-7.
53. Ibid., pp. B-9, B-10.
54. The authors are indebted to Rocco Petrone for this idea: interview of 25 May 1972 and remarks delivered by Petrone to Apollo History Workshop, NASA Hq., 19–21 May 1971.

Chapter 5: Acquiring a Launch Site

1. DDJ, 26 Apr. 1961.
2. Joint Air Force-NASA Hazards Analysis Bd., *Safety and Design Considerations for Static Test and Launch of Large Space Vehicles*, 1 June 1961, part I, "Hazards Analysis," p. I-A-1.
3. Ibid., part I, pp. I-D-3, I-B-1, and I-B-2.
4. AFMTC, NASA, & Pan American, *Preliminary Field Report, Cumberland Island & Vicinity for Nova Launch Facilities*, 13–14 June 1961, pp. 22–23; Hal Taylor, "Big Moon Booster Decisions Looming," *Missiles and Rockets*, 28 Aug. 1961, p. 14.
5. Charles J. Hall to R. P. Dodd, "Land Development for Future Launch Sites," 12 May 1961.
6. Debus, memo for the record, "Land Acquisition Book II," 11 June 1961.
7. NASA, *A Feasible Approach for an Early Manned Lunar Landing* (the Fleming Report), 16 June 1961.
8. Roswell Gilpatric to the Secretaries of the Army, Navy, and Air Force, "National Space Program Facilities Planning," 16 June 1961.
9. *Air Force Systems Command News Review*, Mar. 1963.
10. Debus-Davis Report, reference b.
11. Petrone and Shapiro, memorandum of understanding, "Guideline for Preparation of NASA Manned Lunar Landing Project Report," 7 July 1961.
12. Gibbs interview.
13. Debus-Davis Report, pp. 19, 20.
14. Ibid., p. 20.
15. Ibid.
16. Ibid., pp. D-70, D-71.
17. Ibid., pp. D-19, D-20.
18. Gordon E. Dunn and Banner I. Miller, *Atlantic Hurricanes*, rev. ed. (Baton Rouge: LSU Press, 1960), pp. 266–67.
19. Bramlitt, *History of the Canaveral District*, p. 34.
20. Debus interview, 22 Aug. 1969; Owens interview, 12 Apr. 1972.
21. Debus interview, 16 May 1972.
22. DDJ, 27 July 1961.
23. *Washington Post*, 3 Aug. 1961.
24. Milton Rosen to Hugh Dryden and James Webb, "Selection of a Launch Site for the Manned Lunar Program," attached to a letter from Dryden to Webb, same subj., 18 Aug. 1961.
25. Hugh L. Dryden to Webb, "Selection of a Launch Site for the Manned Lunar Program," 18 Aug. 1961.
26. NASA release 61-189, "Manned Lunar Launch Site Selected," 24 Aug. 1961.
27. Interview with Gibbs by James Covington, 7 Aug. 1969.
28. Petrone interview.
29. "Agreement between DOD and NASA Relating to the Launch Site for the Manned Lunar Landing Program," signed by James Webb and Roswell Gilpatric, 24 Aug. 1961.
30. Ibid., p. 2.
31. NASA-LOD proposal, "Integrated Master Planning, Atlantic Missile and Space Operations Range," 22 Nov. 1961; Minutes of the meeting prepared by Raymond L. Clark, Asst. to the Director.
32. House Committee on Science and Astronautics, Subcommittee on Manned Space Flight, *Hearings, 1963 NASA Authorization*, 87th Cong., 2d sess., pt. 2, pp. 643–55.
33. Lloyd L. Behrendt, comp., *Development and Operation of the Atlantic Missile Range* (Patrick A.F.B., FL, 1963), pp. 63–64, Air Force Eastern Test Range Archives.

34. "Background Information on Agreement between DOD and NASA re: Management of the AMR of DOD and the Merritt Island Launch Area of NASA," prepared by Paul T. Cooper, Brig. Gen., USAF, Mar. 1963, p. 4.
35. Behrendt, *Development of AMR*, pp. 66–67.
36. Debus interview, 22 Aug. 1969; reiterated in an interview 16 May 1972.
37. DDJ, 14 July 1961.
38. Senate, *Amending the National Aeronautics and Space Administration Authorization Act for the Fiscal Year 1962*, report 863 to accompany Senate Bill 2481, 87th Cong., 1st sess., 1 Sept. 1961.
39. Seamans to Lt. Gen. W. K. Wilson, Chief, Corps of Engineers, U.S. Army, 21 Sept. 1961.
40. Senate Committee on Appropriations, 87th Cong., 1st sess., *Hearing, Second Supplemental Appropriation Bill for 1962*, p. 154; Arthur G. Procher, memo for record, "Land Acquisition for NASA Lunar Launch Facility," 26 Sept. 1961.
41. NASA Audit Div., "Review of Management Controls over Contract Modifications Executed by the Corps of Engineers, Launch Operations Center, Cocoa Beach, Florida," report Bu/LO-W 64-7, Washington, DC, 23 Sept. 1963, p. 3.
42. Real Estate Div., U.S. Army Engineer Dist., Jacksonville, FL, "Progress and Status, Real Estate Acquisition and Manned Lunar Landing Program Project," Cape Canaveral, FL, 6 June 1962.
43. Sollohub to Debus, 8 June 1962.
44. *Titusville Star-Advocate*, 17 Feb. 1962.
45. Morgan T. Nealy, Jr., Proj. Mgr., Real Estate Project Off., Titusville, FL, to LOC, 21 Mar. 1963.
46. Telephone interview with Faherty, 1 May 1972. The lady preferred to remain anonymous.
47. "List of Buildings Retained for Interim Use, NASA–Merritt Island Launch Area," in R. J. Pollock's files, KSC Maintenance Div.
48. AFMTC, "Cape Canaveral Missile Test Annex Development Plan," 12 Mar. 1962; AFMTC, "Status Report, NASA–Merritt Island Launch Area Development," Patrick A.F.B., 27 Sept. 1962; James Trainor, "Titan III Plan Awaits DOD Approval," *Missiles and Rockets*, 14 May 1962, p. 35.
49. Shriever to Seamans, 14 Mar. 1962.
50. NASA, "Management Council Meeting of 29 May 1962," corrections, dated 7 Mar. 1962.
51. Debus, memo for record, "Holmes-Shriever-Davis-Debus Meeting, Saturday, 17 Mar. 1962, on Titan III Siting," 19 Mar. 1962.
52. NASA, "Minutes of the Management Council Meeting of 27 Mar. 1962."
53. *Hearings, 1963 NASA Authorization*, pp. 634–35.
54. Ibid., pp. 641–55.
55. Ibid., p. 652.
56. Behrendt, *Development of AMR*, p. 69.
57. Davis to District Engineer, "Real Property Accountability for MLLP Area," in "Background Information and Agreement between DOD and NASA: Re: Management of the AMR of DOD and the Merritt Island Launch Area of NASA," Mar. 1963. Brig. Gen. Paul T. Cooper, USAF, stated that when the question of title arose between the Launch Operations Center and the AF Missile Test Center, the two staff elements agreed to bring the matter to the attention of higher levels by this letter of Gen. Davis. We have seen no NASA document suggesting that this letter arose from a mutual decision, although several NASA communications refer to Davis's letter (Debus to Holmes, 3 Apr. and 4 Apr. 1962; Webb to Gilpatric, 23 May 1962).
58. Petrone interview.
59. Seamans to Webb and Dryden, 13 Apr. 1962.
60. Webb to McNamara, 17 Apr. 1962.

61. *Missiles and Rockets*, 30 Apr. 1962, p. 75.
62. Bell to McNamara, 3 May 1962.
63. Jackson to McNamara, 21 May 1962. The 18 June 1962 issue of *Aviation Week and Space Technology* credited Rep. George Miller with a major role in securing NASA's autonomy on Merritt Island.
64. McNamara to Jackson, 24 May 1962.
65. Debus to Davis, 14 June 1962.
66. Subcommittee of the Senate Committee on Appropriations, *Hearings on H.R. 12711*, 87th Cong., 2d sess., pp. 861–904.
67. P.L. 87-584, *NASA Authorization Act for 1963*, sec. 5.
68. Webb to Gilpatric, 14 Aug. 1962, with enclosure, "Relationships and Responsibilities of the DOD & NASA at the AMR and the Manned Lunar Landing Program Area."
69. Cmdr., AFMTC, to Sec. of Defense, 4 Jan. 1963.
70. "Agreement, the Department of Defense and the National Aeronautics and Space Administration regarding Management of the Atlantic Missile Range of DOD and the Merritt Island Launch Area of NASA," signed by Robert S. McNamara and James E. Webb, 17 Jan. 1963.
71. The interim agreement that implemented the Webb-McNamara Agreement was signed on 10 May 1963 by Davis and Debus. Addenda covered joint instrumentation planning procedures, calibration equipment and services, chemical analysis, security and law enforcement, facilities management, visitor control, and other topics. Under the terms of addendum 5, LC-34, LC-36, LC-37, the Saturn Barge Canal, and other facilities were transferred by AFMTC to NASA-LOC. The 11 separate agreements were understood to be consolidated into a single document. Debus, "AFMTC-LOC Agreements Implementing the Webb-McNamara Agreement of 17 January 1963," 5 June 1963.
72. John P. Lacy to Center Dir., "Status of KSC Land Acquisition," 7 Dec. 1967.
73. Policicchio interview.
74. "List of Buildings to Be Retained for AFMTC Use in the Expanded Cape Area," Col. Colie Houck, 29 June 1962, to Mr. Owens; Joseph Hester, memo for record, "Disposition of House Trailers, Area 1," 5 Mar. 1962.
75. Debus to Peterson, 5 Mar. 1962; Bidgood to Debus, "Real Estate Procurement Policy," 22 Dec. 1961.
76. Senate Committee on Appropriations, *Hearings on H. R. 11038*, 87th Cong., 2d sess., 4 Apr. 1962, pp. 155–56; *Spaceport News* 2 (10 Oct. 1963): 8.
77. "Audit of Merritt Island Purchase," Office, Chief of Engineers, 16 Nov. 1971, in Corps of Engineers Files, per telephone conversation with Joseph Hester, 18 May 1972.
78. *Spaceport News* 11 (1 June 1972): 1, 4; O'Conner interview.
79. "Agreement between National Aeronautics and Space Administration and Bureau of Sport Fisheries and Wildlife for Use of Property at John F. Kennedy Space Center, NASA," 2 June 1972, signed by Willis H. Shapley for NASA and Nathaniel P. Reed for the Dept. of Interior.

Chapter 6: LC-39 Plans Take Shape

1. Logsdon, "NASA's Implementation," p. 22; Ivan D. Ertel and Mary Louise Morse, *The Apollo Spacecraft, A Chronology*, vol. 1 (NASA SP-4009, 1969), pp. 95, 108–109.
2. Logsdon, "NASA's Implementation," p. 34.
3. Ibid., pp. 40–44; Shea interview; Rosen interview, 14 Nov. 1969; Ertel and Morse, *Apollo Chronology*, 1: 118–20, 134.
4. Ertel and Morse, *Apollo Chronology*, 1: 131–34; Akens, *Saturn Chronology*, pp. 33–35.
5. James Grimwood and Barton Hacker, with Peter Vorzimmer, *Project Gemini, A Chronology* (NASA SP-4002, 1969), pp. 2–20.

6. Ertel and Morse, *Apollo Chronology*, 1: 111.
7. Ibid., p. 101.
8. Ibid., pp. 101–104, 121, 128; NASA release 66-15, "Apollo Spacecraft Contract," 21 Jan. 1966.
9. Martin Marietta Corp., *Saturn C-3 Launch Facilities Study Final Report*, vol. 1, *Selection of Optimum Concept*, report ER 12125-1 (Baltimore, Dec. 1969), p. 1.
10. Ibid., pp. vii, 1–11, 70–85; Martin Co., *Special Study Saturn Launch Facilities*, report ER 11996 (Baltimore, 17 Oct. 1961), pp. II-1 through II-5; calendar and schedule of events in O. K. Duren's private papers.
11. LFSEO, LOD, *A Preliminary Study of Launch Facility Requirements for the C-4* (Huntsville, AL, 27 Oct. 1961), p. 38.
12. George von Tiesenhausen, memo for record, "Launch Complex 39," 11 Oct. 1951.
13. Harvey F. Pierce, Maurice H. Connell & Assoc., Inc., to Debus, 21 Nov. 1961, Debus papers.
14. Poppel to Petrone, memo, "Saturn C-3/C-4 Study," 9 Oct. 1961.
15. Biographies, KSC Archives; Owens interview, 21 Nov. 1972.
16. MSFC, *Saturn Mobile (Canal) Concept Flame Deflector and Launcher/Transporter Emplacement Evaluation*, by George Walter, report MIN-LOD-DH-2-62 (Huntsville, AL, Feb. 1962).
17. Poppel to E. House, "Temporary Employment of Naval Architecture Consultant," 22 Dec. 1961; "LFSEO Monthly Progress Report," 15 Feb. 1962, p. 8.
18. Martin, *Saturn C-3 Study*, vol. 3, *Design Criteria for Launch Facilities*, report ER 12125-3, Dec. 1961, pp. 57–84.
19. "LFSEO Monthly Progress Report," 15 Feb. 1962, p. 8.
20. Debus to Petrone, "Transportation Proposals for Complex 39," 30 Jan. 1962; DDJ, 30 Jan. 1962.
21. Duren interview, 29 Mar. 1972; Zeiler interview, 24 Mar. 1972; private papers of Duren.
22. "LOD Weekly Notes," Petrone, 8 Feb. 1962; W. T. Clearman, Acting Sec., Heavy Vehicle Systems Off., memo for record, "Complex 39 Staff Meeting," 12 Mar. 1962, Petrone papers.
23. MSFC, *Appraisal of Transfer Modes for Saturn C-5 Mobile Systems as of 11 June 1962*, by Donald D. Buchanan and George W. Walter, report MIN-LOD-DH-9-62 (Huntsville, AL, 11 June 1962), pp. 5–8; Buchanan interview, 7 Nov. 1972; Walter interview, 7 Nov. 1972.
24. MSFC, *Transporter for Nova Track Design and Stresses*, by William H. Griffith, report NASA-MFSC-LOD-D; MSFC, *Appraisal of Transfer Modes*, p. 5; MSFC, *Saturn Mobile (Rail) Concept: An Examination of Rail Transfer Systems for a Launcher/Transporter*, by George W. Walter, report MIN-LOD-DH-3-62 (Huntsville, AL, 3 Apr. 1962).
25. Maurice H. Connell & Assoc., Inc., *Saturn C-5 Launch Facilities Complex 39: Study of Rail Systems for Vertical Transporter/Launcher Concept* (Huntsville, AL: MSFC, May 1962), p. 7.
26. MSFC, *Appraisal of Transfer Modes*, pp. 9–11; MSFC Weekly Notes, Debus to von Braun, 28 May 1962; Buchanan, memo for record, "Analysis by H. Pierce, 15 May 1962," Buchanan's private papers.
27. LOD Weekly Notes, Zeiler, 15 Feb. 1962; Poppel, Zeiler, Buchanan, and Duren made up the team.
28. Donald Buchanan, memo for record, "Launcher/Transporter Crawler Version," 23 Mar. 1962; Buchanan interview, 28 Nov. 1972; Duren interview, 29 Mar. 1972; MSFC, *Appraisal of Transfer Modes*, pp. 5, 8–9.
29. LOD Weekly Notes, Poppel, 16 May 1962; MSFC, *Appraisal of Transfer Modes*, p. 9; Buchanan, memo for record, "TDY at Bucyrus-Erie, South Milwaukee, Wisconsin," 16 Apr. 1962, Buchanan's private papers; Buchanan interview, 28 Nov. 1972; Buchanan, memo for record, "Analysis by H. Pierce, 15 May 1962," Buchanan's papers.

30. Army Corps of Engineers, Jacksonville Off., "Summary of Opinions Developed by the Jacksonville District Engineering Staff on Mobile Launch Concepts for the Advanced Saturn C-5 Vehicle," June 1962, in Buchanan's papers.
31. MSFC, *Appraisal of Transfer Modes*, p. 11.
32. William T. Clearman, Jr., memo, "Launch Operations Directorate Complex 39 Review," 18 Sept. 1962; E. M. Briel's notes, 12–13 June 1962.
33. Biographies, in KSC Archives; Claybourne interview; Clearman interview, 5 Jan. 1973.
34. Launch Operations Center (hereafter cited as LOC), "Summary of Conference with Members of Manned Space Flight Sub-Committee of House Committee on Science and Astronautics at the NASA Launch Operations Center," 23 Mar. 1962, p. 27.
35. Martin, *Saturn C-3 Study*, 3: 46–52.
36. Clearman, "Complex 39 Staff Meeting," 12 Mar. 1962; Deese to Moser et al., "Preliminary Concepts, Vertical Building," 6 Mar. 1962; Brown Engineering Co., Inc., *An Evaluation of an Enclosed Concept for a C-5 Vertical Assembly Building (VAB)*," 2 Apr. 1962, pp. 7–8; Brown Engineering Co., Inc., *An Evaluation of an Open Concept for a C-5 Vertical Assembly Building*, 2 Apr. 1962, pp. 9–10.
37. Briel's notes, 12–13 June 1962; Brown Engineering Co., "Evolution of the Saturn C-5 Mobile System Vertical Assembly Building," a mimeographed report prepared by E. M. Briel, 7 Sept. 1962.
38. LOC, "Minutes of the Saturn C-5 Launch Operations Working Group Meeting, 18–19 July 1962," 8 Aug. 1962, pp. 2–6.
39. Briel's notes, 31 July 1962; DDJ, 15 Aug. 1962.
40. URSAM, "VAB-LC39: A Report of Meeting with Representatives of LOC, Corps of Engineers and Component Contractors" (Cape Canaveral, FL, 28 Aug. 1962), app. A.
41. Isom G. Rigell, memo, "LC-39 Networks," 4 Sept. 1962; Norman Gerstenzang, memo for record, 5 Sept. 1962; unsigned memo, "Information Required by LO-FEE for LC-39, VAB Criteria," 6 Sept. 1962.
42. Joe J. Koperski, Chief, Engineering Div., Corps of Engineers, to R. P. Dodd, "Back-to-Back vs. In-Line Configuration, Comparisons and Conclusions—Launch Complex 39–Vertical Assembly Building," 21 Sept. 1962.
43. Deese interview, 4 Oct. 1973.
44. NASA, "A Report on Launch Facility Concepts for Advanced Saturn Launch Facilities," by Marvin Redfield, John Hammersmith, and Jay A. Salmonson, 13 Feb. 1962.
45. House, *Hearings: 1963 NASA Authorization*, p. 941.
46. LOC, "Summary of Conference with Members of Manned Space Flight," 23 Mar. 1962, pp. 13–34.
47. Ibid., pp. 21, 30. In an interview with James Frangie on 13 Aug. 1969, Col. Bidgood pointed out that LC-39 provided launch rate flexibility but had limitations in its ability to accommodate different vehicles.
48. NASA, "Minutes of the Management Council Office of Manned Space Flight," 29 May 1962.
49. Ibid., 22 June 1962; DDJ, 15 June 1962.
50. Logsdon, "NASA's Implementation," pp. 56–60.
51. The mode selection story continued several more months as NASA had to defend the choice against strong criticism from the President's Science Advisory Committee. For a lengthier treatment of one of Apollo's most interesting episodes, see Logsdon, "NASA's Implementation."
52. "LOD Weekly Notes," Sendler, 5, 19 July 1962, Bidgood, 5 July, 2, 23 Aug. 1962; DDJ, 15, 21 Aug. 1962.
53. Poppel to Bidgood, "Preliminary Design for a Mobile Arming Tower for Launch Complex 39," 10 Aug. 1962.
54. LOC, "Minutes of the Saturn C-5 Launch Operations Working Group Meeting, 18–19 July 1962," 8 Aug. 1962, pp. 1–5 and app. 9.
55. Redfield interview.

Chapter 7: The Launch Directorate Becomes an Operating Center

1. Gen. Ostrander, "Proposed Organization for Launch Activities at AMR and PMR," 6 July 1961; Debus, "Proposed Organization for Launch Activities at Atlantic Missile Range and Pacific Missile Range Due to Proposed Realignment of NASA Programs," 12 June 1961.
2. Debus to Rees, Dep. Dir. for R&D, MSFC, "Operating Procedures and Responsibilities of MSFC Divisions and Others at Launch Site," 4 Aug. 1961.
3. Ibid.; see also DDJ, 27 July 1961.
4. D. M. Morris, Dep. Dir. of Admin., MSFC, to Albert Siepert, Dir., Office of Business Admin., NASA Hq., 6 June 1961; Harry H. Gorman, Assoc. Dep. Dir. for Admin., MSFC, to Seamans, 26 Sept. 1961.
5. Debus, "A Paper on Launch and Spaceflight Operations," 27 Sept. 1961.
6. Debus, "Analysis of Major Elements Regarding the Functions and Organization of Launch and Spaceflight Operations," 10 Oct. 1961.
7. Concurrence by Wernher von Braun, appended to n. 6 reference.
8. Seamans to Young and Siepert, 13 Oct. 1961; Debus interview, 16 May 1972.
9. Rees to von Braun, "New Organization Proposals for LOD," 17 Oct. 1961; Debus interview, 16 May 1972.
10. Debus interview, 22 Aug. 1969.
11. NASA release 62-53, "Establishment of the Launch Operations Center at AMR and the Pacific Launch Operations Office at PMR," 7 Mar. 1962.
12. Young to Seamans, "Internal Organization of the Launch Operations Center," 29 June 1962.
13. Debus interview, 22 Aug. 1969.
14. "MSFC-LOC Separation Agreement," 8 June 1962, printed in Francis E. Jarrett, Jr., and Robert A. Lindemann, "Historical Origins of NASA's Launch Operations Center to 1 July 1962" (KSC, 1964), app.; NASA release LOC-63-64, 24 Apr. 1963. The transfer of LVOD personnel from MSFC to LOC was completed by 6 May 1963. See James M. Ragusa, "John F. Kennedy Space Center (KSC) NASA Reorganization Policy and Methods," (M.S. thesis, Florida State Univ., Apr. 1968), p. 25.
15. Melton interview.
16. DDJ, 1 Sept. 1961; Clark interview.
17. *Missiles and Rockets,* 6 Nov. 1961, p. 18.
18. Ernest W. Brackett, Dir., Procurement and Supply, to Dir., Off. of Admin., with enclosure, "Establishment of Launch Operations Center," 15 June 1962.
19. Clarence Bidgood, "Facilities Office Memo No. 2," 1 Feb. 1962; *Spaceport News,* 22 Feb. 1963.
20. Debus interview, 22 Aug. 1969.
21. Parker interview, 14 Feb. 1969.
22. By early June 1962, 930 sq m of off-site space had been leased and plans were made to lease 1400 more by 30 July 1962. Robert Heiser to Rachel Pratt, "Notes from von Braun," 11 June 1962; Gordon Harris, Chief of Public Affairs, to Bagnulo, 1 Oct. 1964.
23. Hall to Facilities Program Off., "Justification for Leasing Additional Space in CAC Building," 13 Mar. 1963. During the period of limited office space, thought was given toward acquiring a barge from the Navy to be moored at the Saturn Barge Terminal and used as an office. Hall to Bidgood, "Barge Anchorage," 18 Sept. 1962; Hawkins to Hall, "Trailer Request for LC-34," 1 Mar. 1963; *Spaceport News,* 13 Oct. 1966, p. 6.
24. NASA General Management Instruction (hereafter GMI) 2-2-9.1, "Basic Operating Concepts for the Launch Operations Center at Merritt Island and the Atlantic Missile Range," 10 Jan. 1963.
25. House Committee on Science and Astronautics, Subcommittee on Manned Space Flight, *Hearings: 1964 NASA Authorization,* 88th Cong., 1st sess., pt. 2a, pp. 127, 129.

26. Bidgood interview, 13 Aug. 1969.
27. Seamans to Dir., LOC, "General Responsibilities and Functions of the NASA Center Director," 10 Jan. 1963, with enclosure, "General Responsibilities and Functions of a NASA Center Director," 10 Jan. 1963. These documents were circulated in LOC on 18 Jan. 1963; see Office of the Dir., "General Responsibilities and Functions of a NASA Center Director," 18 Jan. 1963. They were subsequently revised and published as attachments to NASA GMI 2-0-3, "Informational Material on Assignment of Responsibilities in the NASA Organization Structure," 3 June 1963.
28. NASA GMI 2-2-9.1, "Basic Operating Concepts for the Launch Operations Center at Merritt Island and the Atlantic Missile Range," 10 Jan. 1963. A 4 Mar. 1963 plan provided for an "Assistant Director for Program Management" (a title actually adopted in the reorganization of 28 Jan. 1964); a 28 Mar. 1963 plan provided for an "Assistant Director for LOC Programs," as did also a 2 Apr. 1963 plan.
29. The NASA Daytona Beach Operations was established and designated an integral part of LOC in NASA circular 2-2-9, 23 June 1963, which stated that the Manager would "report to the Director, Launch Operations Center, Cocoa Beach, Florida." See also Debus to George Mueller, "Change in Organizational Structure," 13 Dec. 1963; Debus to staff, "LOC Organization Structure," 6 Aug. 1963, and accompanying manual, "LOC Organization Structure," 2 Aug. 1963; LOC Organization Chart, approved by Hugh L. Dryden, 24 Apr. 1963
30. *Spaceport News,* 1 May 1963, p. 6.
31. Bidgood, who had joined LOC on 1 Nov. 1962, organized a Facilities Office by late 1962 and published his first organization chart on 13 Feb. 1963. Bidgood interview, 14 Nov. 1968.
32. "LOC Organization Structure," 2 Aug. 1963.
33. *Spaceport News,* 1 May 1963, p. 6. For Debus's views on the "development operational loop," see his "Analysis of Major Elements Regarding the Functions and Organization of Launch and Spaceflight Operations," 10 Oct. 1961; also "LOC Organizational Structure," 2 Aug. 1963, p. 9.
34. C. C. Parker, "Boards, Committees, Panels, Teams and Working Groups," 25 Sept. 1963.
35. *Spaceport News,* 30 June 1963.
36. Hugo Young, Bryan Silcock, and Peter Dunn, *Journey to Tranquility* (Garden City, NY: Doubleday & Co., 1970), p. 158.
37. Rosholt, *An Administrative History of NASA,* pp. 288–89.
38. *The Washington Post,* 21 Sept. 1963, p. A-10.
39. Thomas's letter and the President's reply appear in Senate Committee of Appropriations, *Hearings: Independent Offices Appropriations, 1964,* 88th Cong., 1st sess., pt. 2, pp. 1616–18.
40. *Fortune,* Nov. 1963, pp. 125–29, 270, 274, 280.
41. *Spaceport News,* 21, 27 Nov. 1963.
42. Executive Order 11129, *Designating Certain Facilities of the National Aeronautics and Space Administration and of the Department of Defense in the State of Florida, as the John F. Kennedy Space Center,* 29 Nov. 1963; NASA announcement 63-283, "Designation of the John F. Kennedy Space Center, NASA," 20 Dec. 1963; message SAF 82841, Sec. of the Air Force to Cmdr., AFSC, Andrews AFB, 7 Jan. 1964; "Decisions on Geographic Names in the United States, Dec. 1962 through December 1963," decision list 6303, U.S. Board on Geographic Names (Washington: Dept. of the Interior, 1964), p. 20.
43. Debus to Mrs. W. L. Stewart, 26 Dec. 1963.
44. Ernest G. Schwiebert, *A History of the U.S. Air Forces Ballistic Missiles,* pp. 130, 201–203, 247; Akens, *Saturn Illustrated Chronology,* pp. 67–68.
45. OMSF, "Management Council Minutes, 29 Oct. 1963," 31 Oct. 1963.

46. Rosholt, *Administrative History,* pp. 289–97. Newell headed the Office of Space Sciences and Applications, Bisplinghoff, the Office of Advanced Research and Technology.
47. NASA release KSC-10-64, 6 Feb. 1964; KSC, "NASA Organization Chart," 28 Jan. 1964; *Spaceport News,* 13 Feb. 1964.
48. "KSC Notes," Petrone to Debus, 29 Oct. 1964.
49. "Kennedy Space Center Apollo Document Tree," approved by Rocco Petrone, 3 Nov. 1965, Joel Kent's private papers.
50. Childers interview, 7 Nov. 1972; Gramer interview, 21 Sept. 1972.

Chapter 8: Funding the Project

1. Lyndon B. Johnson, *The Vantage Point: Perspectives of the Presidency, 1963–69* (New York: Popular Library, 1971), p. 283.
2. Unless otherwise indicated the information on the early budgetary process was gathered from interviews with Robert G. Long, Resources Management Br., KSC; William E. Pearson, Chief, Management Information Control Br., KSC; Alton D. Fryer, Resources Management Div., KSC; and Elizabeth A. Johnson, Financial Management Div., KSC. Much of this information was later formalized in OMSF, *Apollo Program Development Plan,* M-D MA 500, 1 Jan. 1966.
3. The real estate paragraph was usually omitted since land acquisition for the MLLP was provided for under separate CoF documents.
4. Robert C. Seamans, Jr., "Guidelines for Preparation of Detailed Fiscal Year 1964 Budget Estimates—Section I," 30 Aug. 1962, attach. 5, p. 6.
5. "It should be borne in mind, however, that, as a matter of policy, NASA may not choose to exercise its authority to the full extent permitted by law," General Counsel (John A. Johnson) to Paul C. Dembling, "Request from Congressman Karth regarding the 'extent of NASA's authority to reprogram or transfer appropriate funds within the agency,' " 3 Apr. 1963, with encl. NASA was authorized to transfer sums from one budget line item to another to the extent of 5% of the item to which the transfer was to be made, to meet unusual cost variations, provided the total amount authorized was not exceeded. It was also authorized to transfer up to 5% of the CoF appropriation to the RD&O appropriation, and vice versa. Additionally, the NASA Administrator was authorized under certain circumstances to use CoF funds for such things as emergency repairs if of greater urgency than the construction of new facilities.
6. C. C. Parker, Chief, Operations Off., LOD, to J. Martin, "FY-63 C&E Requirement," 19 Dec. 1960.
7. Don R. Ostrander, NASA Hq., to MSFC, "Preliminary Fiscal Year 1963 Budget," 27 Feb. 1961.
8. Akens, *Saturn Illustrated Chronology,* pp. 19, 22–23.
9. *Spaceport News,* 19 Jan. 1967, p. 7.
10. Burke interview.
11. Parker/Greenglass to D'Onofrio, "FY-63 CoF Budget Requirements," 13 Dec. 1961.
12. LOD, "Atlantic Missile Range Fiscal Year 1963 Estimates, Construction of Facilities," undated. See also House, *1963 NASA Authorization,* p. 879; Senate, *NASA Authorization for Fiscal Year 1963*, p. 126; and Senate Committee on Appropriations, 87th Cong., 2d sess., *Hearings: Second Supplemental Appropriation Bill for 1962,* 4 Apr. 1962, p. 149.
13. LOD, "Atlantic Missile Range, Fiscal Year 1963 Estimates, Construction of Facilities," undated (ca. Jan. 1962).

14. House, Manned Space Flight Subcommittee, "Summary of Conference with Members of Manned Space Flight Subcommittee of House Committee on Science and Astronautics at the NASA Launch Operations Center, Cocoa Beach, Fla., 23 Mar. 1962."
15. Ibid., pp. 19–21. The conference record frequently does not identify the speaker.
16. Ibid., p. 21.
17. Ibid.
18. Ibid., pp. 21, 24–27.
19. Ibid., pp. 29–30.
20. Ibid., pp. 55–57.
21. Ibid., p. 58.
22. William E. Lilly, Dir., Program Review and Resources Management Off., OMSF, to LOC, "Advance Approval of FY-63 CoF," 14 Mar. 1962; TWX, C. C. Parker to NASA Hq., "Advance Approval of FY-63 CoF," 21 May 1962.
23. Rosholt, *Administrative History,* pp. 233, 284; *Congress and the Nation, 1945–1964* (Washington: Congressional Quarterly Service, 1965), 1:320.
24. OMSF, "Management Council Minutes," 21 Sept. 1962, item 8; LOC, "FY 1963 CoF Resubmission and Supplemental Program," 8 Sept. 1962.
25. C. C. Parker, Dep. Assoc. Dir. for Admin. Services, LOC, to William E. Lilly, Dir. Program Review and Resources Management, OMSF, 8 Sept. 1962.
26. LOC to NASA Hq., "LOC Fiscal Year 1963 Estimates, Advanced Saturn Complex No. 39," 8 Sept. 1962, with encl.
27. C. C. Parker, LOC, to William E. Lilly, 18 Sept. 1962.
28. *Congress and the Nation, 1945–1964,* pp. 320, 390.
29. Debus to Holmes, 26 Oct. 1962.
30. Project Approval Document, "Construction of Facilities, Advanced Saturn Launch Complex No. 39," code 46-46-990-933-3450, 6 Nov. 1962, p. AMR-63-10; NASA Form 504, "Allotment/Sub-Allotment Authorization," amendment 05, 16 Nov. 1962.
31. Frederick L. Dunlap, Chief, Budget Br., NASA Hq., to Ed Melton, Financial Management Off., LOC, 27 Dec. 1962.
32. NASA Procurement Div., KSC, "Status Summary of Active Contracts as of 31 Mar. 1964," sec. III, "Active Intergovernmental Purchase Orders," p. 3.
33. OMSF directive M-D 9330.01, "Manned Space Flight Program Launch Schedule for Apollo and Saturn Class Vehicles," 15 Oct. 1962. See also MSFC, "Reference Director, MSFC/MSC/OMSF, Flight and Mission Schedule History," 21 Feb. 1963; and OMSF, "Management Council Minutes," 31 July 1962, item 5F.
34. Debus, "Fiscal Year 1964 Preliminary Budget," 8 Mar. 1962; Seamans to Dir., OMSF, "Guidelines for Preparation of Detailed Fiscal Year 1964 Budget Estimates," sec. I, 30 Aug. 1962; LOC, "FY 1964 CoF Program," 1 Nov. 1962.
35. LOC, "FY 1964 CoF Program," 1 Nov. 1962, pp. DF-B 16 through DF-B 18.
36. Ibid.
37. W. F. Barney, Chief, Control Off., MSFC, to Mr. Lada, "Fiscal Year 1963 Spacecraft CoF Projects," 19 Feb. 1962; OMSF, "Management Council Minutes," 28 Aug. 1962.
38. G. Merritt Preston, MSC, AMR Ops., to NASA Hq., Attn: G. M. Low, "Chronology and Background Information on Gemini and Apollo Facilities at AMR," 18 Jan. 1963. This letter was prepared in response to an 11 Jan. 1963 request from Congressman George P. Miller of California, Chairman of the House Committee on Science and Astronautics, for an explanation of the decision to combine some Gemini and Apollo spacecraft facilities on Merritt Island. Combining facilities also made it difficult to extract costs directly chargeable to the Apollo program.
39. Manned Spacecraft Center/Atlantic Missile Range, "FY 1963 Construction of Facilities, Project Documentation," 15 Oct. 1962; Debus to William E. Lilly, 15 Oct. 1962. The use of the project title "Apollo Mission Support Facilities" persisted for several months.
40. LOC, "FY 1964 CoF Program," 1 Nov. 1962.

41. House Subcommittee on Manned Space Flight, 88th Cong., 1st sess., *Hearings: 1964 NASA Authorization,* pt. 2(b), p. 986. In fact, the $432 million figure had been given to the subcommittee at Cape Canaveral on 23 Mar. 1962.
42. Ibid., pp. 987–89.
43. Ibid., pp. 989–90.
44. Ibid., p. 1276.
45. Ibid., pp. 991–94, 1275–83.
46. *Congress and the Nation, 1945–1964,* p. 326.
47. Ibid., pp. 326, 329.

Chapter 9: Apollo Integration

1. Senate Committee on Aeronautical and Space Sciences, *Hearings: NASA Authorization for Fiscal Year 1963*, 87th Cong., 2d sess., 14 June 1962, p. 486; see also House Committee on Science and Astronautics, Subcommittee on Manned Space Flight, *1963 NASA Authorization*, 87th Cong., 2d sess., 26 Mar. 1962, pt. 2, pp. 543–44.
2. "Weekly Notes," Petrone to Debus, 12 Apr., 17 May 1962; "Minutes of the Sixth Meeting of the Management Council of the Office of Manned Space Flight, 29 May 1962." OMSF realized that G.E.'s favored status would offend stage contractors and stipulated in the contract (NASw-410) certain provisions that restricted G.E.'s use of sensitive information.
3. "Agreements Reached at the August Meeting at the Cape Concerning the G.E. Contract," unsigned and undated (Debus and Petrone represented LOC).
4. DDJ, 7 Nov. 1962; OMSF, "Management Council Minutes," 21 Sept., 27 Nov. 1962.
5. House Committee on Science and Astronautics, Subcommittee on Manned Space Flight, *Hearings: 1964 NASA Authorization*, 88th Cong., 1st sess., Mar.–June 1963. The G.E. contract and its ramifications crop up throughout these hearings, particularly in vol. 3, pt. 2(b). The subcommittee hearings at Daytona Beach are contained in app. C to pt. 2(b), pp. 1285–1352.
6. DDJ, 3, 9 July 1963; OMSF, "Management Council Minutes," 27 Aug. 1963.
7. Manned Spacecraft Center, Langley AFB, VA, "Minutes of MSFC-MSC Space Vehicle Board No. 1, 3 Oct. 1961," 7 Nov. 1961.
8. Von Braun to Mueller, "Flight Missions Planning Panel," 30 Dec. 1963; Wagner interview.
9. MSC, "Minutes of MSFC-MSC Space Vehicle Board No. 1, 3 Oct. 1961," 7 Nov. 1961.
10. LOD Weekly Notes, Petrone, 8 Feb. 1962.
11. LOD Weekly Notes, Petrone, 3 May, 21 June 1962, and Poppel, 18 Apr. 1962; "Summary of Launch Operations Panel Activities," prepared by Emil Bertram for Joseph Shea, 18 July 1963.
12. Bertram, "Summary of Launch Operations Panel Activities," 18 July 1963; LOC, "Minutes of Meeting, Apollo-Saturn Launch Operations Panel, 6 Aug. 1963."
13. "Minutes of Systems Review Meeting, Houston, Texas, 10 Jan. 1963," JSC Archives.
14. "Panel Review Board Minutes, 9–10 Aug. 1963"; OMSF, "Management Council Minutes," 28 May 1963.
15. Gruene interview, 19 Nov. 1970. The U.S. Comptroller General ruled these "body-shop" contracts illegal in Mar. 1964. House Committee on Post Office and Civil Service, *Decision of Comptroller General of the United States Regarding Contractor Personnel in Department of Defense*, 89th Cong., 1st sess., report 188, 8 Mar. 1965.
16. Orvil Sparkman, "S-IV Propellant Loading Sequence," 26 Sept. 1961.
17. LOC Weekly Notes, Gruene, 29 Aug., 5 Sept. 1963; Petrone, 10 Oct. 1963.

18. Debus to Shea, Dep. Dir. for Systems, NASA Hq., "Apollo Interface Control Procedures," 24 June 1963; minutes of meeting, "Delineation of Interface Responsibility between Astrionics Division and LOC," 29 May 1962, signed by Poppel and H. J. Fichtner, Chief, Electrical Systems Integration.
19. Debus and von Braun, "Memo of Agreement: MSFC/KSC Relations," 11 Aug. 1964.
20. Debus and von Braun, "Clarification and Implementation Instruction, MSFC/KSC Relations Agreement dated 11 August 1964," 9 Mar. 1965.
21. Bertram, "LIEF Implementation," 3 Apr. 1964; Bertram interview, 15 Nov. 1973.
22. Bertram, "LIEF Implementation," 3 Apr. 1964.
23. Mary Louise Morse and Jean Kernahan Bays, *The Apollo Spacecraft: A Chronology*, vol. 2, *November 8, 1962—September 30, 1964*, NASA SP-4009 (Washington, 1973), pp. iii–vi; Jay Holmes, "Minutes of Special Staff Meeting, Office of Associate Administrator for Manned Space Flight," 31 Jan. 1964; Shea to Phillips, 27 Mar. 1964, Phillips File, NASA Hq. History Off.; Joachim P. Kuettner, Mgr., Saturn Apollo Systems Integrations, MSFC, "Trip Report," 22 Oct. 1962, in Petrone's notes.
24. Memo attached to DDJ, 6 Aug. 1962; Poppel interview, 24 Jan. 1973.
25. Petrone to M. Dell, Apollo Support Off., MSC, 5 Nov. 1962.
26. Debus to Holmes, 14 Nov. 1962 (letter summarizing discussions between the two men on 19 Oct.), Debus papers.
27. Petrone to J. T. Doke, Apollo Project Off., 22 Oct. 1962; Petrone to B. Porter Brown, 13 Nov. 1962.
28. B. Porter Brown, Prelaunch Ops. Div., Ops. Support Off., to Walter Wagner, KSC, "Mission Operations Control Room Information," 3 Feb. 1964; Petrone to Brown, 7 Feb. 1963.
29. KSC Weekly Notes, Petrone, 12 Dec. 1963.
30. KSC Weekly Notes, Petrone, Jan.–May 1965.
31. LOC Weekly Notes, Petrone, 20 Sept. 1962; Petrone to Kuettner, "Weekly Report to MSC," 19 Oct., 20 Nov., 18 Dec. 1962, in Petrone's notes (1962–1964).
32. Debus notes of 20 June 1963, in Petrone's notes; Bertram interview, 28 Sept. 1973; Horn interview; Moore interview; Hand interview.
33. MSFC, *Saturn V Flight Manual, SA 506*, 25 Feb. 1969, sec. 3, 9.
34. KSC, *Apollo/Saturn V Flight Safety Plan, Vehicle AS-501* (1967), pp. 1-1, 2-1, 3-1.
35. Taylor, *Liftoff!* p. 83; Adolf H. Knothe, "Range Safety—Do We Need It?" paper 70-249, American Institute of Aeronautics and Astronautics, Launch Operations Meeting, Cocoa Beach, FL, 2–4 Feb. 1970, p. 2.
36. R. M. Montgomery, "Range Safety of the Eastern Test Range," paper 70-246, American Institute of Aeronautics and Astronautics, Launch Operations Meeting, Cocoa Beach, FL, 2–4 Feb. 1970, p. 2; Arthur Moore to Benson, "Comments on Launch Operations History," 4 Oct. 1974.
37. Debus to Davis, "Range Safety Policies and Procedures," 11 June 1962, with attached letter from Davis to Debus, 10 May 1962, Debus papers.
38. Emil Bertram, memo for record, "Range Safety Information Channels," 30 Mar. 1962, KSC Range Safety Off. Notes; LOC Weekly Notes, Knothe, 3 May 1962; Bertram to Petrone, "Apollo Saturn Range Safety," 7 May 1962, KSC Range Safety Off. Notes; Bertram to Petrone, 9 May 1963, Petrone's notes.
39. AFETR Manual 127-1, *Range Safety Manual*, 1 Sept. 1972, 1:4–6; according to KSC officials the wording on this matter in the current manual is practically unchanged from the manual in force ten years earlier, no copy of which was available.
40. LOC Weekly Notes, Knothe, 9 May, 3 July 1963.
41. Knothe, "Minutes of Meeting on the Use of Liquid Explosives for a Fuel Dispersion System," 12 July 1963, in KSC Range Safety Off. Notes; LOC Weekly Notes, Knothe, 25 July 1963; Christopher C. Kraft, "Range Safety Aspects of the Apollo Program," 5 Aug. 1963.

42. Knothe to attendees, "Minutes of Meeting: Range Safety Aspects of Apollo Program, Held at NASA/LOC on 29 Aug. 1963," 5 Sept. 1963, KSC Range Safety Off. Notes.
43. Kraft, "Aspects of Apollo Range Safety," 1 Nov. 1963.
44. Kraft, "Apollo Range Safety," 11 Dec. 1963, in KSC Range Safety Off. Notes; Hans Gruene, "Apollo Service Module Propellant Dispersion System Interface Disagreement between MSC and KSC/MSFC," 25 Mar. 1964; LOC Weekly Notes, Knothe, 16 Apr. 1964 (marginal note by Debus); George E. Mueller, Assoc. Admin. for Manned Space Flight, to Cmdr., National Range Div., USAF, 18 Sept. 1964.

Chapter 10: Saturn I Launches (1962–1965)

1. LOD, "SA-2 Daily Status Reports," Robert Moser papers, Federal Archives and Records Center, East Point, GA, accession 68A1230, boxes 436257, 436259.
2. MSFC, *Results of the Second Saturn Launch Vehicle Test Flight SA-2*, report MPR-SAT-63-13 (Huntsville, AL, 16 Oct. 1963), pp. 1–5, 24, 49; Speer, *Saturn I Flight Test Evaluation*, pp. 1–6.
3. ARINC Research Corp., *Reliability Study of Saturn SA-3 Pre-Launch Operations*, by Arthur W. Green et al. (Washington, 3 Jan. 1963), pp. 4-7 through 4-11, 7-1.
4. MSFC, *Results of the Third Saturn I Launch Vehicle Test Flight, SA-3*, report MPR-SAT-64-13 (Huntsville, AL, 26 Feb. 1964), pp. 1–8; Speer, "Saturn I Flight Test Evaluation," p. 2; DDJ, 1 Nov. 1962.
5. MSFC, *Results of the Fourth Saturn I Launch Vehicle Test Flight, SA-4*, report MPR-SAT-63-6 (Huntsville, AL, 10 May 1963), pp. 5–7; *Spaceport News*, 9 Apr. 1964, p. 3; Chambers interview.
6. MSFC, *Results of SA-4*, pp. 1–7, 16–17; Speer, "Saturn I Flight Test Evaluation," p. 2.
7. MSFC, *Results of the Saturn I Launch Vehicle Test Flights*, report MPR-SAT-FE-66-9 (Huntsville, AL, 9 Dec. 1966), pp. 26–27.
8. MSFC, *Results of SA-3*, pp. 7–8; MSFC, *Results of SA-4*, p. 7; House Committee on Science and Astronautics, Subcommittee on Manned Space Flight, *Hearings: 1964 NASA Authorization*, 88th Cong., 1st sess., 6 Mar. 1963, pt. 2(a), p. 198; MSFC, *Saturn Monthly Progress Reports* (Jan.–Aug. 1962).
9. Debus to Rees, "Additional Saturn Launch Complex," 29 Jan. 1960.
10. Capt. Arthur G. Porcher, Chief, Facilities Br., Army Test Off., AFMTC, "Additional Launch Facilities for Saturn Type Vehicle," 5 Feb. 1960; Col. Donald Heaton to Gen. Ostrander, "Price Increase in Second Saturn Launch Complex," 12 Feb. 1960.
11. Philip Claybourne, Saturn Project Off., to MFL Br. Chiefs, "Back-Up of Saturn Launch Facilities," 10 May 1960.
12. Harvey F. Pierce to Debus, 26 Feb. 1960.
13. Debus to Zeiler, "Formation of Committee to Review Service Structure Design," 9 Mar. 1960; DDJ, 11, 13 Apr. 1960.
14. Debus, memo for record, "Drift of the Saturn C-2 Vehicle at Launching," 12 July 1960; MSFC, "Summary Report and Recommendations of Saturn Service Structure No. II Design Committee," by Harvey F. Pierce, 12 July 1960, pp. 4–18.
15. MSFC, "Summary Report of Saturn Service Structure Committee," pp. 4–18.
16. Ibid.; LOC, "Concept Development of Saturn Service Structure, No. II," by James Deese, Apr. 1963, pp. 26–27, James Deese papers.
17. DDJ, 29 Aug. 1960.
18. Debus to von Braun, "Hazard Study of Liquid Hydrogen, LO_2 and RP-1," 10 Jan. 1961; DDJ, 9, 11, 13 Jan. 1961.
19. DDJ, 13 Jan. 1961.

20. LFSEO Monthly Progress Report, 12 June 1961, p. 1; Poppel to Parker, "Criteria for VLF 37," 22 Dec. 1960; J. W. Ault, memo for record, "Contract for LC-37 Design," 23 Feb. 1961; Dodd to Corps of Engineers, "Vibroflotation for Complex 37A," 18 Oct. 1961, Debus papers.
21. MSFC, *Saturn Quarterly Progress Report* (July-Sept. 1961), report MPR-SAT-61-11, 1 Dec. 1961, p. 94; MSFC, *Saturn Quarterly Progress Report* (Jan.-Mar. 1962), MPR-SAT-62-3, p. 36; Michael Getler, "Complex 37 Will Dwarf Predecessors," *Missiles and Rockets*, 18 Dec. 1961, pp. 24–25, 47.
22. Getler, "Complex 37," pp. 24–25, 47; "The Biggest Thing on Wheels in the World," prepared by Batten, Barton, Durstine, and Osborne, Inc., Pittsburgh, for U.S. Steel, Jan. 1963.
23. D. E. Eppert, Chief, Construction Div., Canaveral Dist., Corps of Engineers, to James J. Frangie, "List of Saturn Construction Contracts," 12 Sept. 1968, p. 16; *NASA Fifth Semi-Annual Report to Congress*, 1 Oct. 1960 through 30 June 1961, p. 145; Emil Bertram, memo for record, "Apollo-Saturn Subpanel Activities," 15 July 1963, p. 3.
24. MSFC, *Results of Saturn I Launch Vehicle Tests*, pp. 3–5.
25. "Daily Status Reports, LC-37B Wet Test Vehicle," Robert Moser papers; Moser interview, 18 July 1973; Akens, *Saturn Illustrated Chronology*, pp. 58–61.
26. Gruene to Debus, 12 Sept. 1963.
27. D. L. Childs to LVO, S-IV-5 Status Reports #23, 29 Aug., and #33, 11 Sept. 1963; S-IV-5 Daily Log, 21–22 Sept. 1963, Rober Moser papers; LVO, "SA-5 Daily Status Report," 23, 24, 25 Sept. 1953.
28. "SA-5 Daily Status Reports," 11, 14, 17 Oct. 1963; Gruene to Debus, 17 Oct. 1963.
29. LVO, "SA-5 Daily Status Reports," 11 Oct. 1963.
30. Ibid., 22 Oct., 7 Nov. 1963; Gruene to Debus, 31 Oct. 1963; Fannin interview.
31. LVO, "SA-5 Daily Status Reports," 27 Nov. 1963; Corn interview, 23 July 1973; Zeiler interview, 23 July 1973; Pickett interview.
32. LVO, "SA-5 Daily Status Report," 27 Nov., 6, 8, 10, 13 Dec. 1963.
33. Ibid., 23, 27 Dec. 1963, 14, 17, 19 Jan. 1964.
34. Akens, *Saturn Illustrated Chronology*, pp. 72–73; *Cocoa Tribune*, 29 Jan. 1964.
35. MSFC, *Results of the Fifth Saturn I Launch Vehicle Test Flight, SA-5*, report MPR-SAT-FE-64-17 (Huntsville, AL, 22 Sept. 1964), pp. 5–7; *Cocoa Tribune*, 28, 29 Jan. 1964.
36. KSC, "Presentation to the Subcommittee on Manned Space Flight of the House Committee on Science and Astronautics at KSC," 27 Jan. 1964; Sherrer interview.
37. *Cocoa Tribune*, 28 Jan. 1964, p. 2.
38. KSC, "Presentation to the Subcommittee," 27 Jan. 1964.
39. R. P. Eichelberger, "The Saturn Telemetry System," pp. 1–3; KSC, "Technical Progress Report," 24 Jan. 1964; *Spaceport News*, 23 Jan. 1964, p. 2; "Consolidated Instrumentation Plan," pt. IIA of *Firing Test Report, Saturn I SA-5*, 22 Jan. 1964 (TR-4-36), pp. 6, 19, 31.
40. *Spaceport News*, 4 June 1964, p. 1; NASA release 63-268, 23 Jan. 1964; *New York Times*, 27 Jan. 1964; Speer, "Saturn I Flight Test Evaluation," pp. 1–8.
41. *Orlando Sentinel*, 30 Jan. 1964, pp.1, 42.
42. James Grimwood, JSC Historian, supplied information for this section.
43. *Spaceport News*, 4 June 1964, p. 2; NASA, *Astronautics and Aeronautics, 1964*, pp. 70, 126; Sasseen interview, 26 July 1973.
44. *Orlando Sentinel*, 21 May 1964; *Melbourne Daily Times*, 26, 27 May 1964.
45. *Spaceport News*, 4 June 1964, p. 5; MSFC, *Results of the Saturn I Launch Vehicle Test Flights*, p. 23.
46. *Spaceport News*, 13 Aug. 1964, p. 3; Davidson interview.
47. *Cocoa Tribune*, 20 July 1964; *Spaceport News*, 23 July 1964, p. 2; Newall interview.
48. *Cocoa Tribune*, 28 Aug. 1964; *Orlando Sentinel Star*, 8, 9 Sept. 1964; *Miami Herald*, 16 Sept. 1964; *Spaceport News*, 27 Aug., 3, 10, 17 Sept. 1964; *Aviation Week and Space Technology*, 28 Sept. 1964, p. 27.

49. Gen. Samuel Phillips to George Mueller, 14 Jan. 1965; Mueller to Debus, 10 Feb. 1965; weekly notes from Petrone to Debus, 4 Feb. 1965.
50. MSFC, *Results of the Eighth Saturn I Launch Vehicle Test Flight, SA-9*, report MPR-SAT-FE-65-6 (Huntsville, AL, 30 Apr. 1965), p. 14; Akens, *Saturn Illustrated Chronology*, p. 104.
51. Ibid., pp. 9–14; "Pegasus Returning Meteoroid Flux Data," *Aviation Week and Space Technology*, 22 Feb. 1965, p. 28.
52. MSFC, *Results of the Ninth Saturn I Launch Vehicle Test Flight, SA-8*, report MPR-SAT-FE-11 (Huntsville, AL, 27 July 1965), pp. 7–15; "First Industry-Built Saturn I Puts Pegasus-2 in Precise Orbit," *Aviation Week and Space Technology*, 31 May 1965, p. 21.
53. MSFC, *Results of the Tenth Saturn Launch Vehicle Test Flight, SA-10*, report MPR-SAT-FE-65-14 (Huntsville, AL, 24 Sept. 1965), p. 8.

Chapter 11: Ground Plans for Outer-Space Ventures

1. In this section the authors relied extensively on research by William Lockyer, Jr., and James Covington.
2. Col. J. V. Sollohub to Debus, 15 Oct. 1962.
3. "To Design for the Moon Age, Four Firms Work as One Team," *Engineering News–Record* 172 (6 Feb. 1964): 46–48.
4. Alexander interview; Anton Tedesko to Urbahn, Knecht, and Rutledge, 10 Aug. 1962.
5. URSAM, "VAB-LC39: Report of a Meeting with Representatives of LOC, Corps of Engineers and Component Contractors," Cape Canaveral, FL, 28 Aug. 1962.
6. Ibid., pp. 2–6.
7. Wesley Allen, Brown Engineering Co., memo for record, "Meeting with Facilities and MSC," 17 Sept. 1962.
8. Bidgood to Poppel, 26 Sept. 1962.
9. Col. Wm. Alexander, "Report on VAB," undated, p. 5; J. Bing to R. P. Dodd, 7 Nov. 1962; Theodor A. Poppel to Bidgood, 21 Nov. 1962; and Gerstenzang and Carraway to Dodd, 23 Nov. 1962.
10. William D. Alexander, "Vertical Assembly Building—Project Description, Organization, and Procedures," *Civil Engineering* 35 (Jan. 1965): 42–44.
11. Ibid., p. 44.
12. Gerald C. Frewer, "Kennedy Space Center—Assembly Line on a Gigantic Scale," *The Engineering Designer,* May 1967, p. 7.
13. Anton Tedesko, "Base for USA Manned Space Rockets (Structures for Assembly and Launching)," *International Association for Bridge and Structural Engineering Publications* 26 (May 1971): 535; Tedesko, "Design of the Vertical Assembly Building," *Civil Engineering* 35 (Jan. 1965): 45–49.
14. Anton Tedesko, "Space Truss Braces Huge Building for Moon Rocket," *Engineering News–Record* 172 (6 Feb. 1964): 24–27.
15. James H. Deese, "The Problem of Low Level Wind Distribution," paper presented at the Structural Engineers Councils of Florida, First Annual Conference, Tampa, 9 Nov. 1964.
16. Kurt Debus, "Some Design Problems Encountered in Construction of Launch Complex 39," paper given in Darmstadt, Germany, 25 June 1964; R. P. Dodd, "HVAC Temperature Control System for VAB and LCC," with attachment, "VAB HVAC Temperature Control System," 14 July 1963; G. J. Burrus, LCC and Sup. Fac. Sec., memo for record, "LCC Air Conditioning Unit Reliability," 28 July 1965.
17. Debus, "Some Design Problems Encountered," p. 35. *Apollo Launch Complex 39 Facilities Handbook,* issued by the U.S. Army Corps of Engineers, South Atlantic Div.,

p. 14, gives different numbers: height of each door opening, 140 meters; lower door opening 46.32 meters wide and 34.74 meters high; upper door opening 23.16 meters wide and 104.24 meters high.

18. Alexander, "Report on VAB," p. 13; Tedesko, "Design of the Vertical Assembly Building," pp. 48–49; Dodd interview.
19. Philip C. Rutledge, "Vertical Assembly Building—Design of Foundations," *Civil Engineering* 35 (Jan. 1965): 50–52.
20. Alexander, "Vertical Assembly Building—Project Description," pp. 43–44.
21. Stein interview; Bidgood to Clearman, "Design of the Vertical Assembly Building, Advanced Saturn Launch Complex 39," 6 Mar. 1963.
22. Bidgood to Mr. Lenezewski, CE Canaveral Dist., "LC-39 Transfer Aisles," 7 July 1963; Andrew Pickett to Dodd, "Platform Access to S-IC Inter-Tank Area VAB," 27 June 1963; Dodd to Bertram, "LC-39 VAB and LCC," 3 July 1963.
23. "Launch Complex 39," brochure issued by Corps of Engineers for contractors' conference, Oct. 1963, p. 4.
24. R. P. Young, NASA Exec. Off., to Webb, 13 June 1963.
25. M. Menghini, Field Rep., URSAM, memo for record, "Telephone calls to and from Col. Alexander," 31 Oct. 1962; Anton Tedesko, "Assembly and Launch Facilities for the Apollo Program, Merritt Island, Florida: Design of the Structure of the Vertical Assembly Building," paper presented at the ADCE Structural Engineering Conference and Annual Meeting, 19–21 Oct. 1964, p. 10.
26. Stein interview.
27. Brown interview. The NASA–Corps of Engineers movie *The Big Challenge* confirms Brown's testimony.
28. D. T. Brewster to W. W. Kavanaugh, "Minutes of Meeting between M-ASTRA and M-LVOD, 13 Sept. 1962," 31 Oct. 1962; MSFC, "Saturn V Electrical Ground Support Equipment for Launch Complex 39," pp. 1–11.
29. C. Q. Stewart, Mechanical Engineering, memo for record, 1 Aug. 1962.
30. "Minutes of Crawlerway Design Conference," NASA-LOC, Cape Canaveral, 21 Feb. 1963.
31. "Minutes of Crawler Transporter Crawlerway Meeting," LOC E & L Building, 27 Mar. 1963.
32. J. B. Bing, memo for record, 9 July 1963; Bing, memo for record, "Trip Report," 16 Aug. 1963. In line with the complaint of Mr. Bing, the authors found no reference on the part of URSAM people to Giffels and Rossetti in any of the many articles that appeared on URSAM's work on the VAB. In an article in *Engineering News-Record* for 6 Feb. 1964, p. 28, for instance, one of the URSAM principals mentions the companies that constructed the first launch pad and the crawlerway, but not the firm that designed both of them.
33. A. H. Bagnulo to U.S. Army Engineer Dist., Canaveral, 7 Oct. 1963; Lt. Col. Leo J. Miller, Corps of Engineers, Asst. Dist. Engineer, "Construction Coordination Group for NASA-LOC Complex 39," 14 Oct. 1963; Ernst interview.
34. Launch Support Equipment Engineering Div. Monthly Progress Reports, 10 Oct. 1962, 13 Mar. 1964.
35. J. H. Deese, Chief, Facilities Engineering Sec., memo for record, "Engineering Analysis of Launch Pad Diaphragm Construction, Launch Pad 39B," 11 Mar. 1963.
36. "Theoretical Analysis of Surface Temperatures, Flame Trench, Complex 39 A/B, KSC," technical memo 2-62, Mar. 1967, U.S. Army Engineer, Ohio River Div. Laboratories, Cincinnati.
37. Giffels and Rossetti, "Structural Design of Pad Terminal Connection Room and Environmental Control System Buildings," 11 Apr. 1963, sheet 85; Launch Support Equipment Engineering Div., "Preliminary Release Levels for Ground Support Equipment, Launch Complex 39," 19 Dec. 1963.

38. Poppel to Petrone, "Policy Statement and Design Concept for C-5 Propellant Loading Systems," 1 June 1962. A Complex 39 Foundation Prestudy Conference was held on 29 May 1962 at Jacksonvile: C. Q. Stewart, memo for record, "Foundation Design Prestudy Conference, Jacksonville DE, 29 May 1962," 31 May 1962.
39. *NASA Merritt Island Launch Area Master Plan,* vol. 3, pt. 1, *Industrial Area,* sec. 1, "General Site Plan," 22 Mar. 1963. Cf. *John F. Kennedy Space Center, NASA, Master Plan,* pt. 2, *Industrial Area Plans,* sec. 1, "General Site Plan," 25 Oct. 1965.
40. House Committee on Science and Astronautics, *Master Planning of NASA Installations,* House report 167, 89th Cong., 1st sess., 15 Mar. 1965, p. 24.
41. MSC Florida Ops., *Merritt Island Facilities,* undated pamphlet describing facilities funded through FY 1963, 1964, and 1965.
42. Dir., Information Systems, KSC, "Project Development Plan for Launch Instrumentatation," 6 June 1966, p. 2-1.
43. Bruns interview, 22 Aug. 1969.
44. Bidgood to CE Jacksonville, "Central Instrumentation Facility, MILA," 10 May 1963; B. Baker, memo for record, "Siting of the CIF," 22 Aug. 1963.
45. LOC Staff Study, *Concepts for Support Service at the Merritt Island Launch Area,* 6 May 1963.
46. Ibid.; Albert F. Siepert, memo for record, 20 Oct. 1966.
47. "Interim Agreement Implementing the 17 Jan. 1963 Agreement between the Department of Defense and NASA Regarding Management of the Atlantic Missile Range of DOD and the Merritt Island Launch Area of NASA, Part III, Logistic and Administrative Functions," signed 10 May 1963 by Maj. Gen. L. I. Davis, USAF, and Debus; NASA release 63-111, 23 May 1963.
48. C. C. Parker, LOC Asst. Dir. for Admin., to Debus, 23 May, 5, 19 June, 13 Aug., 9 Oct., 14 Nov. 1963, 8 Jan. 1964; LOC release 74-63, 4 Oct. 1963.
49. Parker to Debus, 18 Sept., 17 Oct., 7, 27 Nov. 1963, 2, 16 Jan. 1964; KSC release 56-64, 24 Apr. 1964.

Chapter 12: From Designs to Structures

1. LOC, "Construction Progress Reports," 6 Nov. 1962, p. 5; 27 Nov. 1962, p. 5; 10 July 1963, p. 2.
2. *Spaceport News,* 31 Oct. 1963, p. 4.
3. "Construction Progress Reports," 15 June 1963, p. 7; 13 Sept. 1963, p. 1; D. E. Eppert, Chief, Construction Div., Canaveral Dist., Corps of Engineers, to J. J. Frangie, with attachment: "Tabulation of Contracts Supervised by the Corps of Engineers for Construction of Complexes 34 and 37 as Well as Work on Merritt Island for NASA Facilities" (hereafter cited as Tabulation of Corps of Engineers Contracts).
4. *Spaceport News,* 8 Aug. 1963, pp. 4–5.
5. Ibid., 28 Feb., 8, 15 Aug. 1963, 29 July 1965.
6. Ibid., 15 Aug. 1963; "Launch Operations Progress Report," 26 Aug. 1963.
7. John F. Kennedy, address at Rice University 24 Sept. 1962, *Public Papers of the Presidents,* Washington, 1963, p. 329.
8. NASA Hq., OMSF Instruction MD-M9330.001, 15 Oct. 1962, with enclosure.
9. Facilities Programing Off., Facilities Engineering and Construction Div., LOC, "Summary Project Status Report," 29 Nov. 1963, p. III-1; Tabulation of Corps of Engineers Contracts, Sept. 1968.
10. *Spaceport News,* 16 Jan. 1964.
11. Corps of Engineers, South Atlantic Div., Canaveral Dist., *Apollo Launch Complex 39 Facilities Handbook* (hereafter cited as *LC-39 Facilities Handbook*), pp. 4–5.

12. *Spaceport News,* 16 Jan. 1964, p. 7.
13. "Construction Progress Reports," 16 Aug. 1963, p. 9; 20 Jan. 1964, p. 11; *Spaceport News,* 16 Jan. 1964, 26 Sept. 1963; *LC-39 Facilities Handbook,* p. 15.
14. *Spaceport News,* 16 Jan. 1964.
15. "Summary Project Status Report," 29 Nov. 1963, p. III-2.
16. "LOC Monthly Status Report to the Management Council, Office of Manned Space Flight," presented by Kurt H. Debus, 24 Sept. 1963, p. 15; Summary Project Status Report, 29 Nov. 1963, p. III-2.
17. "Construction Progress Report," 22 Nov. 1963, p. 15.
18. "Summary Project Status Reports," 29 Nov. 1963, p. III-3; 17 Apr. 1964, p. III-2; *Spaceport News,* 19 Sept. 1963, p. 1.
19. "Construction Progress Report," 29 Jan. 1964, p. 20.
20. "Narrative Project Status Report, 1–30 Apr. 1964," p. I-9.
21. "Construction Progress Report," 20 Jan. 1964, pp. 10–11; *LC-39 Facilities Handbook,* pp. 3, 5.
22. "Narrative Project Status Report, 1–28 Feb. 1964," p. I-9; "Summary Project Status Report," 2 Oct. 1964, pp. IV-1, IV-2.
23. *Detailed Construction Schedule, VAB Area Facilities,* 1 May 1964, rev. 30 June 1964, with cover letter from William E. Pearson, Chief, Schedules Off., 27 July 1964 (this series of schedules, revised and issued periodically, will be cited as *Detailed Construction Schedule,* facility, date).
24. "Narrative Project Status Report, 1–30 Apr. 1964," p. I-9.
25. *Apollo/Saturn V MILA Facilities Description,* report K-V-011, p. 1-13.
26. *Spaceport News,* 3, 10, 17 Sept. 1964.
27. Ibid., 3 Sept. 1964; Jones interview.
28. *Spaceport News,* 8 Oct. 1964.
29. "Narrative Project Status Report, 1–31 Dec. 1964," pp. I-4, I-5.
30. Ibid; *LC-39 Facilities Handbook,* pp. 9–10.
31. A. H. Bagnulo, Dir., Facilities Engineering and Construction Div., KSC, "Revised Designations for NASA Facilities," 3 Feb. 1965; George E. Mueller, Assoc. Admin. for Manned Space Flight, to Dir., KSC, "Facility Titles, KSC," with attachment, 8 Sept. 1965.
32. "Narrative Project Status Report," 1–31 Jan. 1965, pp. I-3 through I-5.
33. Ibid., 1–31 Mar. 1965, p. I-2; *Spaceport News,* 1 Apr. 1965.
34. "Construction Progress Report," 27 Nov. 1962.
35. Ibid., 22 Jan., 25 Feb. 1963.
36. MSC Florida Ops., "Description and Justification for Spacecraft Operations and Checkout Building," included in *John F. Kennedy Space Center, NASA Fiscal Year 1963 Estimates, Apollo Mission Support Facilities, Project 7623;* "Manned Spacecraft Center Consolidated Activity Report for 16 Feb.–21 Mar. 1964," p. 78.
37. KSC, "Project Status Report," 1–31 Dec. 1964, p. I-39; Tabulation of Corps of Engineers Contracts, Sept. 1968.
38. Reyes interview, 24 June 1974; Chauvin interview, 24 June 1974.
39. KSC, *Apollo/Saturn V MILA Facilities Description*, K-V-011, p. 3-1.
40. KSC, "Technical Progress Report," 19 Feb. 1964, p. 19.
41. Ling-Temco-Vought, Inc., "Historical Events—Calendar year 1964, Gemini and Apollo Programs and Facilities, Manned Spacecraft Center Florida Operations, Cape Kennedy and Merritt Island," 22 Dec. 1964; Morris interview.
42. *Spaceport News,* 15 Apr. 1965.
43. Ibid.
44. Ibid., 27 May 1965.
45. Ibid., 13 Sept., 26 Dec. 1965.

46. KSC Weekly Notes, Miraglia, 6 July 1964; Parker, 7 June 1964; KSC, *C. O. E., Report of Fatal Accident at LC-39,* signed by Col. W. L. Starnes.

Chapter 13: New Devices for New Deeds

1. D. D. Buchanan, memo for the record, 16 April 1962.
2. LOC, *Procurement Plan,* signed by Kurt Debus, Director, 11 Sept. 1962, p. 2.
3. DDJ, 1, 7 Nov. 1962; Debus to Brackett, 29 Nov. 1962; LOD, "Crawler/Transporter Proposal Conference Attendees, 17 December 1962" (in Buchanan file).
4. R. P. Young, Exec. Off., memo for record, 13 Mar. 1963, in NASA History Office; James E. Webb and Robert C. Seamans, Jr., "Statement of the NASA Administrator on Selection of a Contractor for the Crawler-Transporter," 13 May 1963, ibid.; *Congress and the Nation, 1945–1964* (Washington: Congressional Quarterly Service, 1965), p. 320.
5. "Briefing, Crawler-Transporter Procurement," 5 Feb. 1963, copy in Fred Renaud's private papers.
6. "Decision to Negotiate an Individual Contract under 10 USC 2394 (a) (11)," 5 Dec. 1963; "Determination and Findings for Method of Contracting, Cost-Plus-Incentive Fee," 7 Dec. 1962, both in NASA History Off.
7. W. Kraft, Admin. Asst., Marion Power Shovel Co., to Theodor A. Poppel, 11 Dec. 1963, pp. 1–4, in Fred Renaud's private papers.
8. Ibid., pp. 5–6.
9. Renaud interview, 4 Apr. 1973.
10. *Aviation Week* 84 (20 June 1966): 78.
11. Gramer interview, 19 July 1973.
12. "Fire Alarm System for Crawler-Transporter," 29 Jan. 1965, in Fred Renaud's private papers.
13. Gorman to Petrone, 22 Mar. 1965, including a memo of 29 Feb. (*sic*) 1965, in Fred Renaud's private papers.
14. "Fire Protection Survey and Recommendations," attachment to letter of C. W. Conway to Ronald Worchester, 16 June 1965; copies in Fred Renaud's private papers.
15. *Spaceport News*, 15 Feb. 1968, p. 6.
16. Unless otherwise cited, the descriptive information in this and the following paragraphs concerning the crawlerway and launch pad A facilities is based on the *LC-39 Facilities Handbook,* pp. 35–51.
17. *Spaceport News,* 19 Sept. 1963, p. 1; 17 Oct. 1963, p. 1; Tabulation of Corps of Engineers Contracts, Sept. 1968; "Summary Project Status Report," 29 Nov. 1963, pp. III-2, IV-3; *Technical Progress Report, Second Quarter CY 1965* (TR-194), 30 July 1965, p. 38.
18. *Apollo/Saturn V MILA Facilities Descriptions,* report K-V-011, p. 1-26.
19. Memo, Col. Bagnulo, 3 Feb. 1965.
20. Hahn interview.
21. Boylston interview.
22. Wm. Clearman, "Prototype of Service Arm 6," 30 July 1963, on microfilm in Vehicle Servicing and Accessories Sec. of Design Engineering Off., KSC.
23. "Qualification Test for Cable Retract Sled for Saturn V and Pneumatic Console No. 2," prepared by C. Dyer, Brown Engineering, in Design Engineering Files; photographs in Brad Downs's Design Engineering Files, KSC. The authors are indebted to Mr. Downs for his help in this section.
24. "Weekly Notes," Haworth, 5 Aug.; Clark, 6 Aug. 1964; Gramer interview, 19 Sept. 1972.
25. "Weekly Notes," 19 Aug. 1964.

26. *Technical Progress Report Third and Fourth Quarter CY 1964* (TR-159), 5 Mar. 1965, p. 61; "Saturn V Swing Arm Program Problem," an analytical statement, unsigned and undated. This contract was NAS 10-1751.
27. Rowland interview; R. D. Rowland, Hayes International Corp., to Benson, 25 July 1972.
28. James W. Dalton and Willard Halcomb, Apollo–Saturn V Test and Systems Engineering Off., to Petrone, 28 Oct. 1964; "Saturn V Swing Arm Program Problem," p. 1.
29. William L. Clearman, Jr., Chief, Apollo-Saturn V Test and Systems Engineering Off., to Chief, Launch Equipment Support Sec., Procurement Div., "Contract NAS10-1751, Proposed Changes to Incorporate Revised Drawing Lists," 23 Nov. 1966.
30. James W. Dalton, Apollo-Saturn V Test and Systems Engineering Off., "Minutes of Meeting—Change Review Board—Service Army Contract with International—17 Sept. 1964," 22 Sept. 1964; *Method of Handling Engineering Changes, Contracts NAS10-1751—NAS10-1847.*
31. James W. Dalton to William T. Clearman, "Status of Hayes Service Arm Contract as Result of Sole Source Vendor Items," 2 Dec. 1964.
32. "Saturn V Swing Arm Program Problem," p. 2.
33. Kurt H. Debus to L. F. Jeffers, Hayes International Corp., Birmingham, AL, 5 Nov. 1965.
34. "Saturn V Swing Arm Program Problem," p. 1; "Management Inquiry into the Procurement of Service Arms for Launch Complex 39," pp. 42, 43, 62.
35. "Saturn V Swing Arm Program Problem," p. 1; Gramer interview, 21 Sept. 1972.
36. *Spaceport News,* 28 Oct. 1965, p. 3.
37. *Technical Progress Report Third and Fourth Quarter CY 1964* (TR-159), 5 Mar. 1965, p. 61; Procurement Div., "Status Summary of Active Contracts as of 31 Mar. 1964," sec. II, p. 25.
38. *Technical Progress Report Third Quarter CY 1965* (TR-250), 30 Sept. 1965, pp. 3–18.
39. *Saturn V Launch Support Equipment General Criteria and Description* (SP-4-37-D), rev. 15 Sept. 1964, Launch Support Equipment Engineering Div., pp. 2–62; *Technical Progress Report First Quarter CY 1965,* 26 Apr. 1965, p. 46; KSC Procurement Div., "Status Summary of Active Contracts as of 30 Sept. 1966," sec. II, p. 6.
40. "Apollo/Saturn V MILA Facilities Descriptions," pp. 2-81, 2-82; "Construction Progress Reports," 1 July 1965, p. 4.
41. R. T. Cruden and J. R. Ellis, memo for record, "Ordnance Meeting, LC-39 Arming Tower," 25 Mar. 1963; J. R. Ellis, memo for record, "Ordnance Requirements, Arming Tower LC-39," 26 Mar. 1963; W. T. Clearman, Jr., and James H. Deese, "Meeting at Complex 34 Operations Support Building to Discuss Saturn-V Ordnance Installation Problem," 27 Mar. 1963. The authors are indebted to Francis Jarrett for research on this subject, which is covered more fully in Jarrett and Lindemann, "History of the John F. Kennedy Space Center, NASA, to 1965," typescript.
42. J. R. Ellis, memo for record, "Meeting and Discussions Concerning Arming Tower, LC-39," 9 Apr. 1963; minutes of meeting, "Rust Contract BE-9002, LC-39 Arming Tower, Contract DA-08-123-ENG-(NASA-1752)," 15 Apr. 1963.
43. Off. of the Canaveral Dist. Engineer, "Report on Restudy of Arming Tower to Resolve Dead Load and Wind Load Problems," 20 Dec. 1963, pp. 1–5.
44. Ibid.
45. Vehicle Design Integration Working Group, "Minutes of the Saturn V Common Ordnance Meeting," Huntsville, AL, 10–11 Dec. 1963, pp. 1–4.
46. "Summary Project Status Report," 17 Apr. 1964, p. IV-3; "Development Summary Schedule, Complex 39, 1963"; *Technical Progress Report, Third and Fourth Quarter CY 1964,* 5 Mar. 1965, pp. 45, 58; Tabulation of Corps of Engineers Contracts, Sept. 1968;

Technical Progress Report, Second Quarter CY 1965, 30 July 1965, p. 38; *Technical Progress Report, Third Quarter CY 1965* (TR-250), 30 Sept. 1965, pp. 3–18; *LC-39 Facilities Handbook*, p. 54.
47. *Technical Progress Report, Third and Fourth Quarter CY 1964* (TR-159), 5 Mar. 1965, p. 45.
48. *Technical Progress Report, First Quarter CY 1965* (TR-168), 26 Apr. 1965, p. 33; *Technical Progress Report, Second Quarter CY 1965* (TR-194), 30 July 1965, p. 34.
49. H. D. Brewster and E. G. Hughes, *Lightning Protection for Saturn Launch Complex 39,* report TR-4-28-2-D, 18 Oct. 1963.
50. Ibid., app. A.
51. Ibid., pp. 3-6 through 3-16 and app. A.
52. Ibid., pp. 2-3, 3-3, 3-4, A-1; A. R. Raffaelli, "Introduction to Lightning," report LOC LT1R-2-DE-62-6, 14 Dec. 1962.
53. H. D. Brewster to Lightning Protection Committee, "Minutes of the Third Lightning Protection Committee Meeting, 29 Sept. 1965, at KSC," 20 Oct. 1965, KSC Technical Documents Library.
54. KSC, "Weather Effects on Apollo/Saturn V Operations, Apollo 4 through Apollo 13," report 630-44-0001, 27 July 1970.
55. KSC release 11-66, 21 Jan. 1966.
56. *Building Construction Magazine,* Feb. 1966, p. 29.

Chapter 14: Socio-Economic Problems on the Space Coast

1. Petrone interview, 25 May 1972, pp. 62–68. Petrone discussed the difference between the industrial and construction workers in a sympathetic and understanding way.
2. Senate Committee on Government Operations, Subcommittee on Investigations, *Hearings on Work Stoppages at Missile Bases*, 87th Cong., 2d sess., 25 Apr.–9 June 1961.
3. Ibid., pt. 1, pp. 11–15, 36–46.
4. The John F. Kennedy Space Center Missile Site Labor Relations Committee, "Function Responsibilities and Procedure," p. 1.
5. Glenn M. Parker, "The Missile Site Labor Commission," *ILR Research* 8 (1962): 11.
6. John Miraglia, "Project Stabilization Agreement," pp. 1–2.
7. Senate Committee on Government Operations, Subcommittee on Investigations, *Hearings on Work Stoppages at Missile Bases*, 87th Cong., 1st sess., pt. 2, 4 May 1961, pp. 520 ff.
8. Yates interview.
9. Edward Kiffmeyer, Labor Relations Off., AFETR, "Strike Summary Reports," Patrick Air Force Base, FL, monthly reports from Jan. 1962–July 1965; *History of Air Force Missile Test Center*, vol. 1, 1964, p. 166.
10. *Spaceport News*, 6 Feb. 1964.
11. *Melbourne Daily Times*, 18 Feb. 1964.
12. *Orlando Sentinel*, 5 Feb. 1964.
13. Charles L. Buckley, Jr., Chief, Security Off., memo for record, "FEC Incident, MILA," 1 Feb. 1964.
14. *Cocoa Tribune*, 10 Feb. 1964.
15. *Orlando Sentinel*, 12 Feb. 1964.
16. *Melbourne Daily Times*, 18 Feb. 1964.
17. KSC Weekly Notes, Miraglia, 14 Feb. 1964.
18. *Orlando Sentinel*, 19, 28 Feb. 1964.
19. KSC Weekly Notes, Miraglia, 22 Apr. 1964.
20. Ibid.

21. KSC Weekly Notes, Miraglia, 3 June 1964.
22. KSC Weekly Notes, Miraglia, 11 June 1964; Titusville *Star-Advocate*, 10 June 1964.
23. Paul Styles, Dir., Off. of Labor Relations, memo, 9 June, 1964, copy in files of KSC Security Office; *Miami Herald*, 19 June 1964, p. 2; Gooch interview; Horner interview.
24. KSC Weekly Notes, Miraglia (signed by Oliver E. Kearns), 25 June 1964.
25. Ibid.; *History of the Air Force Missile Test Center*, vol. 1, 1964, pp. 159, 165.
26. KSC Weekly Notes, Miraglia, 16 Sept. 1964.
27. KSC Weekly Notes, Miraglia (signed by Oliver E. Kearns), 23 July, 20 Aug., 2 Sept. 1964.
28. Ibid., 10 Sept. 1964. Several contractor representatives who dealt with labor matters shared Kearns' view of Baxley.
29. Ibid., 16 Sept. 1964.
30. *Cocoa Tribune*, 14 Sept. 1965.
31. KSC Weekly Notes, Kearns, 22 Sept. 1965; *Orlando Sentinel*, 17 Sept. 1965.
32. *Cocoa Tribune*, 4 Oct. 1965.
33. *Time*, 4 July 1969, p. 38.
34. George L. Simpson, Jr., to Webb and Dryden, 25 June 1965.
35. Annie May Hartsfield, Mary Alice Griffin, and Charles M. Grigg, eds., *Summary Report NASA Impact on Brevard County* (Tallahassee: Institute of Social Research, Florida State Univ., 1966), pp. 10–11, table 2, p. 21, citing U.S. census reports.
36. Ibid., pp. 13, 52.
37. Ibid., pp. 104, 106, 107.
38. Ibid., pp. 17, 18, 26, 96.
39. Charles Grigg and Wallace A. Dynes, *Selected Factors in the Deceleration of Social Change in a Rapidly Growing Area* (Tallahassee, 1966), table 3, p. 144.
40. *Spaceport News*, 16 May 1963, p. 6.
41. Ibid., 13 June 1963, p. 3.
42. Ibid., 14 May, 9 July 1964. Sixteen thousand individuals, 56% of the total work force, responded to the questionnaires.
43. Siebeneichen interview.
44. Peter Dodd, *Social Change in Space-Impacted Communities* (Cambridge, MA: The Committee on Space of the American Academy of Arts and Sciences, Aug. 1964), pp. 20–21.
45. KSC Weekly Notes, Miraglia, 22 Apr., 27 May 1964.
46. KSC Weekly Notes, Van Staden, 7 Apr. 1965.
47. *Spaceport News*, 20 Aug. 1964.
48. Ibid., 1 July 1965.
49. *Florida Statistical Abstract, 1969* (Gainesville: Univ. of Florida, 1969), pp. 21, 28.
50. *Spaceport News*, 29 Feb. 1968.
51. *Time*, 4 July 1969, p. 38.
52. Quoted in John G. Rogers, "What Life at Cape Kennedy Does to Marriage," *Parade*, 9 July 1969.
53. Dr. Ronald C. Erbs, M.D., to Faherty, 17 July 1974, in author's personal files.
54. Nazaro interview.

Chapter 15: Putting It All Together: LC-39 Site Activation

1. KSC, "Apollo/Saturn V Facility Activation Plan," 3d Coordination Draft, 30 Dec. 1965; Petrone interview, 17 Sept. 1970; "Presentation of the NASA Oversight Subcommittee, Committee on Science and Astronautics, House of Representatives," 29 Oct. 1968, pp. 38–57.

2. William T. Clearman, Jr., to T. A. Strong, "Lt. Col. Donald R. Scheller, USAF (NASA)," 12 Oct. 1964; Clearman interview, 13 Sept. 1973.
3. KSC, "Minutes of Apollo/Saturn V Site Activation Board Meeting #1," 19 Mar. 1965.
4. KSC, *LC-39 Site Activation Master Schedule (Preliminary), Level A,* rev. 17 Feb. 1967.
5. KSC, "Minutes of Site Activation Board Meeting #1"; "Presentation to the NASA Oversight Subcommittee," 29 Oct. 1968, pp. 43–44.
6. Gruene to Apollo/Saturn V Test Off., "Comments on Site Activation Board Charter," 1 Apr. 1965.
7. Bagnulo to Scheller, "Site Activation Board Charter," 30 Mar. 1965.
8. Clark to Scheller, "Comments on Apollo/Saturn V Site Activation Board," 31 Mar. 1965.
9. KSC, "Minutes of Apollo/Saturn V Site Activation Board Meeting #3," 5 Aug. 1965.
10. Donald R. Scheller, "Management by Exception, Activation of Apollo/Saturn V Launch Complex 39," 15 May 1967; "Presentation to the NASA Oversight Subcommittee," 29 Oct. 1968, pp. 41–42.
11. L. S. Harris, Chief, Site Operation Gp., Engineering and Development Dir., KSC, to Bagnulo, "Activation Projects, LC-39," 23 Sept. 1965, in KSC Engineering and Development Dir. Reading Files, 1965–66; "Minutes of KSC Site Activation Working Group Meeting #1," 3 Dec. 1965; Scheller to SAB, "Apollo/Saturn V SAB Management Meeting Membership," 19 Jan. 1966.
12. Scheller, "Management by Exception"; Clearman interview, 26 Oct. 1973; Fulton interview; Chandler interview.
13. Boeing Atlantic Test Center Management Systems Staff, "ERS Recovery Plan," by A. J. Culver and K. G. Baird, Apr. 1966.
14. Scheller, "Management by Exception"; Murphy interview.
15. Petrone's notes, 14 Feb. 1963; Wagner interview, 21 Sept. 1973; Gassman interview; *NASA Apollo Inter-Center ICD Management Procedure*, report CM-001-001-1B, Jan. 1969, pp. 3–4.
16. Petrone, "KSC Apollo/Saturn Configuration Management Program Directive," 29 Sept. 1965, Management Configuration Off. files.
17. KSC Apollo Program Directive No. 2, 9 Dec. 1965.
18. Gassman interview; Leet interview, 8 Nov. 1973.
19. KSC/MSC, *ICD-IRN Processing,* 6 May 1968; Wagner interview, 21 Sept. 1973.
20. Petrone to Poppel, "VLF-39 Facility Checkout Vehicle Minimum Requirements and Operational Characteristic," 1 Mar. 1962; Poppel (signed by Owens) to Petrone, "Minimum Requirements for Facility Checkout Vehicle for Complex 39," 2 Apr. 1962.
21. Phillips to NASA Manned Space Flight Centers, "Apollo Delivery and Launch Schedules," 16 Feb. 1965; Phillips to NASA Manned Space Flight Centers, "Apollo Schedules and Mission Assignments," 12 Jan. 1965, in Phillips file, NASA Hq. History Off.; KSC, Plans, Program, and Resources Dir., "Verification of '500-F' Schedule Dates Based upon OMSF Approved Apollo Schedule," 10 Mar. 1966.
22. Petrone to Arthur Rudolph, Saturn V Program Manager, MSFC, 11 Jan. 1966.
23. Haggard interview; "Marion Power and Shovel Company, PMSLC Hearing, Miami, Florida, May 1965," contract NASA 10-477, KSC Labor Relations Off.
24. Michael E. Haworth, Jr., NASA Contracting Off., to Marion Power Shovel, "Contract NAS 10-477," 25 Jan. 1965.
25. Clearman to Petrone, Weekly Notes of 25 June 1965.
26. George W. Walter, *Modifications to Bearings for Traction Support Rollers on Crawler-Transporters*, report KSC TR-260-D, 15 Dec. 1965, pp. 1–2; KSC Weekly Notes, Poppel, 29 July 1965; F. Jones, Technical Supervisor, to Richard McCoy, "Contract NAS 10-477, Salvage of Bearings," 26 Oct. 1965.
27. Walter, *Modifications to Bearings*, p. 2; "KSC Press Briefing and Crawler Demonstration," 25 Jan. 1966, pp. 3–4, 12.

28. M. E. Haworth to Patrick Kraft, Treasurer, Marion Power Shovel Co., 15 June 1965.
29. M. E. Haworth to F. Boyle, Pres., Marion Power Shovel Co., "NAS 10-477," 14 Oct. 1965; Gordon Harris, Chief of Public Affairs, to Debus, 30 Sept. 1965.
30. Morgan F. Jones to Poppel, 15 Nov. 1965; Poppel, "Weekly Notes to Debus," 8 Oct. 1965; "Technical Progress Reports," Third Quarter 1965, pp. 3–16; Buchanan interview, 4 Oct. 1974; Walter, *Modifications to Bearings*, p. 12; "KSC Press Briefing and Crawler Demonstration," 25 Jan. 1966, pp. 16–17.
31. KSC Weekly Notes, Poppel, 3, 10 Dec. 1965; Bagnulo, 17 Dec. 1965; *Spaceport News*, 3 Feb. 1966.
32. NASA, *PERT, Program Evaluation and Review Technique, Handbook*, NPC-101, 1 Sept. 1961; "KSC Presentation to the NASA Oversight Subcommittee, Committee on Science and Astronautics," 29 Oct. 1968, pp. 38–58; Potate interview, 6 June 1972.
33. "Minutes of Apollo/Saturn V Site Activation Board Meeting #3," 5 Aug. 1965; KSC Weekly Notes, Petrone, 30 July 1965.
34. L. Steven Harris to Bagnulo, "Site Activation Board Meeting," 28 Oct. 1965, in Engineering and Development Dir. reading files, 1965–1966; Wiley interview, 31 Oct. 1973.
35. Petrone to Phillips, 4 Nov. 1965.
36. Petrone to Phillips, "Proposed SA500F-1/501 Work Around Schedule," 7 Dec. 1965.
37. KSC Weekly Notes, Bagnulo, 17 Nov. 1965; Bagnulo to Debus and others, "Pad A Settlement," 9 Dec. 1965; Roberts interview.
38. "Minutes of Apollo/Saturn Site Activation Board Meeting #14," 6 Jan. 1966; "LC-39 Site Activation Master Schedule Meeting," 17 Jan. 1966.
39. Potter interview; Steven Harris interview; Tom Wills interview; "Reading File of Engineering Division's Site Activation Group," Sept. 1965, Steven Harris's files, KSC.
40. "Minutes of Apollo/Saturn V Site Activation Board Meeting #16," 3 Feb. 1966; KSC, "PERT Analysis Report," 20 Jan. 1966.
41. Clearman, Weekly Notes to Petrone, 7 Jan., 4, 18, 25 Feb., 3 Mar. 1966.
42. Clearman, Weekly Notes to Petrone, 3, 25 Mar., 1, 15 Apr. 1966.
43. "LC-39 Site Activation Status Reports," weekly for March and April 1966; Brewster interview; Hahn interview.
44. "LC-39 Site Activation Status Report," 27 Apr. 1966; Hahn interview.
45. "LC-39 Site Activation Status Reports," weekly reports for Mar. 1966; Weekly Notes, Bagnulo, 25 Mar. 1966; *Spaceport News*, 10, 31 Mar. 1966.
46. Rigell interview; "LC-39 Site Activation Weekly Reports," Mar., Apr. 1966.
47. "LC-39 Site Activation Reports," 13 Apr., 5 May, 3 Aug. 1966.
48. "Minutes of Apollo/Saturn V Site Activation Board Meeting #24," 26 May 1966; *Spaceport News*, 26 May 1966.
49. "Minutes of Apollo/Saturn V Site Activation Board Meeting #25," 16 June 1966; *Spaceport News*, 16 June 1966.
50. "Minutes of Apollo/Saturn V Site Activation Board Meetings" 25, 26, and 27, dated 16, 23 June, 7 July 1966.
51. Barfus interview; Enlow interview; Sparkman interview, 6 Dec. 1973.
52. "Minutes of Apollo/Saturn V Site Activation Board Meeting #27," 7 July 1966.
53. "Minutes of Apollo/Saturn V Site Activation Board Meetings" 27 and 28, dated 7 and 21 July 1966; *Spaceport News*, 4 Aug. 1966.
54. "Minutes of Apollo/Saturn V Site Activation Board Meetings" 28 and 29, 21 July, 4 Aug. 1966.
55. William I. Moore and Raymond J. Arnold, "Failure of Apollo/Saturn V Liquid Oxygen Loading System," 1967 Cryogenic Engineering Conference, 21–23 Aug. 1967, Stanford Univ., CA, paper K-1, in *Advances in Cryogenic Engineering* 13 (1967): 534–44; Boeing Atlantic Test Center, "Technical Report of Complex 39A LOX System Failure, 10 Sept. 1966"; "Fund Board of Inquiry Findings on Failure of LOX Distribution System—19

Aug. 1966," J. G. Shinkle, Chairman; "Presentation to the Congressional Subcommittee on Manned Space Flight," pp. 114–19.

56. Moore and Arnold, "Failure of LOX Loading System," pp. 534–44; "Presentation to the Subcommittee on Manned Space Flight," pp. 114–19.
57. "LC-39 Site Activation Status Reports," weekly for Sept., Oct. 1966.
58. Robert Hotz, *Aviation Week and Space Technology*, 22 Mar. 1965, p. 11.
59. "KSC Presentation to the NASA Oversight Subcommittee," 29 Oct. 1968; Petrone interview, 17 Sept. 1970.

Chapter 16: Automating Launch Operations

1. Sidney Sternberg, "Automated Checkout Equipment—The Apollo Hippocrates," in *Man on the Moon,* ed. Eugene Rabinowitch and Richard Lewis (New York: Basic Books, 1969), pp. 196–97.
2. W. Haeussermann, Dir., Guidance and Control Div., MSFC, memo for record, "Meeting on Saturn Checkout Equipment," 22 July 1960; Paul interview.
3. Debus to Dieter Grau, "Automatic Checkout Committee," 2 Sept. 1960; Richard interview, 12 Dec. 1973. See B. J. Funderburk, *Automation in Saturn I First Stage Checkout* (NASA TN D-4328, Jan. 1968), for story of the Packard Bell 250 and MSFC's early automation efforts.
4. Ludie Richard and Charles O. Brooks, *The Saturn Systems Automation Plan,* MSFC, 15 Sept. 1961, sec. II.
5. Ibid., sec. VII.
6. "Brief Chronological History of the Saturn V Breadboard," attached to *MSFC Automation Plan,* 8 May 1962; Burns interview; Greenfield interview.
7. Jafferis interviews, 19 Dec. 1973, 22 Jan. 1974; Greenfield interview; Whiteside interview, 4 Jan. 1974.
8. Jafferis interview, 19 Dec. 1973; "Description for Use of Saturn Ground Computer on SA-5," draft copy in Jafferis's private papers; *Spaceport News,* 21 May 1964, p. 2.
9. B. E. Duran, "Saturn I/IB Launch Vehicle Operational Status and Experience," read at Aeronautic and Space Engineering and Manufacturing Meeting, New York, 7–11 Oct. 1968, Society of Automotive Engineers reprint 680739; KSC, "Utilization of Saturn/Apollo Control and Checkout System for Prelaunch Checkout and Launch Operations," GP-663, 25 Mar. 1969.
10. W. O. Frost and D. E. Norvell, "Telemetry System Design for Saturn Vehicles," *Proceedings, 1966 International Telemetering Conference,* Los Angeles, 18–20 Oct. 1966, p. 70. See also E. A. Robin, "Development and Utilization of Computer Test Programs for Checkout of Space Vehicles," p. 297; Canaveral Council of Technical Societies, *Proceedings of the Second Space Congress,* Cocoa Beach, FL, 5–7 Apr. 1965, pp. 617, 634; D. M. Schmidt, "Automatic Checkout Systems for Stages of the Saturn V Manned Space Vehicle," *International Convention Record of Electrical and Electronics Engineers* 13 (pt. 4, Mar. 1965), p. 87.
11. Canaveral Council of Technical Societies, *Proceedings of the Second Space Congress,* p. 656; William G. Bodie, "Techniques of Implementing Launch Automation Programs, Saturn IB Space Vehicle System," *Practical Techniques and Applications*, 4: 740. See also *Apollo/Saturn IB Launch Operations Plan AS-203,* KSC document K-IB-021.3, p. 6-8.
12. Duran, "Saturn I/IB Launch Vehicle Operational Status"; KSC, "Utilization of Saturn/Apollo Control and Checkout."
13. Richard Dutton and William Jafferis, "Utilization of Saturn/Apollo Control and Checkout System for Prelaunch Checkout and Launch Operations," paper read at New York

Univ., Project SETE, 24–28 July 1967, pp. 3-34 through 3-43; Medlock interview; Thompson interview.

14. F. Brooks Moore and William Jafferis, "Apollo/Saturn Prelaunch Checkout Display Systems," read at IEEE Conference on Displays, Univ. of Loughborough, England, 7–10 Sept. 1971, pp. 7–9.
15. Ibid., pp. 9, 15–16.
16. Ibid., p. 14.
17. Richard Jenke to Benson, 17 Jan. 1975; Richard Smith interview; Medlock interview; Thompson interview.
18. Dutton and Jafferis, "Utilization of Saturn/Apollo Control and Checkout System," pp. 3-44 through 3-48; Jenke to Benson, 17 Jan. 1975.
19. Jenke to Benson, 17 Jan. 1975; Medlock and Thompson interviews.
20. Jenke to Benson, 17 Jan. 1975; Medlock and Thompson interviews.
21. Fridtjof Speer, Chairman, Saturn System Evaluation Working Gp., MSFC, to LOD Dir., "Justification for Early Delivery of the Saturn Blockhouse Records and Sequence Records," 19 Sept. 1961, Debus reading file.
22. R. W. Bivans, G. D. Matthews, and F. T. Innes, "A Scanning and Digitizing System for Multiple Asynchronous Telemetry Data Sources," read at National Telemetry Conference, Los Angeles, June 1964, p. 1, G. D. Matthews's private papers.
23. Bruns interview, 3 Jan. 1974; Bobby Griffin and G. D. Matthews, *The Real-Time Telemetry Data Processing Effort at the Launch Operations Center,* MTP-LVO-63-2, MSFC, p. 1.
24. Griffin and Matthews, *Real-Time Telemetry Data,* pp. 9–11; Bruns interview, 3 Jan. 1974; Corbett, Hughes, and Jelen interviews.
25. Griffin and Matthews, *Real-Time Telemetry Data,* pp. 3–6; George Matthews interview.
26. Griffin and Matthews, *Real-Time Telemetry Data,* pp. 14–17; Bivans, Matthews, and Innes, "Scanning and Digitizing"; George Matthews interview; LOC Weekly Notes, Sendler to Debus, 23 Aug., 15 Nov. 1962.
27. KSC Computation Br., "Scientific Computation Support of Saturn/Apollo Vehicle, SA-7," TR-103-2, 3 Dec. 1964.
28. Joralan interview, 3 Jan. 1974; LOC Weekly Notes, Sendler to Debus, 22 Mar. 1962.
29. LOC Weekly Notes, Gruene to Debus, 21 Feb. 1962.
30. Raymond Clark, Asst. LOC Dir., to Col. Max Carey, "Request for Additional Data on NASA Telemetry Requirements," 6 Oct. 1962; Debus to Davis, "The AFMTC Launch Area Telemetry System Plan, 28 September 1962," 18 Oct. 1962.
31. Telephone directory, Project Mercury Field Ops., STG, Cape Canaveral, FL, Sept. 1961; "Patent Application on ACE, NASA Case No. 8012," encl. to letter, James O. Harrel to Harold G. Johnson, 20 Jan. 1967, Johnson's private papers; Walton interview, 17 Dec. 1970. The Cape launch team first appeared as Preflight Operations Division on a Sept. 1962 MSC organization chart. Earlier it was called Mercury Field Operations or MSC's Atlantic Missile Range Operations.
32. Parsons interview; Preston interview, 22 Jan. 1974.
33. Parsons interview; Walton interview, 17 Dec. 1970; Preston interview, 22 Jan. 1974; W. E. Parsons, Head, Flight Instrumentation Sec., to C. W. Frick, Head, Apollo Project Off., MSC, "Implementation Plan for Apollo SPACE System," 26 June 1962, Johnson's private papers; "PACE-S/C History," compiled by Harold Johnson ca. 1963, Johnson's private papers.
34. Parsons interview; Harold Johnson interview; "PACE-S/C History"; Walton interview, 17 Dec. 1970.
35. Tom S. Walton, MSC Florida Ops., *Experimental Station Implementation and Planning,* 18 Dec. 1964; Walton interview, 17 Dec. 1970.
36. "PACE-S/C History"; Parsons interview; Walton interview, 23 Jan. 1974.
37. Parsons interview; Norwalk interview; Walton interview, 23 Jan. 1974.

38. Parsons interview.
39. Page interview; *Spaceport News,* 6 Jan. 1966; Apollo Support Dept., General Electric Co., *ACE-S/C, Acceptance Checkout Equipment, Spacecraft,* Daytona Beach, FL, undated.
40. Apollo Support Dept., General Electric Co., *ACE-S/C;* James O. Hassell to Harold G. Johnson, "Patent Application on ACE," 20 Jan. 1967, with encl. 1, "Patent Application on ACE, NASA Case No. 8012," Johnson's private papers; Harold Johnson interview.
41. Moore and Jafferis, "Apollo/Saturn Prelaunch Checkout," pp. 4–7.

Chapter 17: Launching the Saturn IB

1. Debus to Dep. Assoc. Admin. for Manned Space Flight, "Saturn I/IB Pad Utilization," 13 Nov. 1963; T. F. Goldcamp, memo for record, "Modification of LC-34 for Saturn IB," 12 Dec. 1963.
2. OMSF, *Mission Operation Report, Apollo/Saturn Flight Mission AS-201,* NASA report M-932-66-01, pp. 14–17; KSC Weekly Notes, Poppel, 1 July 1965; NASA release 66-32, *Apollo/Saturn 201 Press Kit,* 17 Feb. 1966, pp. 41–43.
3. OMSF, *Mission Operation Report, AS-201*, pp. 14–17; NASA, *AS-201 Press Kit*, pp. 41–43.
4. KSC Weekly Notes, Bagnulo, 5 Aug. 1965; *Spaceport News,* 26 Aug. 1965; KSC, "Daily Status Report, AS-201," 27 Dec. 1965–Jan. 1966.
5. KSC Weekly Notes, Hans Gruene, 26 Aug. 1965.
6. Akens, *Saturn Illustrated Chronology,* p. 117; KSC Weekly Notes, Gruene, 26 Aug. 1965.
7. KSC Weekly Notes, Gruene, 17 Sept., 1 Oct. 1965.
8. KSC Weekly Notes, Gruene, 17, 24 Sept., 1 Oct. 1965; Petrone, 7 Oct. 1965.
9. *Brevard Sentinel,* 20 Feb. 1966; KSC release 17-66, 16 Feb. 1966; *Spaceport News,* 18 Feb., 3 Mar. 1966.
10. MSFC, *Saturn IB Vehicle Handbook,* vol. 1, *Vehicle Description* (prepared by Chrysler Corp. Space Div.), 25 July 1966, p. II-7 (S-IB stage data summary); MSFC, *Saturn-Apollo Space Vehicle Summary, AS-201,* p. 21; NASA, *AS-201 Press Kit,* pp. 37–38; KSC Weekly Notes, Von Staden, 19 Aug. 1965.
11. MSFC, *Saturn-Apollo Space Vehicle Summary, AS-201;* Akens, *Saturn Illustrated Chronology,* p. 121; KSC Weekly Notes, John J. Williams, 28 Oct. 1965; NASA, *AS-201 Press Kit,* pp. 22, 39–40.
12. KSC Weekly Notes, Preston, 30 Sept. 1965; Petrone, 30 Sept. 1965.
13. KSC Weekly Notes, Petrone, 7, 28 Oct. 1965; Gruene, 8, 15, 22, 29 Oct. 1965.
14. KSC Weekly Notes, John J. Williams, 28 Oct., 10, 18 Nov. 1965; Preston, 10, 18 Nov. 1965.
15. KSC Weekly Notes, Gruene, 10 Nov. 1965.
16. KSC Weekly Notes, Gruene, 26 Nov., 3 Dec. 1965.
17. KSC, "Daily Status Report, AS-201," 8–23 Dec. 1965; KSC Weekly Report, John J. Williams to Debus, 6 Jan. 1966.
18. KSC, "Daily Status Report, AS-201," 8–23 Dec. 1965; KSC Weekly Report, Gruene to Debus, 10 Dec. 1965.
19. KSC, "Daily Status Report, AS-201," 27 Dec. 1965–7 Jan. 1966; KSC Weekly Reports, Gruene to Debus, 7 Jan. 1966; Williams to Debus, 6 Jan. 1966.
20. *Miami Herald,* 13 Jan. 1966.
21. Carlson interview, 16 Dec. 1970.
22. Bryan interview.
23. KSC, "Daily Status Report, AS-201," 24 Jan.–18 Feb. 1966; KSC Weekly Notes, Gruene to Debus, 15 Oct. 1965; KSC, *Apollo/Saturn IB Launch Plan, AS-201,* 27 Oct. 1965.

24. Phillips to Petrone, TWX, 17 Feb. 1966, Phillips chronological files.
25. KSC, Launch Vehicle Operations, "Problems in AS-201 Checkout," 11 Mar. 1966.
26. Donnelly interview, 17 Nov. 1970.
27. Gruene interview, 19 Nov. 1970.
28. KSC, *Apollo/Saturn IB Ground Systems Evaluation Report, AS-201,* Apr. 1966.
29. Melvyn Savage, Apollo Test Dir., to Phillips, Apollo Program Dir., "A/S 201 Hold," 3 Mar. 1966.
30. Gruene interview, 19 Nov. 1970.
31. Savage to Phillips, "A/S 201 Hold," 3 Mar. 1966.
32. *Brevard Sentinel,* 20 Feb. 1966; NASA, *Apollo/Saturn 201 Press Kit,* pp. 6–8; KSC, *AS-201 Ground Systems Evaluation Report,* p. iii.
33. Carlson interview, 16 Dec. 1970.
34. KSC, *AS-201 Ground Systems Evaluation Report,* p. iii; NASA, *Sixteenth Semi-Annual Report to Congress,* 1 July–31 Dec. 1966, p. 58; Weekly Notes, E. P. Bertram to Petrone, 3 Mar. 1966.
35. Debus to KSC Management Board, 17 Jan. 1966; Siepert to Debus, "Approach and Status of KSC Task Force on Management Appraisal," 1 Mar. 1966. The research for this portion on KSC's 1966 reorganization was done by Robert Lindemann and Frank Jarrett.
36. KSC, draft briefing memo, "Proposed KSC Reorganization," n.d., p. 3; NASA announcement, "Approval of Revised KSC Organizational Structure," 29 Apr. 1966.
37. KSC, draft briefing memo, "Proposed KSC Reorganization"; KSC, "Approval of Revised KSC Organizational Structure," KSC release 123-66, 29 Apr. 1966.
38. Akens, *Saturn Illustrated Chronology,* p. 138; *Spaceport News,* 30 June, 7 July 1966; NASA, *Apollo/Saturn 203 Press Kit*, 21 June 1966, pp. 2–3, 18–19; KSC, "Daily Status Report, AS-203," 6–15 Apr. 1966.
39. KSC, "Daily Status Report, AS-203," 19 Apr.–31 May 1966.
40. Guy Thomas to Chief, NASA Requirements Br., 1 June 1966, in Rocco Petrone's notes, 1966.
41. NASA, *Apollo/Saturn, AS-203, Post-Launch Report No. 1,* 22 July 1966; Akens, *Saturn Illustrated Chronology,* p. 144; *Spaceport News,* 7 July 1966.
42. KSC, "Daily Status Reports, AS-202," 28 Feb.–22 Aug. 1966, in particular see 30 Mar., 14, 27 Apr., 22 June, 5, 15, 29 July, 8, 15 Aug.; Sasseen interview, 4 Feb. 1974.
43. *Spaceport News,* 18 Aug. 1966.
44. NASA, *Sixteenth Semiannual Report to Congress, 1 July–31 Dec. 1966,* pp. 47–48; NASA release 66-213, 25 Aug. 1966.

Chapter 18: The Fire That Seared the Spaceport

1. Senate Committee on Aeronautical and Space Sciences, *Report on Apollo 204 Accident,* report 956, 90th Cong., 2d sess., 30 Jan. 1968, pp. 3–7.
2. Idem, *Apollo Accident: Hearings,* 90th Cong., 1st sess., pt. 1, pp. 13–54. Dr. Charles A. Berry, chief of medical programs at MSC, introduced and discussed Dr. E. Roth's four-part report, "The Selection of Space-Cabin Atmosphere."
3. Frank J. Handel, "Gaseous Environments during Space Missions," *Journal of Space Craft and Rockets* 1 (July–Aug. 1964): 361.
4. *Report of Apollo 204 Review Board to the Administrator, NASA,* 5 Apr. 1967, app. D, panel 2, pp. D-2-25, D-2-26.
5. *Science Journal* 2 (Feb. 1966): 83.
6. *Space/Aeronautics* 45 (Feb. 1966): 26, 28, 32.
7. Gen. Samuel Phillips, Apollo Program Dir., to John Leland Atwood, Pres., North American Aviation, "NASA Review Team Report," 19 Dec. 1965.

8. Ibid., p. 1.
9. Ibid., p. 66.
10. Senate Committee on Aeronautical and Space Sciences, *Report on Apollo 204 Accident,* pt. 4, p. 318.
11. House Subcommittee on NASA Oversight of the Committee on Science and Astronautics, *Investigation into Apollo 204 Accident: Hearings,* 90th Cong., 1st sess., 1: 404.
12. Ibid., p. 450.
13. *Report of Apollo 204 Review Board,* p. 4-1.
14. "Daily Status Report, AS-204," 29 Aug. 1966; unless otherwise noted, the material in this section is based on these reports between 29 Aug. 1966 and 26 Jan. 1967.
15. *Report of Apollo 204 Review Board,* p. 4-1.
16. Ibid., pp. 4-1, 4-2.
17. Chauvin interview, 6 June 1974.
18. *Report of Apollo 204 Review Board,* p. 4-2.
19. Chauvin and Reyes interviews, 6–7 June 1974.
20. Ibid.
21. *Report of Apollo 204 Review Board,* p. 4-2.
22. Ibid.
23. Notes by M. Mogilevsky, signed, undated, relative to his conversation with Thomas R. Baron, 12–13 Dec. 1966, in files of Frank Childers, KSC.
24. Statement of Frank Childers, 9 Feb. 1967, submitted at the request of the KSC Director, copy in files of Childers.
25. John H. Brooks, Chief, NASA Regional Inspections Off., to Kurt Debus, "Thomas Ronald Baron, North American Aviation Employee," 3 Feb. 1967.
26. Ibid.
27. Hansel interview.
28. Brooks to Debus, 3 Feb. 1967.
29. *Orlando Sentinel*, 6 Feb. 1967. John Hansel said later than North American had ample reason for firing Baron, because he had violated procedural requirements that brought automatic dismissal. Hansel interview.
30. Brooks to Debus, 3 Feb. 1967.
31. Ibid.
32. Titusville *Star-Advocate,* 7 Feb. 1967.
33. Childers interview.
34. Reyes interview, 19 Jan. 1973.
35. House Subcommittee on NASA Oversight of the Committee on Science and Astronautics, *Investigation into Apollo 204 Accident: Hearings,* 90th Cong., 1st sess., 1: 498 ff.
36. Erlend A. Kennan and Edmund H. Harvey, Jr., *Mission to the Moon* (New York: William Morrow and Co., Inc., 1969), pp. 115–16, 147n. This book is highly critical of NASA and the space program, with special emphasis on the 204 fire.
37. Chauvin and Reyes interviews, 6–7 Jun. 1974.
38. *Report of Apollo 204 Review Board,* app. D, panel 7, p. D-7-12.
39. Ibid., app. B, p. B-142, testimony of Clarence Chauvin.
40. Ibid., p. B-145, testimony of William Schick.
41. *Report of Apollo 204 Review Board,* app. D, panel 7, p. D-7-13.
42. Ibid., pp. D-7-4, D-7-5.
43. Ibid., app. B, pp. B-153, B-154, testimony of Gary W. Propst; p. B-159, testimony of A. R. Caswell.
44. Ibid., p. B-91, testimony of Bruce W. Davis.
45. Ibid., p. B-39, testimony of D. O. Babbitt.
46. *Report of Apollo 204 Review Board,* app. D, panel 11, p. D-11-36. At least one member of the Pan American Fire Department, James A. Burch, testified that he had arrived in

time to help open the hatch—even though he admitted the trip to the gantry took from five to six minutes and ascent on the slow elevator consumed two minutes more. Ibid., app. B, p. B-177.

47. *Time,* 10 Feb. 1967, p. 19.
48. *Newsweek,* 13 Feb. 1967, pp. 96–97.
49. *The Sunday Star,* Washington, 21 May 1967.
50. Quoted in *Today,* 14 Apr. 1967; 14 May 1967.
51. *New York Times*, 4 Apr. 1967.
52. H. Bliss, "NASA's in the Cold, Cold Ground," *ATCHE Journal* 13 (May 1967): 419.
53. Lyndon B. Johnson, *The Vantage Point: Perspectives of the Presidency, 1963–1969* (New York: Popular Library, 1971), p. 284.
54. House Subcommittee on NASA Oversight of the Committee on Science and Astronautics, *Investigation into Apollo 204 Accident: Hearings,* 90th Cong., 1st sess., 1: 207.
55. Announcement of Dr. Kurt H. Debus, 3 Feb. 1967, "KSC Cooperation with the Apollo 204 Investigation."
56. *Time,* 10 Feb. 1967, reported rumors of lengthy suffering that preceded the astronauts' deaths. The autopsy disproved these charges.
57. *Aviation Week and Space Technology,* 13 Feb. 1967, p. 33.
58. *Time,* 14 Apr. 1967.
59. *Report of Apollo 204 Review Board,* 6 Apr. 1967, pp. 5-1, 5-2.
60. Ibid., p. 5-9.
61. Ibid., pp. 6-1, 6-2, 6-3.
62. Atkins interview, 29 May 1974.
63. *Report of Apollo 204 Review Board,* pp. 6-2, 6-3.
64. Ibid., p. 6-3.
65. House Subcommittee on NASA Oversight of the Committee on Science and Astronautics, *Investigation into Apollo 204 Accident: Hearings,* 90th Cong., 1st sess., 1: 81.
66. Atkins interview, 5 Sept. 1973.
67. *Report of Apollo 204 Review Board,* app. B, pp. B-39 through B-146.
68. Senate Committee on Aeronautical and Space Sciences, *Apollo Accident: Hearings,* pts. 1, 2.
69. Ibid., pt. 4, p. 365; House Subcommittee on NASA Oversight of the Committee on Science and Astronautics, *Investigation into Apollo 204 Accident: Hearings,* 90th Cong., 1st sess., 1: 13.
70. Senate Commitee on Aeronautical and Space Sciences, *Apollo Accident: Hearings,* 90th Cong., 1st sess., pt. 6, p. 541.
71. Ibid., pt. 2, p. 127; pt. 5, pp. 416–17; House Subcommittee on NASA Oversight of the Committee on Science and Astronautics, *Investigation into Apollo 204 Accident: Hearings,* 90th Cong., 1st sess., 1: 265.
72. House Subcommittee on NASA Oversight of the Committee on Science and Astronautics, *Investigation into Apollo 204 Accident: Hearings,* 90th Cong., 1st sess., 1: 386–87.
73. Ibid., pp. 390–91.
74. Ibid., p. 391.
75. Ibid., 1: 460–80, 501.
76. Senate Committee on Aeronautical and Space Sciences, *Report on Apollo 204 Accident,* report 956, 90th Cong., 2d sess., p. 7; Senate Committee on Aeronautical and Space Sciences, *Apollo Accident: Hearings,* 90th Cong., 1st sess., pt. 4, p. 319. "Some early tendency to shift blame for the fire upon North American Aviation," Tom Alexander wrote in *Fortune,* July 1969, p. 117, "was gradually supplanted by NASA's admission that the fire was largely its own management's failure. NASA had overlooked and thereby in effect approved an inherent fault in design, namely the locking up of men in a capsule full of inflammable materials in an atmosphere of pure oxygen at sixteen pounds per

square inch of pressure. NASA, after all, had more experience in the design and operation of space hardware than any other organization and was, therefore, more to blame than North American if the hardware worked badly."

In 1972, however, North American Rockwell Corp., North American Aviation, Inc., Rockwell Standard Corp., and Rockwell Standard Co. settled out of court with the widows of the three astronauts who charged the spacecraft builders with negligence. The widows of White and Chaffee each received $150000, the widow of Grissom $300000. *Washington Post,* 11 Nov. 1972.

77. Senate Committee on Aeronautical and Space Sciences, *Apollo Accident: Hearings,* 90th Cong., 1st sess., pt. 5, pp. 397, 428.
78. Senate Committee on Aeronautical and Space Sciences, *Report on Apollo 204 Accident,* report 956, 90th Cong., 1st sess., pp. 11, 20.
79. "New Hatch Slashes Apollo Egress Time," *Aviation Week and Space Technology,* 15 May 1967, p. 26.
80. William J. Normyle, "NASA Details Sweeping Apollo Revisions," *Aviation Week and Space Technology,* 15 May 1967, p. 24.
81. George E. Mueller, "Apollo Actions in Preparation for the Next Manned Flight," *Astronautics and Aeronautics* 5 (Aug. 1967): 28–33; "Records of Spacecraft Testing, July 1968," in files of R. E. Reyes, Preflight Operations Br., KSC.
82. Normyle, "NASA Details," p. 25; Reyes interview, 30 Oct. 1973; Atkins interview, 5 Nov. 1973. Actually the official reports to Debus during 1966 show no written reports from the Safety Office. Atkins must have reported orally at irregular intervals.
83. Mueller, "Apollo Actions," p. 33.
84. House Special Studies Subcommittee of the Committee on Government Operations, *Investigation of the Boeing-TIE Contract: Hearings,* 90th Cong., 2d sess., pp. 3–9.
85. Ibid., pp. 10, 13–14, 24.
86. "Technical Integration and Evaluation Contract," NASw 1650, Statement of Work, 15 June 1967.
87. Wagner interview; "Boeing-TIE Goals and Accomplishments," copy in file of Walter Wagner, KSC.

Chapter 19: Apollo 4: The Trial Run

1. OMSF, *Apollo Program Flight Summary Report, Apollo Missions AS-201 through Apollo 8,* pp. 13–17; MSFC, *Technical Information Summary, AS-501, Apollo Saturn V Flight Vehicle,* R-ASTR-S-67-65, 15 Sept. 1967.
2. "NASA Announces Changes in Saturn Missions," NASA release 63-246, 30 Oct. 1963.
3. Dir., Apollo Program, "Clarification of Apollo Saturn IB and V Flight Mission Designations," 12 Apr. 1965.
4. MSFC, *Technical Information Summary, AS-501,* pp. 24–75.
5. OMSF, *Apollo Program Directive No. 4D,* 1 July 1966; *No. 4E,* 22 Sept. 1966; *No. 4F,* 30 Nov. 1966; Proffitt interview; NASA, *Sixteenth Semiannual Report to Congress, 1 July–31 Dec. 1966,* pp. 49, 51–52. See also NASA, *Seventeenth Semiannual Report to Congress, 1 Jan.–30 June 1967,* p. 11, for information on the spacer.
6. KSC, "LC-39 Site Activation Status Report," 14 Sept. 1966; *Spaceport News,* 15 Sept. 1966; KSC, "Apollo 4 (AS-501) Daily Status Reports," Sept.–6 Oct. 1966.
7. KSC, "Apollo 4 Daily Status Reports," 29 Nov., 1, 2, 6, 7, 13 Dec. 1966.
8. Ibid., Dec. 1966 and Jan. 1967; KSC, "Program Milestone Data—Apollo," 15 July 1971.
9. KSC, "Apollo 4 Daily Status Reports," Feb.–Mar. 1967 (see 16 Mar. for number of wiring discrepancies in spacecraft); NASA, *Seventeenth Semiannual Report to Congress, 1 Jan.–30 June 1967,* pp. 11–12.

10. *Spaceport News,* 3 Mar. 1966; Fowler interview.
11. KSC, "Apollo 4 Daily Status Reports," Feb. 1967; *Spaceport News,* 15 Feb. 1968.
12. KSC, "LC-39 Site Activation Status Report," 19, 26 Apr. 1967.
13. NASA release 67-132, summarized in *Astronautics and Aeronautics, 1967,* p. 164; KSC, "Apollo 4 Daily Status Reports," May–Aug. 1967; *Spaceport News,* 31 Aug. 1967.
14. KSC, *Catalog of Launch Vehicle Tests, Saturn V, Apollo/Saturn V, Revision 1,* 15 June 1966, GP-244.
15. Ibid., p. 1-27; Carlson interview, 5 Sept. 1974.
16. KSC, *Catalog of Launch Vehicle Tests, Saturn V,* p. 2-18.
17. Ibid., pp. 1-18, 1-28.
18. KSC, "Apollo 11 (AS-506) Daily Status Report," 25 Mar. 1969.
19. *Catalog of Launch Vehicle Tests, Saturn V,* pp. 7-1 through 7-13.
20. Ibid., pp. 9-1 through 9-43; KSC, *Apollo/Saturn Program Development/Operations Plan,* 2: 3-90 and 3-93 provide a comparison of the two tests and their objectives.
21. KSC, *Catalog of Launch Vehicle Tests, Saturn V,* pp. 9-1 through 9-43.
22. Ibid., pp. 9-21 and 9-25.
23. Ibid., p. 9-3.
24. Ibid.; Donnelly interview, 19 June 1974.
25. Harris interview; Carlson interview, 5 Sept. 1974.
26. KSC, "Program Milestone Data, Apollo," 6 June 1973.
27. KSC, "Apollo 4 Daily Status Reports," Sept. 1967; Donnelly interview, 19 June 1974.
28. KSC, "Apollo 4 Daily Status Reports," 27 Sept.–13 Oct. 1967; Richard S. Lewis, *Appointment on the Moon: The Inside Story of America's Space Venture* (New York: Viking Press, 1968), p. 406.
29. Donnelly interview, 19 Nov. 1970.
30. NASA release 67-274 and *Baltimore Sun,* 26 Oct. 1967, p. A6, summarized in *Astronautics and Aeronautics, 1967,* p. 319.
31. Phillips to Mueller, memo for record, 5 Sept. 1968; KSC, "Apollo 4 Daily Status Reports," 19–20 Oct. 1967.
32. *Spaceport News,* 23 Nov. 1967.
33. Ibid.
34. Ibid.
35. *KSC Information and Protocol Operations Plan—Apollo 4 Mission,* pp. 1–4.
36. Ibid.
37. J. E. Ballou, GAO Area Audit Mgr., Atlanta, to L. Melton, KSC, 30 Mar. 1965; *Aviation Week* 89 (23 Sept. 1968): 74; H. L. DeLung, Acting Regional Mgr., GAO, Atlanta, to Debus, 20 Aug. 1965. The authors recognize their reliance on thorough researches of Maj. James J. Frangie in this section.
38. Debus to DeLung, 15 Oct. 1965.
39. R. J. Madison, Mgr., GAO Regional Off., Atlanta, to Debus, 6 July 1966. See also p. 16 of draft report, Comptroller General of the U.S., "Review of Launch Complex 39 Facilities for the Saturn V Vehicle, John F. Kennedy Space Center, Florida, NASA," undated, accompanying letter of Clerio P. Pin, Assoc. Dir., GAO, to Webb, 8 June 1967 (hereafter cited as "GAO Draft Report, KSC, 1967"); Mueller to William Parker, Asst. Dir., Civil Accounting and Auditing Div., GAO, Washington, 16 Aug. 1966.
40. Melton, memo for record, 29 Aug. 1966; Malcolm S. Stringer, memo for record, "GAO Review of Justification for Redundant and Duplicate Launch Complex 39 Facilities, 31 Oct. 1967."
41. Raymond Einhorn, Dir. of Audits, to Debus, "GAO Draft Report on Review of Launch Complex 39 Facilities for the Saturn V Vehicle, KSC, NASA," 13 June 1967; "GAO Draft Report, KSC, 1967," p. 24.
42. Debus to Mueller, 19 July 1967, in NASA Hq. History Office; Einhorn to Debus, 15 Sept. 1967.

43. Raymond Middleton to KSC Liaison Representative with GAO, 30 Oct. 1967; Melton, memo for record, 19 Oct. 1967; Debus to Mueller, 2 Nov. 1967, in NASA Hq. History Office; Harold B. Finger, Assoc. Admin. for Organization and Management, to Clerio P. Pin, Assoc. Dir., GAO, Washington, 24 Jan. 1968, with enclosures, pp. 10–12, NASA Hq. History Office.
44. Pin to Finger, 27 Mar. 1968, in NASA Hq. History Office.
45. Stringer interview; Finger to Pin, 24 Jan. 1968, NASA Hq. History Office.
46. House Committee on Science and Astronautics, *1964 NASA Authorization,* 88th Cong., 1st sess., 1963, pt. 1, p. 5; pt. 2, pp. 126–27.

Chapter 20: Man on Apollo

1. NASA, OMSF, *Apollo Program Directive No. 4H,* 3 Nov. 1967.
2. KSC, "Apollo 5 Daily Status Reports," 3 Mar.–14 Apr. 1967.
3. KSC release 1-68, 3 Jan. 1968; Widick interview.
4. "Apollo 5 Daily Status Reports," 23 June, 14–21 Aug., 19 Nov., 22 Dec. 1967, 19 Jan. 1968.
5. Statement of Rocco Petrone and Gen. S. Phillips, "Apollo 5 Post-Launch Press Conference," 22 Jan. 1968; NASA, OMSF, *Apollo Program Flight Summary Report, Apollo Missions AS-201 through Apollo 8,* Jan. 1969, pp. 20–22.
6. "LC-39 Site Activation Status Report," 5, 26 Apr. 1967; KSC, "Apollo 6 (AS-502) Daily Status Report," 22 Mar. 1967.
7. "Apollo 6 Daily Status Report," Mar. 1967.
8. Ibid., May–June 1967.
9. Ibid., July–Aug. 1967.
10. Ibid., 12 Sept.–Oct. 1967.
11. Ibid., Dec. 1967.
12. Ibid., 28 Dec. 1967–11 Jan. 1968.
13. Ibid., 15–16, 30 Jan., 6–9 Feb. 1968; *Spaceport News,* 15 Feb. 1968.
14. Phillips to Mueller, 5 Sept. 1968; "Apollo 6 Daily Status Report," 8–21 Mar. 1968.
15. "Apollo 6 Daily Status Report," 25 Mar. 1968; Phillips to Mueller, 5 Sept. 1968; KSC, *Apollo/Saturn V Ground Systems Evaluation Report, AS-502,* KSC document 140-44-0010, pp. 2-2, 4-1; NASA, "Apollo 6 Pre-Launch Press Conference," Cocoa Beach, 3 Apr. 1968, pp. 3–4, 7–10.
16. House Committee on Government Operations, *Hearing: Investigation of the Boeing-TIE Contract,* 90th Congress, 2d sess., 15 July 1968, p. 10.
17. NASA Hq., "Apollo 6 SLA Problem and Resolution," 17 Dec. 1968; NASA, *Nineteenth Semiannual Report to Congress, 1 Jan.–30 June 1968,* p. 19.
18. NASA, *Nineteenth Semiannual Report,* pp. 8–18.
19. NASA, "Apollo 6 Post-Launch Press Conference," LC-39 Press Site, 4 Apr. 1968, pp. 3–5.
20. *Spaceport News,* 11 Apr. 1968.
21. Erlend A. Kennan and Edmund H. Harvey, Jr., *Mission to the Moon,* pp. 284–85.
22. NASA, *Astronautics and Aeronautics, 1968,* pp. 83, 119–20; NASA, *Nineteenth Semiannual Report,* p. 19.
23. Phillips to Mueller, 5 Sept. 1968; Erich E. Goerner, "LOX Prevalve to Prevent POGO Effect on Saturn V," *Space/Aeronautics,* Dec. 1968, p. 72; House Committee on Science and Astronautics, *Hearings on 1970 NASA Authorization,* 91st Cong., 1st sess., pt. 2, pp. 27–29.
24. MSC, *Apollo Spacecraft Program Quarterly Status Report,* no. 25, 30 Sept. 1968, pp. 1–9.

25. *Spaceport News,* 29 Feb. 1968.
26. Robert Sherrod, "The Selling of the Astronauts," *Columbia Journalism Review,* May–June 1973, pp. 17–25.
27. Ibid., pp. 16–17. When former astronaut John Glenn entered the Ohio senatorial primary in the spring of 1964, news broke that he and the other original astronauts had financial interests in Cape Colony Inn in Cocoa Beach. Profitability of the Inn was obviously related to the space program.
28. One of the astronauts was so fearful of heights that he hesitated to cross the catwalk at the 31st floor of the VAB, so the ground crew covered the grating on the swing arm with boards whenever he crossed to the spacecraft.
29. *Spaceport News,* 4 Oct. 1968.
30. Wendt interview.
31. Neil Armstrong et al., *First on the Moon* (Boston: Little, Brown and Co., 1970), non-documented, interesting account of Apollo 11, previous Apollos, and the astronauts and their families, based on interviews.
32. *Spaceport News,* 28 Mar. 1968.
33. Ibid.
34. *Kennedy Space Center Story* (1971 ed.), pp. 227–28.
35. NASA, OMSF, "Apollo Program Directive 4H," 3 Nov. 1967; KSC, "Apollo 7 (AS-205) Daily Status Reports," 11 May–3 June 1968.
36. "Apollo 7 Daily Status Reports," June–July 1968; KSC, "Minutes of Apollo Launch Operations Committee (ALOC) Meetings," 13 June, 11 July, 1 Aug. 1968; *Spaceport News,* 1 Aug. 1968.
37. "Minutes of ALOC Meetings," 1, 15, 29 Aug. 1968; *Spaceport News,* 29 Aug., 12 Sept. 1968; Ragusa interview.
38. NASA, OMSF, *Mission Operation Report: Apollo 7 (AS-205) Mission,* 30 Sept. 1968.
39. Ibid.
40. Wernher von Braun and Frederick L. Ordway III, *History of Rocketry and Space Travel* (New York: Thomas V. Crowell and Co., 1969), pp. 226–27.
41. *Spaceport News,* 24 Oct. 1968.
42. Ibid.
43. KSC release 22-68, 29 Jan. 1968; NASA, *Nineteenth Semiannual Report to Congress,* 19 Jan.–30 June 1968, p. 13; Roderick O. Middleton, KSC Apollo Program Mgr., to Samuel C. Phillips, TWX, 7 Mar. 1968; MSFC, *Saturn V Launch Vehicle Flight Evaluation Report—AS-503*, p. 3-1.
44. Harold B. Finger to Mueller, 1 May 1968.
45. Ibid.
46. MSFC, *Saturn V Launch Vehicle Flight Evaluation Report—AS-503,* pp. 3-1, 3-2; KSC, "Apollo 8 Daily Status Reports," 29 Apr.–6 May 1968; unsigned document on working schedule for manned (CSM-103) and unmanned (BP-30) AS-503 missions, 21 Apr. 1968; Phillips to Debus, TWX, 29 Apr. 1968.
47. KSC, "Apollo 8 Daily Status Reports," 8, 10, 17, 31 May 1968.
48. Ibid., 14, 17 June 1968.
49. Proffitt interview, 1 Dec. 1970.
50. NASA, *Astronautics and Aeronautics, 1968,* p. 191.
51. George M. Low to C. H. Bolender and K. S. Kleinknecht, "Chuck Mathews Review of KSC Activities," 14 Sept. 1968, Apollo discussion papers, JSC Historical Archives.
52. Debus to Mueller, 16 July 1968.
53. Phillips to Debus, 24 Aug. 1968.
54. KSC, "Apollo 8 Daily Status Reports," 10 June–22 July 1968.
55. Ibid., 22 July–4 Aug. 1968; George M. Low, memo for record, 19 Aug. 1968, in Apollo discussion papers, JSC Historical Archives.

56. KSC, "Apollo 8 Daily Status Reports," 28 June–25 July 1968.
57. Ibid., 12–16 Aug. 1968; Low, memo for record, 19 Aug. 1968; "Transcript of News Conference on Apollo Program Changes," 19 Aug. 1968.
58. Phillips to Debus, 10, 20 Aug. 1968.
59. KSC, *Apollo/Saturn V Launch Operations Test and Checkout Requirements, AS-503,* document K-V-051-01/3, p. 6-1.
60. KSC, "Apollo 8 Daily Status Report," 16 Sept. 1968.
61. NASA, "Transcript of Saturn AS-503 Delta Design Certification Review," 19 Sept. 1968; KSC, "Apollo 8 Daily Status Reports," 8–10 Oct. 1968.
62. NASA, *Astronautics and Aeronautics, 1968,* p. 266.
63. Ibid., p. 278.
64. KSC, "Apollo 8 Daily Status Reports," 15, 19 Nov. 1968; KSC, *Apollo/Saturn V Launch Mission Rules, Apollo 8 (AS-503/CSM 103),* document K-V-05.10/3, pp. 1-2 through 1-29.
65. KSC, "Apollo 8 Daily Status Reports," 5–11 Dec. 1968.
66. NASA, *Astronautics and Aeronautics, 1968,* p. 313.
67. John N. Wilford, "Final Countdown On for Moon Shot Tomorrow," *New York Times,* 20 Dec. 1968.
68. "Apollo 8 Onboard Voice Transcription, As Recorded on the Spacecraft Onboard Recorder (Data Storage Equipment)," MSC, Jan. 1969, tape 58-4.
69. *Kennedy Space Center Story* (1971 ed.), pp. 101–102.
70. *Boeing Atlantic Test Center News* 6 (13 Jan. 1969): 1.
71. MSC, *Apollo 8 Technical Debriefing,* 2 Jan. 1969, p. 139.
72. *Spaceport News,* 16 Jan. 1969; Reyes interview and Chauvin interview, June 1973.
73. "Briefing by KSC-NASA for the Congressional Subcommittee on Manned Space Flight," 28 Feb. 1969, pp. 4–5.

Chapter 21: Success

1. MSC, *Apollo Spacecraft Program Quarterly Status Report No. 25*, 30 Sept. 1968, p. 28.
2. NASA, *Current News*, 20 December 1968, p. 1; NASA, *Astronautics and Aeronautics, 1969*, p. 16.
3. NASA, *Astronautics and Aeronautics, 1967,* pp. 330–31; Phillips to Debus, 19 Aug. 1968.
4. KSC, "Apollo 9 (AS-504) Daily Status Reports," May–Oct. 1968.
5. Ibid., 8–26 Nov. 1968.
6. Ibid., 2–31 Dec. 1968.
7. Titusville *Star-Advocate*, 16 Dec. 1968.
8. Renaud interview, 16 May 1973.
9. Titusville *Star-Advocate*, 16 Dec. 1968.
10. Ibid.
11. Ibid.
12. Ibid.
13. KSC, "Apollo 9 (AS-504) Daily Status Reports," 3–9 Jan. 1969.
14. Ibid., 10–11 Feb. 1969; KSC release 33–69, 4 Feb. 1969.
15. Ibid., 12, 20 Feb. 1969; KSC, *Apollo/Saturn V Test and Checkout Plan, AS-504 and All Subsequent Missions*, pp. 4-9 through 4-12; Widick interview, 15 Dec. 1970; KSC, *Apollo/Saturn V Space Vehicle Countdown Demonstration Test (Apollo 9)*, p. vi.
16. KSC, "Apollo 9 Daily Status Report," 27–28 Feb. 1969; NASA, "Apollo 9 Postponement News Conference," 27 Feb. 1969, CST 12:05, pp. 9A/1 through 9C/2; KSC, *Apollo/Saturn V Space Vehicle Countdown (Apollo 9)*, pp. v–vi.

17. KSC, *Apollo/Saturn V Ground Systems Evaluation Report, Apollo 9*, p. 2-1.
18. KSC, "Briefing for the Subcommittee on Manned Space Flight, Committee on Science and Astronautics, House of Representatives," 28 Feb. 1969, pp. 44–70.
19. Youmans interview, 5 Feb. 1971; Proffitt interview, 1 Dec. 1970; George Low to C. H. Bolender and K. S. Kleinknecht, "Chuck Mathews Review of KSC Activities," 14 Sept. 1968, JSC Archives, Apollo activity file.
20. KSC, *Apollo/Saturn V Launch Operations Test and Checkout Requirements, AS-504 and All Subsequent Missions*, document K-V-051-01, p. 1-1; Proffitt interview, 1 Dec. 1970.
21. Chart included in folder with John M. Marshall's interview with Henry C. Paul at KSC, 9 Dec. 1970, in KSC Historian's Office, illustrates graphically the growth of automation overall and of Atoll in particular for the period both before and subsequent to AS-504.
22. R. B. Johansen, "Developments in On-Board and Ground Checkout Systems," American Institute of Aeronautics and Astronautics, Cocoa Beach, FL, 2–4 Feb. 1970, AIAA paper 70-245, pp. 3–5; James E. Rorex and Robert P. Eichelberger, "Digital Data Acquisition System in Saturn V," in *Proceedings of the Second Space Congress*, 5–7 Apr. 1965, Cocoa Beach, FL, sponsored by Canaveral Council of Technical Societies, pp. 632–49; Debus, "Launching the Moon Rocket," p. 25.
23. W. V. George and C. A. Stinson, "An Automated Telemetry Checkout Station for the Saturn V Systems," *NTC/66: Proceedings, National Telemetering Conference*, Boston, 10–12 May 1966, p. 117.
24. NASA *Proceedings of the Apollo Unified S-Band Technical Conference*, p. 248; Edmund F. O'Conner, "Launch Vehicles for the Apollo Program," pp. 165–66; Walyer O. Frost, "SS-FM: A Telemetry Technique for Wide-Band Data," Institute of Radio Engineers, *Transactions on Space Electronics and Telemetry*, SET-2 (Dec. 1962), p. 289, notes the first use of SS-FM telemetry on the SA-2 flight.
25. NASA, *Proceedings of the Apollo Unified S-Band Technical Conference*, p. 248.
26. Samuel C. Phillips to MSFC, 28 June 1965; Adolf H. Knothe, "Range Safety—Do We Need It?" American Institute of Aeronautics and Astronautics, Launch Operations Meeting, Cocoa Beach, FL, Feb. 1970, p. 3.
27. "Tracking and Data Acquisition," *Spaceflight* 11 (June 1969): 190; NASA, *Proceedings of the Apollo Unified S-Band Technical Conference*, p. 3.
28. "How We Will Communicate with Astronauts on the Moon," *Space World*, Jan. 1969, pp. 33, 35.
29. "Tracking and Data Acquisition," p. 190; NASA, *Sixteenth Semiannual Report to Congress*, 1 July–31 Dec. 1966, pp. 167–68.
30. Frank Leary, "Support Net for Manned Space Flight," *Space/Aeronautics*, Dec. 1966, pp. 71–72. KSC, "Apollo/Saturn V Launch Operations Plan," AS-501/502, pp. 7-18 through 7-20 contains a general description of the LIEF and ALDS systems and their relationship to each other and to operations at KSC.
31. Edmond C. Buckley to Mueller, 26 Aug. 1964.
32. *Space Daily*, 28 Jan. 1966, p. 177.
33. John F. Mason, "Modernizing the Missile Range: Part 1," *Electronics*, Feb. 1965, pp. 94–95.
34. This conclusion is derived from the chart shown in NASA, *Proceedings of the Apollo Unified S-Band Technical Conference*, p. 296; NASA, *Sixteenth Semiannual Report to Congress, 1 July–31 Dec. 1966*, p. 165.
35. Twigg interview.
36. KSC, "Apollo 11 (AS-506) Daily Status Reports," Jan.–Apr. 1969; KSC, *Kennedy Space Center Story*, 1971 ed., p. 119.
37. NASA, *Apollo Flight Summary Report*, pp. 82–83.
38. KSC, "Apollo Program Milestone Data," 15 July 1973; KSC, "Apollo 11 Daily Status Reports," 20 May–4 July 1969.

39. KSC, *Kennedy Space Center Story*, 1971, pp. 121–25; *Spaceport News*, 23 July 1969.
40. *Kennedy Space Center Story*, 1971, pp. 222–23.
41. Ibid., p. 124; KSC, "Apollo Countdown Document, C-07."
42. *Spaceport News*, 23 July 1969.
43. Ibid.
44. Ibid., 30 July 1969.
45. Anne M. Lindbergh, *Earthshine* (Harcourt Brace Jovanovich: New York, 1969), pp. 42–43.

Chapter 22: A Slower Pace: Apollo 12–14

1. KSC, "AS-507 Daily Status Reports."
2. NASA, *Apollo 12 Press Kit*, pp. 1–2; Lt. Gen. Samuel Phillips, transcript of news conference at MSC, 24 July 1969, summarized in *Astronautics and Aeronautics, 1969*, p. 243; KSC, "AS-507 Daily Status Reports," June–July 1969.
3. *Armed Forces Journal*, 27 Sept. 1969, p. 8; NASA releases 69-151, 10 Nov. 1969; 70-4, 8 Jan. 1970; *Spaceport News*, 28 Aug., 11 Sept., 4 Dec. 1969.
4. *Spaceport News*, 28 Aug., 11 Sept. 1969, 18 June 1970.
5. KSC, "AS-507 Daily Status Reports"; Chauvin interview, 2 Apr. 1974; *Washington Post*, 30 Oct. 1969, p. A8.
6. KSC, "AS-507 Daily Status Report," 13 Nov. 1969; *Washington Post*, 13 Nov. 1969, p. A1; *Spaceport News*, 20 Nov. 1969; Sieck interview; KSC, "Apollo 12 (AS-507) Quick Look Assessment Report," 26 Nov. 1969; NASA, "Pre-Launch Press Conference," KSC and MSC, 13 Nov. 1969, pp. 6B-3, 6B-4.
7. NASA, *Apollo 12 Press Kit*, pp. 43–45; Widick interview, 23 May 1974.
8. "Launch VIP List Headed by Nixon," *Orlando Sentinel*, 13 Nov. 1969; NASA, *Analysis of Apollo 12 Lightning Incident*, MSC-01540, Feb. 1970, p. 12; Manned Spacecraft Center, "Apollo 12 Technical Crew Debriefing," 1 Dec. 1969, p. 2-1.
9. MSC, "Apollo 12 Technical Crew Debriefing," 1 Dec. 1969, p. 2-1.
10. NASA, "Apollo 12 Post Launch Briefing," 14 Nov. 1969, at KSC, p. 8A/2.
11. Ibid.; Kapryan interview.
12. MSC, "Apollo 12 Technical Crew Debriefing," p. 3-2.
13. NASA, "Apollo 12 Mission Commentary," p. 15/1.
14. NASA, "Apollo 12 Post Launch Briefing," pp. 8A-3, 8D-2.
15. Richard M. Nixon, "Remarks to NASA Personnel at the Kennedy Space Center," 14 Nov. 1969, *Public Papers of the Presidents, 1969* (Washington, 1971), p. 936.
16. NASA, "Apollo 12 Mission Commentary," summarized in *Astronautics and Aeronautics, 1969*, pp. 372–78.
17. *Spaceport News*, 1 Jan. 1970.
18. Ibid.
19. NASA, *Analysis of Apollo 12 Lightning Incident*; for one contribution from the scientific community see app. B, M. Brook, C. R. Holmes, and C. B. Moore, "Exploration of Some Hazards to Naval Equipment and Operations beneath Electrified Clouds."
20. KSC, *Launch Mission Rules Apollo 13 (SA-508/CSM 109.LM-7)*, 17 Feb. 1970, p. 1-17.
21. NASA, *Analysis of Apollo 12 Lightning Incident*, pp. 29, 36.
22. KSC, "Apollo 13 (AS-508) Daily Status Report," July 1969–Mar. 1970; KSC, "Proceedings of Manned Space Flight Subcommittee Hearings at Kennedy Space Center, 10 Apr. 1970," pp. 20–26; Moser interview, 17 Apr. 1974.
23. KSC Board of Investigation, "Investigation of Circumstances Surrounding Incident Resulting in Destruction by Fire of Three Motor Vehicles in Vicinity of Perimeter Fence on Pad A of LC-39 on 3/25/70," see part II, 2A and 2B, for narrative of events and committee recommendations; "Transcript of Proceedings of Manned Space Flight Subcommittee at KSC, 10 Apr. 1970," pp. 31–32; Corn interview, 22 Apr. 1974.

24. KSC, "Apollo 13 (AS-508) Daily Status Report," 27, 30, 31 Mar. 1970; NASA, *Report of the Apollo 13 Review Board*, 15 June 1970, pp. 4-21 through 4-23; Lamberth interview.
25. NASA, "Apollo 13 Status Report," A13-1, 9:30 a.m., 5 Apr. 1970; NASA, "Apollo 13 Change of Shift Briefing," 13 Apr. 1970, 2:30 p.m., p. 20A/1; KSC, "Apollo 13 (AS-508) Daily Status Report," 6 Apr. 1970; Lamberth interview; KSC Weekly Report, Kapryan, 2 Apr. 1970.
26. KSC releases 41-70, 6 Mar.; 43-70, 11 Mar.; 153-70, 23 Mar. 1970; *Spaceport News*, 12, 26 Mar., 23 Apr. 1970; KSC Weekly Report, Kapryan, 9 Apr. 1970.
27. KSC, Apollo News Center, "Apollo 13 Status Reports," 1–4, 6–7, 9–10 Apr. 1970; NASA, "Apollo 13 Medical Status Briefing #1," 6 Apr. 1970, 6:50 p.m., p. 8c/1; NASA, "Apollo 13 Medical Status Briefing #2," 8 Apr. 1970, 6:46 p.m.
28. *Baltimore Sun*, 10 Apr. 1970. The $800000 represented overtime pay for workers at KSC and the cost of the recovery force for the Pacific Ocean splashdown.
29. MSC, "Apollo 13 Prelaunch Press Conference," at KSC, 10 Apr. 1970, 2:10 p.m., pp. 12B/1–12B/4.
30. McCafferty interview, 28 Jan. 1971.
31. MSC, "Apollo 13 Prelaunch Press Conference," pp. 12A/2 and 12B/2.
32. KSC, "Apollo 13 (AS-508) Post Launch Report," 24 Apr. 1970; KSC, "Apollo 13 (AS-508) Flight Summary."
33. MSC, "Apollo 13 Mission Commentary," 13 Apr. 1970, CST 8:34 p.m. GET 55:11, pp. 165/1 through 168/1.
34. Ibid., p. 196/1.
35. NASA, *Report of Apollo 13 Review Board,* pp. 4-25 through 4-46.
36. Ibid., pp. 4-46 through 4-48; NASA, *Current News*, 11, 17 Apr. 1970; McCafferty interview, 28 Jan. 1971; KSC, *Kennedy Space Center Story*, 1971, pp. 152–54.
37. *Washington Post*, 17 Apr. 1970.
38. NASA, *Report of Apollo 13 Review Board*, p. 5-1.
39. Ibid., pp. 4-17 through 4-23, 5-1 through 5-9.
40. Ibid., preface.
41. NASA, "Transcript of Press Conference at KSC on 9 Nov. 1970," quoted in *Astronautics and Aeronautics, 1970*, p. 364; *New York Times*, 10 Nov. 1970, p. 33; for changes in Apollo 14 launch dates see Apollo Program Directives 4K, 4M, 4N; also *Astronautics and Aeronautics, 1970*, pp. 7, 205, 218.
42. KSC, "Apollo 14 (AS-509) Daily Status Reports."
43. NASA, *Apollo 14 Press Kit*, 8 Jan. 1971, p. 93; Humphrey interview.
44. KSC, "Apollo 14 (AS-509) Daily Status Reports"; OMSF, "Apollo Program Weekly Status Reports," June–Aug. 1970.
45. KSC, "Apollo 14 (AS-509) Daily Status Reports"; KSC, "Apollo 14 Post Launch Report," 16 Feb. 1971.
46. *Spaceport News*, 3 Dec. 1970; *Washington Post*, 12 Jan. 1971; *Houston Chronicle*, 18 Jan. 1971.
47. McCafferty interview, 28 Jan. 1971.
48. KSC, "Apollo 14 Post Launch Report"; *Spaceport News*, 11 Feb. 1971.
49. *New York Times*, 15 Nov. 1969, quoted in *Astronautics and Aeronautics, 1969*, p. 380.
50. NASA, *Astronautics and Aeronautics, 1967*, pp. 17, 337; *1968*, pp. 19, 241; *NASA Historical Pocket Statistics*, Jan. 1974, pp. D-2 through D-7.
51. OMSF, "Apollo Program Directive 4-K, Subject: Apollo Program Schedule and Hardware Planning Guidelines and Requirements," 10 July 1969, pp. 5-7 and 10-11.
52. Lee A. DuBridge, testimony on NASA FY 70 Authorization before the Senate Committee on Aeronautical and Space Sciences, 9 May 1969, quoted in *Astronautics and Aeronautics, 1969*, p. 134.
53. *New York Times*, 10 Aug. 1969, p. 44; see *Lunar Exploration: Strategy for Research 1969-1975*, published by the National Academy of Sciences, National Research Council,

Space Science Board, for further evidence of attitudes in the scientific community.

54. *Washington Post*, 5 Oct. 1969.
55. NASA, *America's Next Decades in Space, A Report for the Space Task Group*, Sept. 1969.
56. NASA, "Apollo Program Directive 4-M," 16 Mar. 1970, with cover sheet from Mathews to Debus, 6 Apr. 1970; George Low, Dep. Admin., NASA, "Fiscal Year 1971 Budget Briefing for Community Leaders," KSC, 2 Feb. 1970; transcript of NASA news conference, Washington, D.C., 2 Sept. 1970, summarized in *Astronautics and Aeronautics, 1970*, pp. 284–85.
57. Debus, "Briefing for Community Leaders on FY-1970 Budget at the Kennedy Space Center," 30 Apr. 1960; Miles Ross, Dep. Dir., KSC, "Briefing at Breakfast Meeting of Brevard County Chamber of Commerce," 25 Sept. 1973; Kaufman interview.
58. *New York Times*, 26 Oct. 1969, p. F15; *Orlando Sentinel*, 13 Nov. 1969, p. 2A; *Miami Herald*, 11 Oct. 1970, p. H14; Charles Johnson interview.
59. *Philadelphia Evening Bulletin*, 17 Nov. 1970.
60. Kapryan discussed the question of morale at the Apollo 12 prelaunch briefing, 13 Nov. 1969; see the minutes, p. PC/6E/2.

Chapter 23: Extended Lunar Exploration: Apollo 15–17

1. NASA, *Apollo 15 Press Kit*, 15 July 1971, pp. 1–8; *Astronautics and Aeronautics, 1970*, pp. 284–85; *Time*, 9 Aug. 1971, pp. 10–15.
2. NASA, *Apollo 15 Press Kit*, pp. 5, 60–69.
3. Ibid., pp. 94–100, 134–44; Petrone, Apollo Program Dir., to Manned Space Flight Centers, "6/12/70 Weight and Performance Review Agreements and Actions," 7 July 1970.
4. NASA, *Apollo 15 Press Kit*, p. 133.
5. KSC, "Apollo 15 Post Launch Report," 12 Aug. 1971, pp. 1-1 and 1-2; OMSF, "Apollo Program Weekly Status Reports," June–Sept. 1970; KSC, "Apollo 15 (AS-510) Daily Status Reports," May 1970–Jan. 1971.
6. KSC, "Apollo 15 Daily Status Reports," Jan.–May 1971; Jackie Smith interview.
7. Chauvin interview, 23 May 1974.
8. NASA, *Apollo 15 Press Kit*, pp. 61–69; KSC, "Apollo 15 Daily Status Reports," Jan.–May 1971; KSC, "Apollo 15 Post Launch Report," pp. 1–2; Edwin Johnson interview.
9. NASA, *Apollo 15 Press Kit*, pp. 78–82.
10. KSC release 41-71, "LRV Flight Model Delivery," 10 Mar. 1971; NASA, *Apollo 15 Press Kit*, 15 July 1971, pp. 77–97; *Time*, 9 Aug. 1971; Arthur Scholz to Benson, 18 Oct. 1974. The authors found several different costs cited for the rover. The $12.9 million price is *Time*'s figure and reflects the total project cost of $38+ million divided by three flight vehicle rovers. The cost per vehicle drops if the training rover is included, or if the $13 million R&D costs are excluded. Goldsmith interview.
11. NASA, *Apollo 15 Press Kit*, pp. 77–97; Widick interview, 23 May 1974; Reyes interview, 6 June 1974; Carothers interview; Scholz interview; "Apollo 15 Daily Status Reports," 15 Mar.–25 Apr. 1971.
12. *Washington Post*, 5 May 1971; *New York Times*, 5 May 1971; *Spaceport News*, 6 May 1971.
13. KSC, "Apollo 15 Post Launch Report," p. 1-2; KSC, "Apollo 15 Daily Status Reports," Feb.–Mar. 1971; Collner interview.
14. KSC, "Apollo 15 Daily Status Reports," 29 Mar. 1971; Cochran interview.
15. KSC, "Apollo 15 Daily Status Reports," 29 Mar.–8 Apr. 1971; Lang interview.
16. *Houston Post*, 3 Apr. 1971; *Los Angeles Times*, 3 Apr. 1971.

17. KSC, "Apollo 15 Post Launch Report," pp. 1-2, 1-3; KSC, "Apollo 15 Daily Status Reports," June–July 1971.
18. NASA, "Apollo 15 Mission Commentary," 26 July 1971, 4:11 GET, 12:43 CDT, p. 41/2.
19. KSC, "Apollo 16 (AS-511) Daily Status Reports," Aug.–Nov. 1971; KSC, "Apollo 16 (AS-511) Post Launch Report," 2 May 1972.
20. KSC, "Apollo 16 Post Launch Report"; Crawford interview; Hangartner interview.
21. Ely interview; KSC, S-IC Flight Control, "Test Problem Report," 3 Dec. 1971; KSC, "Apollo 16 Post Launch Report."
22. *New York Times*, 8, 10 Jan. 1972; KSC, "Apollo 16 Post Launch Report."
23. Moxley interview.
24. *New York Times*, 28, 31 Jan. 1972; *Spaceport News*, 27 Jan. 1972; KSC, "Apollo 16 Daily Status Reports," Jan.–Feb. 1972; KSC, "Apollo 16 Post Launch Report."
25. Moxley interview; KSC, "Apollo 16 Daily Status Reports," Feb. 1972.
26. NASA, *Apollo 16 Press Kit*, 22 Mar. 1972.
27. Montgomery interview; McKnight interview.
28. KSC, "Apollo 16 Post Launch Report"; *Spaceport News*, 9 Mar. 1974.
29. *Spaceport News*, 23 Mar. 1972; KSC release 62-72, "Apollo 16 Crewmen Outline Lunar Mission for Spaceport Launch Team," 16 Mar. 1972.
30. KSC, "Apollo 16 Daily Status Reports"; KSC, "Apollo 16 Post Launch Report."
31. *Miami Herald*, 3 Oct. 1972, p. B1; *Baltimore Sun*, 21 Nov. 1972; *Spaceport News*, 21 Sept.–14 Dec. 1972; *Time*, 18 Dec. 1972.
32. KSC, "Apollo 17 Post Launch Report," 19 Dec. 1972, pp. 5-1, 5-2.
33. NASA, *Apollo 17 Press Kit*, 14 Nov. 1972, pp. 56–61.
34. Jackie Smith interview.
35. KSC, "Apollo 17 Post Launch Report," pp. 5-2 to 5-4; *Miami Herald*, 29 Aug. 1972, p. 1.
36. *Lawrence* (Kansas) *Daily Journal-World*, 30 Sept. 1972; *Huntsville Times*, 1 Oct. 1972; NASA, *Astronautics and Aeronautics, 1972*, p. 330; KSC, "Apollo 17 Daily Status Reports," 29 Sept. 1972.
37. *Wall Street Journal*, 13 Nov. 1972, p. 12.
38. *Los Angeles Times*, 27 Nov. 1972; *Spaceport News*, 16 Nov. 1972.
39. NASA, "Apollo 17 Commentary"; KSC, "Apollo 17 Post Launch Report," 19 Dec. 1972.
40. NASA, "Apollo 17 Commentary."
41. *Spaceport News*, 14 Dec. 1972.
42. NASA, "Apollo 17 Commentary"; NASA, "Apollo 17 Post Launch Press Conference"; NASA Apollo Program Dir., "Apollo 17 Mission (AS-512) Post Mission Operation Report No. 1," 19 Dec. 1972, pp. 3–4; *Spaceport News*, 14 Dec. 1972.
43. NASA, "Apollo 17 Flight Summary."

Bibliography

While a great number of books have been written about the Apollo program, there has been no previous history of the launch facilities and operations. *Liftoff*, by L. B. Taylor, Jr. (see *Books*, below) provides a lively journalistic account of the spaceport in the 1960s. Unfortunately, the book ends before Apollo reached its goal. William R. Shelton's *Countdown: The Story of Cape Canaveral* is an entertaining eyewitness account of launch operations at the Cape during the 1950s. Gordon Harris's *Selling Uncle Sam* recalls Apollo events as seen from the Office of Public Affairs at KSC. Michael Collins's *Carrying the Fire* contains an astronaut's views of the Gemini and Apollo programs. The Apollo 11 astronaut set himself a high goal—writing a book without a dull or confusing passage—and then accomplished it. His treatment of technical problems is to be envied. While there is no balanced account of the AS-204 fire, the near tragedy of Apollo 13 is well covered in Henry Cooper's *Thirteen: The Flight That Failed*. A good general account of the Apollo program is John Noble Wilford's *We Reach the Moon*.

The reader will find a wealth of detailed information about Apollo launch facilities and operations in journal articles and conference papers. The popular aerospace magazines (*Missiles and Rockets, Aviation Week and Space Technology*, *Space/Aeronautics*, and *Astronautics and Aeronautics*) trace the progress of the Apollo program. Numerous scientific and engineering journals contain articles by members of the launch team. An even better source for technical exposition is the papers prepared for conferences such as the annual Space Congress held in Cocoa Beach, Florida. The proceedings for most of these conferences were printed.

NASA's dealings with Congress are revealed in thousands of pages of briefings, testimony, and hearings. The agency's *Semiannual Report to Congress* (1958–1969) provides a detailed account of the progress toward a manned lunar landing. At the annual budget hearings, top NASA officials made similar statements and answered numerous questions about specific activities. Special committee hearings at KSC regarding launch operations appear as appendixes in the annual hearings or as special congressional reports.

The authors relied on *Astronautics and Aeronautics* as a basic guide to aerospace events of the 1960s. NASA's History Office has compiled these annual chronologies since 1961; the first two years the work appeared as a report to the House Committee on Science and Astronautics and subsequently as a NASA special publication. Although there are several thousand entries in each volume, the series is well indexed. Another helpful source is *Current News*, a compilation of newspaper articles about NASA activities prepared by the agency's Office of Public Affairs. The authors obtained information on specific missions from surprisingly detailed NASA press kits (e.g., the Apollo 8 press kit is 105 pages), mission summaries, and the transcripts of press conferences. The publications are available in the KSC archives and other NASA installations.

KSC's public affairs publications proved very helpful. *The Kennedy Space Center Story* (1969, 1972, 1974) is a well-written informative account of events at the space center since the early 1960s. The first edition attempted no historical evaluation and ignored unpleasant events, such as the AS-204 fire. Hundreds of KSC news releases about the Apollo program provided interesting sidelights for the history. The Center's newspaper, *Spaceport News*, prepared under the direction of the Public Affairs Office, served a similar function. Distinctly a house organ, the paper avoided controversy, but was, nevertheless, useful for background and specific facts.

Three unpublished works, prepared at KSC, blazed a research path for the authors. Frank Jarrett and Robert Lindemann's "History of the John F. Kennedy Space Center, NASA (Origins through December 1965)" provides a detailed, carefully researched account of early center history. Even more helpful are the unpublished manuscripts of James Covington, James Frangie, and William Lockyer (Apollo Launch Facilities) and George Bittle and John Marshall (Apollo Launch Operations). Both manuscripts are in the KSC historical archives.

Concerning primary sources, the General Accounting Office's criticisms notwithstanding, the authors found an overabundance of source material at the Kennedy Space Center. Documents on the AS-204 fire, alone, occupy more than 60 large cartons in the KSC records-holding area. The card catalog in the center's documents department references several thousand studies and procedures for the Apollo/Saturn. Fortunately, KSC's records retrieval and library systems provide quick access to documents.

Dr. Kurt Debus's "Daily Journal" (1959–1963) and the weekly reports rendered to him by the KSC staff (1962–1972) were key sources of information. These documents are located in the center director's office at KSC. The authors found other valuable data in Debus's correspondence files, in storage at the Federal Records Center in Atlanta. While the originals can be retrieved through KSC's records management office, the letters used for this history have been reproduced for the KSC archives. Rocco Petrone's program office was another rich source of reports, memoranda, and letters. Some carbon copies are on file in the KSC archives, but the bulk of this material has been retired to Atlanta. Similar documents from other KSC sources, numbering in the thousands, have been collected by the KSC staff during the past ten years.

The progress of design and construction of the three Saturn launch complexes is reflected in a series of reports: Saturn Monthly and Quarterly Progress Reports (published at Huntsville with a section on the Cape), the Monthly Progress Reports of the Launch Facilities Support Equipment Office (mainly about ground support equipment), and Construction Progress Reports and Project Status Reports on LC-39. These documents are available in the KSC historical archives along with the minutes of the Site Activation Board meetings and the Site Activation Status Reports. Important documents for the launch operations include the minutes of the Apollo Launch Operations Committee, the daily status reports for Apollo missions, Apollo/Saturn V test procedures, and the postlaunch reports. The daily status reports and the test procedures for the Saturn I launches were secured from Robert Moser's papers in the Federal Records Center at Atlanta.

A number of documents in the KSC archives concern the center's relations with other members of the Apollo team. The minutes of the Management Council Meetings relate important discussions while Brainerd Holmes was head of the Office of Manned Space Flight. Other sets of minutes from 1961–1963 cover the activities of the Launch Operations Panel and the Panel Review Board. Management instructions from the Headquarters and KSC program offices are contained in the Apollo Program Directives. The offices established the crucial scheduling dates in the series of directives referred to as "dash 4"; the frequent revisions chart the vicissitudes of the Apollo program from 1965 through 1972. Most of these directives are available in the historical archives or the documents department. The researcher may wish to make use of other documents in the archives including mission flight manuals and safety plans, interface control documents, and the Apollo Program Development Plans prepared by the Office of Manned Space Flight.

Interviews with participants were among the most valuable sources of information. Whenever possible, the authors evaluated the objectiveness and accuracy of an interview against other accounts of the same events. A list of the interviews is included in the bibliography. The transcripts are available in the KSC archives.

Books

Akens, David S. *Historical Origins of the George C. Marshall Space Flight Center*. Huntsville, AL: Marshall Space Flight Center, 1960.

———. *Saturn Illustrated Chronology*, 5th ed. Huntsville, AL: Marshall Space Flight Center, 1971.

Aldrin, Edwin E., with Wayne Wargo. *Return to Earth*. New York: Random House, 1973.

Alexander, Tom. *Project Apollo: Man to the Moon*. New York: Harper and Row, 1964.

Armstrong, Neil, Michael Collins, and Edwin E. Aldrin, Jr., written with Gene Farmer and Dora Jane Hambilin. *First on the Moon*. Epilogue by Arthur C. Clarke. Boston: Little, Brown and Co., 1970.

Behrendt, Lloyd L., comp. *Development and Operation of the Atlantic Missile Range*, History of the Air Force Missile Test Center, vol. 2. Patrick Air Force Base, FL: Air Force Missile Test Center, 1963.

Bergaust, Erik. *Murder on Pad 34*. New York: G. P. Putnam's Sons, 1968.

Booker, P. J., G. C. Frewer, and G. K. C. Pardoe. *Project Apollo: Way to the Moon*. American Elsevier Publishing Co., 1969.

Bramlitt, E. R. *History of Canaveral District*, South Atlantic Division, U.S. Corps of Engineers, 1971.

Cantafio, Leopold J., ed. *Range Instrumentation*. Englewood Cliffs, NJ: Prentice-Hall, 1967.

Collins, Michael. *Carrying the Fire: An Astronaut's Journeys*, with a foreword by Charles A. Lindbergh. New York: Farrar, Straus, and Giroux, 1974.

Cooper, Henry S. F., Jr. *Apollo on the Moon*. New York: Dial Press, 1969.

———. *Thirteen: The Flight That Failed*. New York: Dial Press, 1973.

Emme, Eugene, ed. *The History of Rocket Technology*. Detroit: Wayne State Press, 1964.

———. *A History of Space Flight*. New York: Holt, Rinehart and Winston, 1965.

Ertel, Ivan D., and Mary Louise Morse. *The Apollo Spacecraft: A Chronology*, vol. 1, *Through November 7, 1962*. NASA SP-4009. Washington, 1969.

Etzioni, Amitoi. *The Moon-Doggle: Domestic and International Implications of the Space Race*. Garden City, NY: Doubleday & Co., 1964.

Green, Constance McLaughlin, and Milton Lomask. *Vanguard: A History*. NASA SP-4202. Washington, 1970.

Grey, Jerry, and Vivian Grey, eds. *Space Flight Report to the Nation*. New York: Basic Books, 1962.

Grimwood, James M., and Barton C. Hacker, with Peter Vorzimmer. *Project Gemini Technology and Operations: A Chronology*. NASA SP-4002. Washington, 1969.

Harris, Gordon L. *Selling Uncle Sam*. Hicksville, NY: Exposition Press, 1976.

Holmes, Jay. *America on the Moon: The Enterprise of the Sixties*. Philadelphia: J. B. Lippincott Co., 1962.

Huzel, Dieter K. *Peenemuende to Canaveral*. Englewood Cliffs, NJ: Prentice-Hall 1962.

Johnson, Lyndon B. *The Vantage Point: Perspectives of the Presidency, 1963–1969*. New York: Popular Library, 1971.

Kennan, Erlend A., and Edmund H. Harvey, Jr. *Mission to the Moon: A Critical Examination of NASA and the Space Program*. New York: William Morrow and Co., 1969.

Klee, Ernest, and Otto Mark. *Birth of the Missile: The Secrets of Peenemuende*. New York: E. P. Dutton and Co., 1963.

Koelle, Heinz H., ed. *Handbook of Astronautical Engineering*. New York: McGraw-Hill Book Co., 1961.

Lay, Beirne, Jr. *Earthbound Astronauts: The Builders of Apollo-Saturn*. Englewood Cliffs, NJ: Prentice-Hall, 1971.

Lewis, Richard S. *Appointment on the Moon: The Inside Story of America's Space Venture*. New York: Viking Press, 1968.

Ley, Willy. *Rockets, Missiles, and Men in Space.* New York: Viking Press, 1968.

Logsdon, John M. *The Decision to Go to the Moon: Project Apollo and the National Interest.* Cambridge: MIT Press, 1970.

McGovern, James. *Crossbow and Overcast.* New York: William Morrow & Co., 1964.

Medaris, John B. *Countdown for Decision*. New York: G. P. Putnam's Sons, 1960.

Morse, Mary Louise, and Jean Kernahan Bays. *The Apollo Spacecraft: A Chronology*, vol. 2, *November 8, 1962–September 30, 1964*. NASA SP-4009. Washington, 1973.

Nieburg, H. L. *In the Name of Science*. Chicago: Quadrangle Books, 1966.

Ordway, Frederick L., III, ed. *Advances in Space Science and Technology*, vol. 6. New York: Academic Press, 1964.

Rabinowitch, Eugene, and Richard Lewis, eds. *Man on the Moon: The Impact on Science, Technology and International Cooperation*. New York: Basic Books, 1969.

Rosholt, Robert L. *An Administrative History of NASA, 1958–1963*. NASA SP-4101. Washington, 1966.

Sänger, Eugene. *Space Flight: Countdown for the Future*, trans. Karl Frucht. New York: McGraw-Hill Book Co., 1965.

Schwiebert, Ernest G. *A History of the U.S. Air Force Ballistic Missiles*. Washington: Frederick A. Prager, Publishers, 1965.

Shelton, William R. *Countdown: The Story of Cape Canaveral*. Boston: Little, Brown and Co., 1960.

———. *American Space Exploration: The First Decade*. Boston: Little, Brown and Co., 1967.

Swenson, Loyd S., Jr., James M. Grimwood, and Charles C. Alexander. *This New Ocean: A History of Project Mercury*. NASA SP-4201. Washington, 1966.

Sorenson, Theodore C. *Kennedy*. New York: Harper & Row, 1965.

Taylor, L. B., Jr. *Liftoff! The Story of America's Spaceport*. New York: E. P. Dutton & Co., 1968.

Von Braun, Wernher, and Frederick L. Ordway III. *History of Rocketry and Space Travel.* New York: Thomas V. Crowell and Co., 1966, 1969, 1975.

———. *Space Frontier.* New York: Holt, Rinehart, and Winston, 1967.

Wilford, John N. *We Reach the Moon.* New York: Bantam Books, 1964.

Webb, James E. *Space Age Management: The Large-Scale Approach*. New York: McGraw-Hill Book Co., 1969.

Young, Hugo, Bryan Silcock, and Peter Dunn. *Journey to Tranquility.* Garden City, NY: Doubleday & Co., 1970.

Journal Articles

Alelyunas, Paul. "Checkout: Man's Changing Role." *Space/Aeronautics* 44 (Dec. 1965): 66–73.

Alexander, George. "Cape Canaveral to Expand for Lunar Task." *Aviation Week*, 31 July 1961, p. 28.

———. "Telemetry Data Confirms Saturn Success." *Aviation Week and Space Technology*, 6 Nov. 1961, pp. 30–32.

———. "Inquiry Focuses on Electrical Systems." *Aviation Week and Space Technology*, 6 Feb. 1967, pp. 30–34.

Alexander, Tom. "The Unexpected Payoff of Project Apollo." *Fortune* 80 (July 1969): 114–117.

Alexander, William D. "Vertical Assembly Building—Project Description, Organization, and Procedures." *Civil Engineering* 35 (Jan. 1965): 42–44.

"Apollo 15: The Most Perilous Journey." *Time*, 9 Aug. 1971, pp. 10–15.

"Apollo: Giant Equipment Problems," *Missiles and Rockets*, 18 Sept. 1961, p. 19.

Bleymaier, Joseph S. "ITL and Titan III." *Astronautics and Aerospace Engineering* 1 (Mar. 1963): 33–36.

Bliss, H. "NASA's in the Cold, Cold Ground." *ATCHE Journal* 13 (May 1967): 419.

Campbell, John B. "What Happened to Apollo." *Space/Aeronautics* 48 (Aug. 1967): 54–70.

Cerquettini, C. "Sprayable Polyurethane Foam Insulation, Saturn II Booster." *SAMPE Journal* 5 (June-July 1969): 28–29.

"Death on Old Shaky." *Time*, 27 Jan. 1961, pp. 15–16.

Debus, Kurt H. "Launching the Moon Rocket." *Astronautics and Aerospace Engineering* 1 (Mar. 1963): 20–32.

————. "Saturn Launch Complex." *Ordnance* 46 (Jan.-Feb. 1962): 522–23.

"First Industry-Built Saturn I Puts Pegasus-2 in Precise Orbit." *Aviation Week and Space Technology*, 31 May 1965, p. 21.

Fisher, Allen C., Jr. "Cape Canaveral's 6000-Mile Shooting Gallery." *National Geographic* 116 (Oct. 1959).

Fleming, William A. "Launch Operations Challenge." *Astronautics* 6 (June 1961): 20–23.

Frewer, Gerald C. "Kennedy Space Center—Assembly Line on a Gigantic Scale." *The Engineering Designer*, May 1967, p. 7.

————. "The Crawler Transporter for Project Apollo." *The Designing Engineer*, July 1967, p. 15.

Getler, Michael. "Apollo: Was It Worth It?" *Space/Aeronautics* 3 (Sept. 1969): 48.

————. "Complex 37 Will Dwarf Predecessors." *Missiles and Rockets*, 18 Dec. 1961, p. 24.

Goerner, Erich E. "LOX Prevalve to Prevent POGO Effect on Saturn V." *Space/Aeronautics* 50 (Dec. 1968): 72–74.

Heaton, Donald H. "Approaches to Rendezvous." *Astronautics* 7 (Apr. 1962): 24.

Hendel, Frank J. "Gaseous Environments during Space Missions." *Journal of Spacecraft and Rockets* 1 (July-Aug. 1964): 353–64.

Holmes, D. Brainerd. "Man in Space—A Challenge to Engineers." *Challenge* 1 (Spring 1963): 28.

Houbolt, John C. "Lunar-Orbit Rendezvous and Manned Lunar Landing." *Astronautics* 7 (Apr. 1962): 26.

"How Soon the Moon?" *Time*, 14 Apr. 1967, pp. 86–87.

"Inquest on Apollo." *Time*, 10 Feb. 1967, pp. 18–19.

Knothe, Adolf H. "Range Safety—A Necessary Evil." *Aerospace Engineering* 20 (June 1961): 20.

Kolcum, Edward H. "S-1 Award Puts Chrysler in Space Field." *Aviation Week and Space Technology* 75 (27 Nov. 1961): 22.

Kovit, Bernard. "The Saturns." *Space/Aeronautics* 42 (Aug. 1964): 40–52.

"Launch Complex 39 Built Specifically for Saturn V." *Space Age News* 12 (Aug. 1969): 92.

"Launch Vehicles." *Spaceflight* 11 (Mar. 1969): 74.

Leary, Frank. "Support Net for Manned Space Flight." *Space/Aeronautics* 46 (Dec. 1966): 68–80.

Lewis, Richard S. "The Kennedy Effect." *Bulletin of the Atomic Scientists* 24 (Mar. 1968): 2.

"Life in the Space Age." *Time*, 4 July 1969, pp. 38–39.

Mason, John F. "Modernizing the Missile Range: Part I." *Electronics* 22 (Feb. 1965): 94–105.

Mast, Larry T. "Automatic Test and Checkout in Missile and Space Systems." *Astronautics and Aerospace Engineering* 1 (Mar. 1963): 41–44.

McGuire, Frank G. "Kapustin Yar Serves as Russia's Cape Canaveral." *Missiles and Rockets* 3 (Feb. 1958): 61–62.

McMillan, Brockway. "The Military Role in Space." *Astronautics* 7 (Oct. 1962): 18–21.

Means, Paul. "Group Taking on Another Vital Role." *Missiles and Rockets*, 14 Mar. 1960, pp. 22–24.

"Men of the Year." *Time*, 3 Jan. 1969, pp. 9–16.

Mendelbaum, Leonard. "Apollo: How the United States Decided to Go to the Moon." *Science*, 14 Feb. 1969, pp. 649–54.

Moore, W. I., and R. J. Arnold. "Failures of Apollo/Saturn V Liquid Oxygen Loading System." *Advances in Cryogenic Engineering* 13 (1966): 534–44.

Norcross, J. S., and Berl W. Martin. "Air Force Eastern Test Range UHF Telemetry Status Report." *Telemetry Journal* 4 (Apr./May 1969): 25–31.

Normyle, William J. "NASA Details Sweeping Apollo Revisions." *Aviation Week and Space Technology*, 15 May 1967, pp. 24–26.

Parker, Glenn M. "The Missile Site Labor Commission." *ILR Research* 8 (1962): 11.

Parker, P. J. "Apollo 9 Tests Lunar Module." *Spaceflight* 11 (July 1969): 230–33.

"Pegasus Returning Meteoroid Flux Data." *Aviation Week and Space Technology*, 22 Feb. 1965, p. 28.

Rogers, John G. "What Life at Cape Kennedy Does to Marriage." *Parade*, 9 July 1969.

Rosen, Milton W. "Big Rockets." *International Science and Technology*, Dec. 1962, pp. 66–71.

Rutledge, Philip C. "Vertical Assembly Building—Design of Foundations." *Civil Engineering* 35 (Jan. 1965): 50–52.

"SA-9 Launch." *Aviation Week and Space Technology*, 22 Feb. 1965, p. 27.

"Saturnalia at Canaveral." *Newsweek*, 6 Nov. 1961.

"Saturn Flight Specs. Manned Shot Plan." *Aviation Week and Space Technology*, 30 Apr. 1962, p. 32.

"Saturn's Success." *Time*, 3 Nov. 1961.

Sherrod, Robert. "The Selling of the Astronauts." *Columbia Journalism Review*, May/June 1973, pp. 17–25.

Smith, Richard A. "Canaveral, Industry's Trial by Fire." *Fortune* 65 (June 1962): 135.

———. "Now It's an Agonizing Reappraisal of the Moon Race." *Fortune* 68 (Nov. 1963): 124–29, 270–80.

Sloan, James E., and Jack F. Underwood. "Systems Checkout for Apollo." *Astronautics and Aerospace Engineering* 1 (Mar. 1963): 37–40.

Taylor, Hal. "Big Moon Booster Decisions Looming." *Missiles and Rockets*, 28 Aug. 1961, pp. 14–15.

Tedesko, Anton. "Base for USA Manned Space Rockets (Structures for Assembly and Launching)." *International Association for Bridge and Structural Engineering Publications* 26 (1967): 535.

———. "Space Truss Braces Huge Building for Moon Rocket." *Engineering News-Record*, 6 Feb. 1964, pp. 24–27.

———. "Design of the Vertical Assembly Building." *Civil Engineering* 35 (Jan. 1965): 45–49.

"To Design for the Moon Age, Four Firms Work as One Team." *Engineering News-Record*, 6 Feb. 1964, pp. 46–48.

"To Strive, to Seek, to Find, and Not to Yield. . . ." *Time*, 3 Feb. 1967, pp. 13–16.

"To the Moon." *Time*, 18 July 1969, pp. 20–31.

"Tracking and Data Acquisition." *Spaceflight* 11 (June 1969): 190.

Trainer, James. "Titan III Plans Await DOD Approval." *Missiles and Rockets*, 14 May 1962, pp. 35–36.

Vonbun, Friedrich O. "Ground Tracking of Apollo." *Astronautics and Aeronautics* 4 (May 1966): 104–15.

von Braun, Wernher. "Exploring the Space Sea." *Ordnance* 49 (July–Aug. 1964): 50.

von Tiesenhausen, Georg. "Saturn Ground Support and Operations." *Astronautics* 5 (Dec. 1960): 30.

"Washington Roundup." *Aviation Week and Space Technology*, 18 June 1962, p. 25.

Congressional Documents

House Committee on Government Operations, Subcommittee. *Investigation of the Boeing-TIE Contract*. Hearing, 90th Cong., 2d sess., 15 July 1968.

House Committee on Post Office and Civil Service, Subcommittee on Manpower. *Decision of Comptroller General of the United States Regarding Contractor Technical Personnel*. H. Rept. 188, 89th Cong., 1st sess., 18 Mar. 1965.

House Committee on Science and Astronautics. *Apollo 13 Accident*. Hearing, 91st Cong., 2d sess., 16 June 1970.

———. *Cape Canaveral: The Hope of the Free World*. Print, 87th Cong., 2d sess., 24 May 1962.

———. *Equatorial Launch Sites—Mobile Sea Launch Capability*. H. Rept. 710, 87th Cong., 1st sess., 12 July 1961.

———. *Management and Operation of the Atlantic Missile Range*. Print, 86th Cong., 2d sess., 5 July 1960.

———. *Master Planning of NASA Installations*. H. Rept. 167, 89th Cong., 1st sess., 15 Mar. 1965.

———. *1962 NASA Authorization*. Hearings, pt. 1, 87th Cong., 1st sess., 13 Mar.–17 Apr. 1961.

———. *Report on Cape Canaveral Inspection*. Print, 86th Cong., 2d sess., 27 June 1960.

———. *Space, Missiles, and the Nation*. Rept. 2092 pursuant to H. Res. 133, 86th Cong., 2d sess., 5 July 1960.

———. *Transfer of the Development Operations Division of the Army Ballistic Missile Agency to the National Aeronautics and Space Administration*. Hearing on H.J. Res. 567, 86th Cong., 2d sess., 3 Feb. 1960.

House Committee on Science and Astronautics, Subcommittee on Manned Space Flight. *1968 NASA Authorization*. Hearings on H. R. 4450, H. R. 6470, pt. 2, 90th Cong., 1st sess., 14–21 Mar. 1967.

———. *1964 NASA Authorization*. Hearings on H. R. 5466, pts. 2a, 2b, 88th Cong., 1st sess., 6 Mar.–6 June 1963.

———. *1963 NASA Authorization*. Hearings on H. R. 10100, pt. 2, 87th Cong., 2d sess., 6 Mar.–10 Apr. 1962.

House Committee on Science and Astronautics, Subcommittee on NASA Oversight. *Apollo Program Pace and Progress*. Print, 90th Cong., 1st sess., 17 Mar. 1967.

———. *Engineering Management of Design and Construction of Facilities of NASA*. Print, 91st Cong., 1st sess., 11 Aug. 1969.

———. *Investigation into Apollo 204 Accident*. Hearings, 2 vols., 90th Cong., 1st sess., 10 Apr.–10 May 1967.

———. *NASA-DOD Relationship*. Print, 88th Cong., 1st sess., 26 Mar. 1964.

———. *Pacing Systems of the Apollo Program*. Print, 89th Cong., 1st sess., 15 Oct. 1965.

Senate Committee on Aeronautical and Space Sciences. *Amending the National Aeronautics and Space Administration Authorization Act for the Fiscal Year 1962*. S. Rept. 863 to accompany S. 2481, 87th Cong., 1st sess., 1 Sept. 1961.

———. *Apollo Accident*. Hearings, pts. 1–6, 90th Cong., 1st sess., 7 Feb.–9 May 1967.

———. *Apollo 204 Accident*. S. Rept. 956, 90th Cong., 2d sess., 30 Jan. 1968.

———. *NASA Authorization for Fiscal Year 1963*. Hearings on H. R. 11737, 87th Cong., 2d sess., 13–15 June 1962.

———. *NASA Authorization for Fiscal Year 1965*. Hearings on S. 2446, pt. 2, 88th Cong., 2d sess., 4–18 Mar. 1964.

———. *NASA Authorization for Fiscal Year 1966*. Hearings on S. 927, pt. 2, 89th Cong., 1st sess., 22–30 Mar. 1965.

Senate Committee on Aeronautical and Space Sciences, NASA Authorization Subcommittee. *NASA Authorization for Fiscal Year 1961*. Hearings on H. R. 10809, pt. 1, 86th Cong., 2d sess., 28–30 Mar. 1960.

Senate Committee on Appropriations, Subcommittee. *Independent Offices Appropriations, 1964*. Hearing on H. R. 8747, pt. 2, 88th Cong., 1st sess., 18 Oct. 1963.

Senate Committee on Government Operations, Permanent Subcommittee on Investigations. *Work Stoppage at Missile Bases*. Hearings pursuant to S. R. 69, pts. 1, 2, 87th Cong., 1st sess., 25 Apr.–9 June 1961.

———. *Work Stoppage at Missile Bases,* S. Rept. 1312, 87th Cong., 2d sess., 29 Mar. 1962.

Conference Papers

Aden, R. M. "Electrical Support Equipment for the Saturn V Launch Vehicle System." *Proceedings of 2d Space Congress*, Cocoa Beach, FL, 5–7 Apr. 1965.

Clements, J. S. "S-1C Stage Instrumentation Checkout Concepts at KSC." *Proceedings of 3d Space Congress*, Cocoa Beach, FL, 7–10 March 1966.

Debus, Kurt. "Some Design Problems Encountered in Construction of Launch Complex 39." Hermann Oberth Gesellschaft, Darmstadt, Germany, 25 June 1964.

———. "Trends and Problems in Instrumentation and Operations in NASA's Future Space Efforts." AFMTC-AFESD-AFCEA Symposium, Patrick A.F.B., FL, 6 Mar. 1962.

———. "Launch Operations for Saturn V/Apollo." 10th Annual Meeting of the American Astronautical Society, New York, 6 May 1964.

Deese, James H. "The Problem of Low Level Wind Distribution." Structural Engineers Councils of Florida, Tampa, FL, 9 Nov. 1964.

Duran, B. E. "Saturn I/IB Launch Vehicle Operational Status and Experience." Paper 680739, Society of Automotive Engineers, Los Angeles, 7–11 Oct. 1968.

Eudy, Glenn. "Saturn V Mechanical Ground Support Equipment." *Proceedings of 2d Space Congress*, Cocoa Beach, FL, 5–7 Apr. 1965.

George, W. V., and C. A. Stinson. "An Automated Telemetry Checkout Station for the Saturn V Systems." 1966 National Telemetering Conference, Boston, MA, 12 May 1966.

Goff, H. C., and J. M. Schabacker. "Apollo Spacecraft Integrated Checkout Planning." *Proceedings of 3d Space Congress*, Cocoa Beach, FL, 7–10 Mar. 1966.

Hope, J. R., and C. J. Neumann. "Probability of Tropical Cyclone Induced Winds at Cape Kennedy." *Proceedings of 5th Space Congress*, Cocoa Beach, FL, 11–14 Mar. 1968.

Jafferis, William. "Prelaunch and Launch Checkout Operations—Uprated Saturn and Saturn V Vehicles." Project SETE, New York, 24–28 July 1967.

Johansen, R. B. "Developments in On-Board and Ground Checkout Systems." AIAA paper 70-245, American Institute of Aeronautics and Astronautics Launch Operations Meeting, Cocoa Beach, FL, 2–4 Feb. 1970.

Knothe, Adolf H. "Range Safety—Do We Need It?" AIAA paper 70-249, American Institute of Aeronautics and Astronautics Launch Operations Meeting, Cocoa Beach, FL, 2–4 Feb. 1970.

Marshall, John M. "The Mobile Concept and Automated Checkout of the Apollo/Saturn V Space Vehicle." 22d International Astronautical Congress, Brussels, Sept. 1971.

Montgomery, R. M. "Range Safety of the Eastern Test Range." AIAA paper 70-246, AIAA Launch Operations Meeting, Cocoa Beach, FL, 2–4 Feb. 1970.

Moore, F. Brooks, and William Jafferis. "Apollo/Saturn Prelaunch Checkout Display Systems." IEEE Conference on Displays, Univ. of Loughborough, England, 7–10 Sept. 1971.

Petrone, Rocco. "Apollo/Saturn V Launch Operations." AIAA Third Annual Meeting, Boston, 29 Nov.–2 Dec. 1966.

———. "Ground Support Equipment and Launch Installations at John F. Kennedy Space Center, NASA, for the Manned Lunar Landing Program." 15th International Astronautical Congress, Warsaw, 1964.

———. "Saturn V/Apollo Launch Operations Plan." AIAA Space Flight Testing Conference, Cocoa Beach, FL, 18–20 Mar. 1963.

Richard, Ludie G. "Saturn V System Philosophies." *Proceedings of 2d Space Congress*, Cocoa Beach, FL, 5-7 Apr. 1965.

Robin, E. A. "Development and Utilization of Computer Test Programs for Checkout of Space Vehicle." *Proceedings of 4th Space Congress*, Cocoa Beach, FL, 3-6 Apr. 1967.

Rorex, James E., and Robert P. Eichelberger. "Digital Data Acquisition System in Saturn V." *Proceedings of 2d Space Congress*, Cocoa Beach, FL, 5-7 Apr. 1965.

Rudolph, Arthur. "Operational Experience with the Saturn V." AIAA paper 68-1003, AIAA 5th Annual Meeting, Philadelphia, 21-24 Oct. 1968.

Salvador, G., and R. W. Eddy. "Saturn IB Stage Launch Operations." *Proceedings of 5th Space Congress*, Cocoa Beach, FL, 11-14 Mar. 1968.

Speer, F. A. "Saturn I Flight Test Evaluation." AIAA paper 64-322, American Institute of Aeronautics and Astronautics, Washington, July 1964.

Taylor, G. H. "Operational Television System for Launch Complex 39 at the John F. Kennedy Space Center." *Proceedings of 5th Space Congress*, Cocoa Beach, FL, 11-14 Mar. 1968.

Taylor, T., Jr. "System Considerations for Establishing Prelaunch Checkout Effectiveness." *Proceedings of 2d Space Congress*, Cocoa Beach, FL, 5-7 Apr. 1965.

Tedesko, Anton. "Assembly and Launch Facilities for the Apollo Program, Merritt Island, Florida: Design of the Structure of the Vertical Assembly Building." ASCE Structural Engineering Conference, 21 Oct. 1964.

Thilges, J. N. "Range Safety, A Thorn in the Flesh." *Proceedings of 3d Space Congress*, Cocoa Beach, FL, 7-10 Mar. 1966.

von Tiesenhausen, Georg. "Ground Equipment to Support the Saturn Vehicle." American Rocket Society, Washington, Dec. 1960.

Technical Reports

Army Ballistic Missile Agency:

Dodd, R. P., and J. H. Deese. *Juno V (Saturn) Heavy Missile Launch Facility, 1st Phase Request, 2nd Phase Estimate.* Atlantic Missile Range, 14 Feb. 1959.

Hall, Charles J. *A Committee Study of Blast Potentials at the Saturn Launch Site*, Rep. No. DHM-TM-9-60. Redstone Arsenal, AL, Feb. 1960.

Hamilton, J. S., J. L. Fuller, and P. F. Keyes. *Juno V Transportation Feasibility Study*, Rep. No. DLMT-TM-58-58. Redstone Arsenal, AL, 5 Jan. 1959.

Koelle, H. H., F. L. Williams, W. G. Huber, and R. C. Callaway, Jr. *Juno V Space Vehicle Development Program (Phase I), Booster Feasibility Demonstration,* Rep. DSP-TM-10-58. Redstone Arsenal, AL, 13 Oct. 1958.

———. *Juno V Space Vehicle Development Program (Status Report)*, Rep. No. DSP-TM-11-58. Redstone Arsenal, AL, 15 Nov. 1958.

Organizational Manual, Army Ballistic Missile Agency, Redstone Arsenal, AL, 5 Mar. 1959.

Army Corps of Engineers:

Apollo Launch Complex 39 Facilities Handbook. South Atlantic Division, Canaveral District, undated.

Army Ordnance Missile Command:

Koelle, H. H., F. L. Williams, and W. G. Huber. *Saturn Systems Study*, Rep. No. DSP-TM-1-59. Redstone Arsenal, AL, 13 Mar. 1959.

Project Horizon: A. U.S. Army Study for the Establishment of a Lunar Military Outpost. Redstone Arsenal, AL, 8 June 1959.

Saturn Systems Study II, Rep. No. DSP-TM-13-59. Redstone Arsenal, AL, 13 Nov. 1959.

Air Force:

Air Force Eastern Test Range. *Range Safety Manual*. AFETR Manual 127-1. Patrick A.F.B., FL, 1 Sept. 1972.

Joint Air Force/NASA Hazards Analysis Board. *Safety and Design Considerations for Static Test and Launch of Large Space Vehicles*. Patrick Air Force Base, FL; Air Force Missile Test Center, 1 June 1961.

Headquarters, NASA:

Ad Hoc Task Group. *A Feasible Approach for an Early Manned Lunar Landing* (Fleming Committee Report). Washington, D.C., 16 June 1961.

Allen, William H., ed. *Dictionary of Technical Terms for Aerospace Use*. Washington, D.C.: GPO, 1965.

America's Next Decades in Space, A Report for the Space Task Group. Washington, D.C., Sept. 1969.

Apollo Inter-Center ICD Management Procedure, CM-001, 001-1B. Washington, D.C., Jan. 1969.

Logsdon, John M. *NASA's Implementation of the Lunar Landing Decision*, HHN-81 (comment edition). Washington, D.C., Aug. 1969.

NASA-DOD. *Joint Report on Facilities and Resources Required at Launch Site to Support NASA Manned Lunar Landing Program* (Debus-Davis Report). Cape Canaveral, FL, 31 July 1961.

NASA Special Committee on Space Technology. *Recommendations Regarding a National Civil Space Program* (Stever Committee Report). 28 Oct. 1958.

Office of Manned Space Flight. *Apollo Configuration Management Manual*, NPC 500-1. Washington, D.C., 18 May 1964.

———. *Apollo Program Flight Summary Report*, Apollo Missions AS-201 through Apollo 8. Washington, D.C., Jan. 1969.

———. *Apollo Program Development Plan*, Rep. No. M-D MA500. Washington, D.C., 1 Jan. 1966.

PERT: Program Evaluation and Review Technique, Handbook, NPC-101. Washington, D.C., 1 Sept. 1961.

"A Plan for Manned Lunar Landing" (Low Committee Report). Washington, D.C., 7 Feb. 1961.

Report of Apollo 204 Review Board to the Administrator, NASA, 5 Apr. 1967, 1 vol. plus 14 vols. of appendixes and a set of looseleaf color photographs.

Report of the Apollo 13 Review Board. Washington, D.C., 15 June 1970.

Report to the Administrator, NASA, on Saturn Development Plan, by Saturn Vehicle Team (Silverstein Committee Report). Washington, D.C., 15 Dec. 1959.

Kennedy Space Center:

Apollo/Saturn V Launch Operations Test and Checkout Requirements, AS-504 and All Subsequent Missions, K-V-051-01. KSC, FL, 1968.

Apollo/Saturn V MILA Facilities Descriptions, K-V-011, Coordination Draft. KSC, FL, 30 June 1965.

Brewster, H. D., and W. G. Hughes. *Lightning Protection for Saturn Launch Complex 39*, KSC, FL, 18 Oct. 1963.

Catalog of Launch Vehicle Tests, Saturn V, Apollo/Saturn V, Revision 1, GP-244. KSC, FL, 15 June 1966.

Hanna, George V. *Chronology of Work Stoppage and Related Events, KSC/AFETR through July 1965*. KSC, FL, Oct. 1965.

Index of KSC Apollo Tree Documents and other KSC Generated Documents in the KSC Library, GP-856. KSC, FL, 1970.

Jarrett, Francis E., Jr., and Robert A. Lindemann. *Historical Origins of NASA's Launch Operations Center to July 1, 1962,* KHM-1. Cocoa Beach, FL, Apr. 1964.

KSC Apollo/Saturn Program Development/Operations Plan, 100-39-0001, 2 vols. KSC, FL, 10 Oct. 1965.

Launch Support Equipment Engineering Division. *Development of Design Criteria for Saturn V Flame Deflector*, TR-174-D. KSC, FL, 1 June 1965.

———. *Saturn V Launch Support Equipment General Criteria and Description*, SP-4-37-D. KSC, FL, 15 Sept. 1964.

———. *Saturn V Electrical Ground Support Equipment for Launch Complex 39*, SP-96-D. KSC, FL, 21 Dec. 1964.

———. *Saturn V Launch Support Equipment General Criteria and Description*, SP-4-37-D, Revision. KSC, FL, 15 Sept. 1964.

Launch Vehicle Checkout Automation and Programming Office. *Apollo (Saturn) Automated Checkout*. KSC, FL, 23 Aug. 1974.

McMurran, W. R., ed. "The Evolution of Electronic Tracking, Optical, Telemetry and Command Systems at the Kennedy Space Center." KSC, FL, 17 Apr. 1973.

Public Affairs Office. *Kennedy Space Center Story*. KSC, FL, Dec. 1971.

Saturn Launch Vehicle Checkout Automation Development Plan, KSC-100-39-0007. Cocoa Beach, FL, 8 Aug. 1966.

Scheller, Donald R. "Management by Exception, Activation of Apollo/Saturn V Launch Complex 39." KSC, FL, 15 May 1967.

A Selective List of Acronyms and Abbreviations, GP-589 Revised. KSC, FL, July 1972.

Walter, George W. *Modifications to Bearing for Traction Support Rollers on Crawler-Transporters*. KSC, FL, 15 Dec. 1965.

"Weather Effects on Apollo/Saturn V Operations, Apollo 4 through Apollo 13," 630-44-0001. KSC, FL, 27 July 1970.

Manned Spacecraft Center:

Analysis of Apollo 12 Lightning Incident, MSC-01540. Houston, TX, Feb. 1970.

Atlantic Missile Range Operations: Facilities, 1959–1964. Houston, TX, 15 Apr. 1963.

John F. Kennedy Space Center, NASA Fiscal Year 1963 Estimates, Apollo Mission Support Facilities. Cape Canaveral, FL; Florida Operations, 1963.

Launch Operations Center:

Concepts for Support Service at the Merritt Island Launch Area. Cape Canaveral, FL, 6 May 1963.

Criteria for Design Pad "A" Launch Complex 39. Cape Canaveral, FL, 19 Dec. 1962.

Criteria for Launch Complex 39, Crawler Transfer System and Utilities. Cape Canaveral, FL, 5 Sept. 1962.

Deese, James. "Concept Development of Saturn Service Structure No. II." Cape Canaveral, FL, Apr. 1963.

Saturn C-5 Facilities Evaluation for Complex 39, LTR-1-2. Cape Canaveral, FL, 10 Sept. 1962.

Raffaelli, A. R. *Introduction to Lightning*, LT1R-2-DE-62-6. Cape Canaveral, FL, 14 Dec. 1962.

Marshall Space Flight Center:

Buchanan, Donald D., and George W. Walter. *Appraisal of Transfer Modes for Saturn C-5 Mobile Systems*, Rep. No. MIN-LOD-DH-9-62. Huntsville, AL, 11 June 1962.

Catalog of Systems Tests for Saturn S-1 Stage. Huntsville, AL, 1 Sept. 1961.

Chrysler Corporation Space Division. *Saturn IB Vehicle Handbook*. Huntsville, AL, 25 July 1966.

Duren, O. K. *Interim Report on Future Saturn Launch Site*, Rep. No. MIN-LOD-DL-1-61. Huntsville, AL, 10 May 1961.

Gwinn, Ralph T., and Kenneth J. Dean. "*Consolidated Instrumentation Plan for Saturn Vehicle SA-1*," Rep. No. MTP-LOD-61-36.2a. Huntsville, AL, 25 Oct. 1961.

Launch Countdown, Saturn Vehicle SA-1, Rep. No. MTP-LOD-61-35.2. Redstone Arsenal, AL, 30 Oct. 1961.

Launch Facilities and Support Equipment Office, Launch Operations Directorate. *A Preliminary Study of Launch Facility Requirements for the C-4 Space Vehicle*. Huntsville, AL, Oct. 1961.

———. *Project Saturn, C-1, C-2 Comparison*. Rep. No. M-MS-G-113-60. Huntsville, AL, 21 Jan. 1961.

MSFC Automation Plan. Huntsville, AL, 8 May 1962 (revised, 1 June 1964).

Results of the Saturn I Launch Vehicle Test Flights, Rep. No. MPR-SAT-FE-66-9. Huntsville, AL, 1 Aug. 1961.

Richard, Ludie G., and Charles O. Brooks. *The Saturn Systems Automation Plan*. Huntsville, AL, 15 Sept. 1961.

SA-1 Vehicle Data Book, Rep. No. MTP-MS and M-E-61-3. Huntsville, AL, 26 June 1961.

Software for IU 201 at MSFC, SA-201 at KSC, SA-501 at KSC, Rep. No. R-DIR-64-1. Huntsville, AL, 1 Dec. 1964.

Sparks, Owen L., comp. *Preliminary Concepts of Launch Facilities for Manned Lunar Landing Program*, Rep. No. MIN-LOD-DL-3-61. Huntsville, AL, 1 Aug. 1961.

———. *Preliminary Feasibility Study on Offshore Launch Facilities for Space Vehicles*, Rep. No. IN-LOD-DL-1-60 Interim. Huntsville, AL, 29 July 1960.

Technical Information Summary, AS-501, Apollo Saturn V Flight Vehicle, R-ASTR-S-67-65. Huntsville, AL, 15 Sept. 1967.

Walter, George W. *Saturn Mobile (Canal) Concept Flame Deflector and Launcher/Transporter Emplacement Evaluation*, Rep. No. MIN-LOD-2-62. Huntsville, AL, Feb. 1962.

———. *Saturn Mobile (Rail) Concept: An Examination of Rail Transfer Systems for a Launcher/Transporter*, Rep. No. MIN-LOD-DH-3-62. Huntsville, AL, 3 Apr. 1962.

Contractor:

Beech Aircraft Corporation. *Saturn C-5 Propellant Transportation Optimization Study, ER-13539*. Boulder, CO, 25 June 1962.

Boeing Atlantic Test Center. *Launch Complex 39 GSE Systems Descriptions*. KSC, FL, 3 Aug. 1965.

Brown Engineering Co. *An Evaluation of an Enclosed Concept for a C-5 Vertical Assembly Building*. Huntsville, AL, 2 Apr. 1962.

———. *An Evaluation of an Open Concept for a C-5 Vertical Assembly Building*. Huntsville, AL, 2 Apr. 1962.

———. *Fixed Pad Concept of Launch Complex 39 for the Saturn C-5 Vehicle*. Huntsville, AL, 28 Sept. 1962.

Chrysler Corp. Space Division. *Saturn I/IB Automation Orientation, HSE-R 115*. Huntsville, AL, undated.

Culver, A. J., and K. G. Baird. "ERS Recovery Plan." KSC, FL: Boeing Atlantic Test Center Management Systems Staff, Apr. 1966.

General Electric Co. Apollo Support Department. *Systems Description of Saturn V Launch Vehicle Ground Electrical Support Equipment at Vehicle Launch Facility 39-1*. Daytona Beach, FL, 27 Sept. 1965.

Green, Arthur W., Lewis E. Williamson, Robert P. Dell, and Reed B. Jenkins. *Reliability Study of Saturn SA-3 Pre-Launch Operations*. Washington, D.C.: ARINC Research Corporation, 3 Jan. 1963.

Hartsfield, Annie May, Mary Alice Griffin, and Charles M. Grigg, eds. *Summary Report NASA Impact on Brevard County*. Tallahassee, FL: Institute of Social Research, Florida State University, 1966.

The Martin Co. *Rescue and Escape Systems from Tall Structures (RESTS).* Denver, CO, Oct. 1963.

———. *Saturn C-2 Operational Modes Study, Summary Report.* Baltimore, MD, June 1961.

———. *Saturn C-3 Launch Facilities Study*, Rep. No. ER 12125, 3 vols. Baltimore, MD, Dec. 1961.

Maurice H. Connell and Associates. *Alternate "I" Heavy Missile Service Structure.* Miami, FL, 24 Apr. 1959.

———. *Heavy Missile Launch Facility Criteria.* Miami, FL, 15 Mar. 1959.

———. *Heavy Missile Service Structure Criteria.* Miami, FL, undated.

———. *Siting Study and Recommendation, Saturn Staging Building and Service Structure, Complex 34 AFMTC.* Miami, FL, Jan. 1960.

———. *Saturn C-5 Launch Facilities, Study of Rail Systems for Vertical Transporter/ Launcher Concept.* Huntsville, AL, May 1962.

Sutherland, L. C. *Sonic and Vibration Environments for Ground Facilities—A Design Manual*, report WR 68-2. Wyle Laboratories, 1968.

TRW Space Technology Laboratories. *A Study of the KSC Safety Program.* Cape Canaveral, FL, May 1965.

Interviews

Aden, Robert, MSFC Astrionics, Electrical, by Benson, 30 Oct. 1974.

Alexander, William, Washington, D.C., by James Frangie, 8 Aug. 1969.

Atkins, John R., KSC Safety Off., by Faherty, 5 Sept., 5 Nov. 1973, 29 May 1974.

Barfus, Armond, KSC Support Ops., Development Testing Lab., by Benson, 5 Dec. 1973.

Bertram, Emil, KSC Launch Ops., Requirements and Resources Off., by Benson, 28 Sept.,* 15 Nov. 1973.*

Bidgood, Clarence, Washington, D.C., by Frangie, 14 Nov. 1968, 13 Aug. 1969.

Black, Dugald, KSC Dep. Dir., Support Ops., by Benson, 28 Feb. 1974.

Bobik, Joseph, KSC Spacecraft Ops., Quality Surveillance Div., by Benson and Faherty, 26 June 1974.

Boylston, Clifford, Brown Engineering Co., Huntsville, AL, by Benson, 21 July 1972.

Brewster, Heyward, KSC Design Engineering, Design Documentation Br., by Benson, 30 Nov. 1973.*

Brown, Joseph Andrew, KSC Design Engineering, Architectural Sec., by Faherty, 25 Sept. 1973.

Bruns, Rudolf, KSC Information Systems, Computer Systems Div., by Frangie, 22 Aug. 1969; by Benson, 3 Jan. 1974.

Bryan, Frank, KSC Launch Vehicle Ops. Engineering Staff, by Benson, 19 Dec. 1973.

Buchanan, Donald, KSC Design Engineering, by Frangie, 5 Sept. 1969; by Benson, 22 Sept., 7 Nov., 28 Nov. 1972; 4 Oct. 1974.

Burke, J. F., Chief, KSC Saturn/Apollo Facilities Br., by Frangie, 23 Apr. 1969.

Carlson, Norman, KSC Launch Vehicle Ops., Test Ops. Br., by John Marshall, 16 Dec. 1970; by Benson, 5 Sept. 1974.*

Carothers, Ralph Dale, KSC Spacecraft Ops., Preflight Ops. Br., by Benson, 14 June 1974.*

Chambers, Milton, KSC Launch Vehicle Ops., Gyro and Stabilizer Systems Br., by Benson, 19 Aug. 1974.*

*Indicates telephone interview.

Chandler, William, KSC Launch Vehicle Ops., Electrical Systems Br. (LCC), by Benson, 20 Nov. 1973.*

Chauvin, Clarence, KSC Spacecraft Ops. CSM Test Staff, by Benson, 2 Apr.,* 23 May, 6–7 June 1974; by Faherty, 24 June 1974.

Childers, Frank, KSC Information Systems, Quality Surveillance Off., by Faherty, 7 Nov. 1972, 18 Mar. 1973.

Clark, Raymond, KSC Design Engineering, by Benson, 30 June 1972.

Claybourne, John P., KSC Sciences and Applications Project Off., Earth Resources, by Benson, 5 Jan. 1973.

Clearman, William T., Jr., Cocoa Beach, FL, by Benson, 5 Jan., 13 Sept.,* 26 Oct. 1973.*

Cochran, Harold, KSC Spacecraft Ops., Communications and R. F. Sec., by Benson, 7 June 1974.*

Collner, Joseph D., KSC Spacecraft Ops., Communications and R. F. Sec., by Benson, 7 June 1974.

Corbett, Belzoni A., Jr., KSC Information Systems, by Benson, 11 Jan. 1974.

Corn, Graydon F., KSC Launch Vehicle Ops., Propellants Br., by Benson, 23 July 1973,* 22 Apr. 1974.*

Crawford, Harvey, KSC Spacecraft Ops., Environmental Control and Cryogenics Sec., by Benson, 28 June 1974.*

Crunk, Henry, KSC Launch Vehicle Ops., Mechanical and Propulsion Systems, by Benson, 12 Apr. 1973.*

Davidson, James, KSC Launch Vehicle Ops., Electrical Systems, by Benson, 31 July 1973.*

Davis, Edwin, KSC Design Engineering, Launch Accessories, by Benson, 18 Jan. 1973.*

Debus, Kurt H., KSC Center Dir., by James Covington, 22 Aug. 1969; by Benson and Faherty, 16 May 1972.

Deese, James, KSC Design Engineering, Systems Analysis, by Benson, 10 May 1972, 4 Oct. 1973; by Faherty, 16 Mar. 1973.

Dodd, Richard P., KSC Design Engineering, Project Integration Off., by Benson, 13 Aug. 1974.

Donnelly, Paul, KSC Assoc. Dir. for Ops., by John Marshall, 19 Nov. 1970; by Benson and Faherty, 19 June 1974.*

Duggan, Orton L., KSC Apollo Off., by James Grimwood, 13 Nov. 1969.

Duren, Olin K., MSFC Astronautical Lab., Materials, by Benson, 29 Mar., 16 May 1972,* 1 Nov. 1974.*

Edwards, Marion, KSC Launch Vehicle Ops., Launch Instrumentation, by Benson, 7 May 1973.

Ely, George, KSC Launch Vehicle Ops., Flight Control, by Benson, 8 July 1974.

Enlow, Roger, KSC Technical Support, by Benson, 5 Dec. 1973.*

Ernst, Lloyd, KSC Design Engineering, LC-39 Area Management Br., by Faherty, 8 Nov. 1973.

Fannin, Edward, KSC Launch Vehicle Ops., Mechanical and Propulsion, by Benson, 23 July 1973.*

Finn, James E., Design Engineering, Cables and Special Power Sec., by Faherty, 16 Jan. 1973.

Fiorenza, Vincent, GE Space Div., Apollo and Ground Systems, KSC, by Benson, 25 Apr. 1974.

Foster, Leroy, General Electric, Daytona Beach, FL, by Benson, 30 Apr. 1974.

Fowler, Calvin, GE Space Div., Apollo and Ground Systems, KSC, by Benson, 25 Apr. 1974.

Fulton, James, KSC Design Engineering, by Benson, 26 Oct. 1973.*

Gassman, Marvin, Apollo-Skylab Program Off., Configuration Management Br., by Benson, 8 Nov. 1973.

Gibbs, Asa, Satellite Beach, FL, by James Covington, 7 Aug. 1969.

Glaser, William, KSC Launch Vehicle Ops., Telemetry, by Benson, 7 May 1973.

Goldsmith, James, MSFC Procurement Off., Saturn Div., by Benson, 6 Nov. 1974.*

Gooch, Harold, KSC Administration, Labor Relations, by Faherty, 1 July 1974.

Gramer, Russell, KSC Installation Support, Quality Surveillance Div., by Faherty, 19, 21 Sept. 1972, 19 July 1973.

Greenfield, Terry, KSC Design Engineering, Digital Electronics, by Benson, 3 Jan. 1974.

Gruene, Hans, KSC Dir. of Launch Vehicle Ops., by John Marshall, 19 Nov. 1970; by Benson and Faherty, 10 May 1972.

Haggard, Ken M., Lockheed, Personnel Industrial Relations, by Faherty, 16 July 1973.

Hahn, Richard, KSC Design Engineering, Analysis, by Benson, 3 Dec. 1973.

Hand, Larry, KSC Design Engineering, Communications, by Benson, 2 Oct. 1973.

Hangartner, James, KSC Spacecraft Ops., Mechanical Systems, by Benson, 3 July 1974.*

Hansel, John, KSC Quality Control, by Faherty, 11 July 1973.

Harris, Gordon, KSC Chief, Public Affairs, by Benson, 12 Apr. 1974.

Harris, Steven, KSC Design Engineering, Field Engineering Off., by Benson, 24 Oct. 1973.*

Henschel, Charles F., KSC/NASA Test Ops. Off., by John Marshall, 17 Nov. 1970.

Horn, Frank W., Jr., KSC Apollo-Skylab Programs, by Benson, 28 Sept. 1973.

Horner, William J., Jr., KSC Security Off., by Faherty, 1 July 1974.

Huffman, Bobby R., KSC Launch Vehicle Ops., Launch Instrumentation Systems Div., by Benson, 7 May 1973.

Hughes, R. Bradley, KSC Information Systems Engineering Application, by Benson, 11 Jan. 1974.

Humphrey, John T., KSC Launch Vehicle Ops., Propulsion and Vehicle Mechanical, by Benson, 2 Aug. 1974.*

Jafferis, William, KSC Launch Vehicle Ops., Systems Engineering, by Benson, 19 Dec. 1973, 22 Jan. 1974.*

Jelen, Wilfred G., KSC Information Systems Data, by Benson, 15 Jan. 1974.

Jenke, Richard, Huntsville, AL, by Benson, 29 Oct. 1974.

Johnson, Charles, Florida Dept. of Commerce, Employment Service, Cocoa, FL, by Benson, 8 Apr. 1974.*

Johnson, Edwin C., KSC Spacecraft Ops., CSM and Payloads Project Engineering, by Benson, 10 June 1974.*

Johnson, Harold G., KSC Support Ops., Planning and Contract, by Benson, 17 Dec. 1973.

Jones, James, Information and Measurement Systems, Test Analysis Sec., by Faherty, 8 Apr. 1973.

Joralan, Albert, KSC Design Engineering Data Systems, by Benson, 3 Jan. 1974.

Kapryan, Walter, KSC Dir. of Launch Ops., by Benson, 25 Apr. 1974.

Kaufman, James R., KSC Administration, Manpower Utilization, by Benson, 27 Mar. 1974.*

Lamberth, Horace, KSC Spacecraft Ops., Fluid Systems, by Benson, 24 Apr. 1974.*

Lang, J. Robert, KSC Spacecraft Ops., Environmental Control and Cryogenic, by Benson, 7 June 1974.*

Lealman, Roy, KSC Launch Vehicle Ops., Electrical G and C Systems, by Benson and Faherty, 27 June 1974.

Leet, Joel, KSC Shuttle Project Planning, by Benson, 8 Nov. 1973, 13 Feb. 1974.

Lloyd, Russell, KSC Support Ops., by Benson, 13 Feb. 1974.

Lowell, Albert, General Electric, Daytona Beach, FL, by Benson, 16 Apr. 1974.

McCafferty, Riley, Johnson Space Center, Crew Training and Simulation, by Ivan Ertel, 28 Jan. 1971; by Benson, 20 Feb. 1974.

McKnight, James N., KSC Spacecraft Ops., Preflight, by Benson, 26 June 1974.*

Malley, George, Chief Counsel, Langley Research Center, by Faherty, 6 Nov. 1973.*

Marsh, Thomas, KSC Launch Vehicle Ops., Propulsion and Vehicle, by Benson, 7 May 1973.

Mathews, Edward, JSC Manager, Space Shuttle Systems Integration, by Benson, 19 Feb. 1974.

Matthews, George D., KSC Information Systems, Telemetry, by Benson, 11 Jan. 1974.

Medlock, Joe, KSC Launch Vehicle Ops., Checkout, Automation and Programming Off., by Benson, 7 Oct. 1974.

Melton, Lewis, KSC Administration, Resources and Financial Management, by Covington, 18 Feb. 1969.

Montgomery, Ann, KSC Spacecraft Ops., Preflight Ops., by Benson, 26 June 1974.

Moore, Robert T., KSC Information Systems, Planning and Technical Support, by Benson, 27 Sept. 1973.

Morris, Owen, JSC Manager, Apollo Spacecraft Program Off., by Benson, Grimwood, and Courtney Brooks, 20 Dec. 1972.

Moser, Robert, KSC Test Planning Off., by Benson, 30 Mar., 18 July 1973,* 17 Apr. 1974.*

Moxley, Paul, KSC Spacecraft Ops., Propulsion Systems, by Benson, 19 July 1974.

Mrazek, William, MSFC Assoc. Dir. for Engineering, by Benson, 2 Aug. 1972.*

Murphy, John, KSC Apollo-Skylab Program Off., Launch Vehicle and Workshop Br., by Benson, 9 Nov. 1973.

Nazaro, Ron, IBM, KSC, by Faherty, 8 July 1974.

Newall, Robert, KSC Launch Vehicle Ops., S-IC Systems, by Benson, 31 July 1973.*

Norwalk, William, Hqs. Auditing Div., by Benson, 23 Jan. 1974.*

O'Conner, H. J., Mgr., Wildlife Refuge, by Faherty, 31 May 1972.

O'Hara, Alfred D., KSC Launch Vehicle Ops. Management, by Benson, 14 Jan. 1974.*

Owens, Lester, KSC Design Engineering Systems, by Benson, 12 Apr., 21 Nov. 1972.

Page, George F., KSC Spacecraft Ops., by Benson, 28 Jan. 1974.*

Pantoliano, Thomas, KSC Launch Vehicle Ops., Mechanical and Propulsion, by Benson, 18 Apr. 1973.*

Parker, Clarence C., KSC Installation Support, by James L. Frangie, 14 Feb. 1969; by Benson and Faherty, 2 Feb. 1972.

Parsons, Walter, KSC Design Engineering Systems, by Benson, 21 Jan. 1974.

Paul, Henry C., KSC, Chief, Checkout Automation and Programming Off., by John Marshall, 9 Dec. 1970.

Petrone, Rocco, NASA Hq., Apollo Program Dir., by Eugene Emme and Tom Ray, 17 Sept. 1970; by Benson, Emme, and Faherty, 25 May 1972.

Pickett, Andrew, KSC Shuttle Projects Off., by Benson, 26 July 1973.*

Policicchio, Mrs. Caroline, KSC, by James Covington, 4 Aug. 1969.

Poppel, Theodor, KSC Design Engineering, Field Engineering Off., by Benson, 12 Jan. 1973,* 24 Jan. 1973.

Porcher, Arthur G., KSC Design Engineering, by Benson, 28 Apr. 1972.*

Potate, John, NASA Office of Manned Space Flight, by Benson, Faherty, and Ray, 25 May 1972.

Potter, John, KSC Design Engineering, Field Engineering Off., by Benson, 24 Oct. 1973.

Preston, G. Merritt, KSC Manager, Shuttle Project, by Benson and Faherty, 12 Dec. 1973, by Benson, 22 Jan. 1974.*

Proffitt, Richard C., KSC Spacecraft Ops., Launch Complex 39, by John Marshall, 1 Dec. 1970; by Faherty, 20 June 1974.

Ragusa, James, KSC, Off. of the Dep. Director, by Benson, 11 Sept. 1974.*

Redfield, Marvin, NASA Hqs., Advanced Development, by Benson, Faherty, and Ray, 25 May 1972.

Reyes, Raul Ernest, KSC Spacecraft Ops. Preflight, by Faherty, 19 Jan., 3 June, 30 Oct. 1973, 24 June 1974.

Renaud, Fred, KSC, Bendix Launch Support, by Faherty, 4 Apr., 16 May 1973.

Richard, Ludie, MSFC Dep. Dir., Science and Engineering, by Benson, 12 Dec. 1973,* 30 Oct. 1974.

Rigell, Isom, KSC Launch Vehicle Ops., by Benson, 3 Dec. 1973.*

Roberts, John T., KSC Design Engineering, Utilities Sec., by Benson, 1 Nov. 1973.*

Rosen, Milton W., NASA Hqs., by Barton Hacker and Eugene Emme, 14 Nov. 1969; by Benson and Faherty, 25 May 1972.

Rowland, R. D., Asst. to the President, Hays International Corp., Birmingham, AL, by Benson, 25 July 1972.*

Russell, Labrada, KSC Installation Support, Librarian, by Benson, 15 Mar. 1973.

Sasseen, George T., KSC Spacecraft Ops., Engineering, by Benson, 26 July 1973,* 4 Feb. 1974*; by Faherty, 8 July 1974.

Scholz, Arthur, KSC, Boeing Aerospace Co., Field Ops. and Support, by Benson, 18 June 1974.

Seully, Edward J., McDonnell-Douglas Astronautics Co., by Faherty, 20 Apr. 1973.

Shea, Joseph, at Washington, D.C., by Eugene Emme, 6 May 1970.

Sherrer, Leroy, KSC Launch Vehicle Ops. Contractor Technical Management, by Benson, 25 July 1973.

Siebeneichen, Paul, KSC Community Relations Off., by Faherty, 29 Jan. 1973.*

Sieck, Robert, KSC Spacecraft Ops., Shuttle Project, by Benson, 4 Apr. 1974.*

Smith, Jackie E., KSC Spacecraft Ops., Experiments, by Benson, 4 June 1974.

Smith, Richard G., MSFC, Manager Saturn Program Off., by Benson, 29 Oct. 1974.

Sparkman, Orval, KSC Design Engineering, Mechanical Design, by Benson, 15 Dec. 1972, 13 June 1974.*

Sparks, Owen L., MSFC, Performance and Flight Mechanics, by Benson, 31 Mar. 1972.

Stringer, M. S., KSC Internal Review Staff, by James Frangie, 19 Dec. 1968.

Stein, Martin, URSAM Project Architect for LCC, by Covington, 8 Aug. 1969.

Thompson, John, KSC Launch Vehicle Ops., Checkout, Automation and Programming Off., by Benson, 7 Oct. 1974.

Twigg, John M., KSC Launch Vehicle Ops., Skylab and Space Shuttle, by John Marshall, 23 Nov. 1970.

von Tiesenhausen, Georg, by Benson, 29 Mar. 1972, 20 July 1973.

Wagner, Walter, KSC Apollo-Skylab Programs, Configuration Management, by Faherty, 7 Aug. 1973; by Benson, 21 Sept. 1973.

Walter, George, KSC Design Engineering, Structures, by Benson, 7 Nov. 1972, 26 Jan. 1973.

Walton, Thomas, KSC Design Engineering, LPS Systems, by John Marshall, 17 Dec. 1970; by Benson, 23 Jan. 1974.*

Wasileski, Chester, KSC Design Engineering, Facilities and Systems, by Benson, 14 Sept., 14 Dec. 1972.

Wedding, Michael A., Chief, Checkout Equipment Br., Automation—Spacecraft, by John Marshall, 11 Dec. 1970.

Wendt, F. Gunter, North American Rockwell Test Management, by Faherty, 18 June 1973.

White, James, KSC Design Engineering, Electrical and Electronic Design, by Benson, 9 May 1973.

Whiteside, Carl, KSC Launch Vehicle Ops., Electrical G and C, by Benson, 4 Jan.,* 29 Aug. 1974.

Widick, Herman K., KSC Spacecraft Ops., LM and Skylab Test, by John Marshall, 15 Dec. 1970; by Benson, 23 May 1974.

Wiley, Alfred N., KSC Spacecraft Ops., by Benson, 31 Oct. 1973, 13 Feb. 1974.

Williams, Francis L., NASA Hq. Off. of Analysis and Evaluation, by Benson, 6 Apr. 1972.*

Williams, Grady, KSC Dep. Dir. for Design Engineering, by Benson, 29 Mar. 1973.

Wills, Tom, KSC Design Engineering, Mechanical Design, by Benson, 28 Nov. 1973.

Wojtalik, Fred, MSFC Astrionics Lab., Guidance and Control, by Benson, 30 Oct. 1974.

Yates, Maj. Gen. Donald N. (USAF, Ret.), by Faherty, 17 Sept. 1973.

Youmans, Randell E., KSC Launch Vehicle Ops., Test Ops., by John Marshall, 5 Feb. 1971.

Zeiler, Albert, KSC Design Engineering, Mechanical Design, by F. E. Jarrett and W. Lockyer, 11 Aug. 1970; by Benson, 24 Mar., 11 July, 24 Aug. 1972, 23 July 1973*; by Faherty, 24 Aug. 1972.

INDEX